Trigonometric Identities

Pythagorean Identities: $\sin^2\theta + \cos^2\theta = 1$

$$1 + \tan^2\theta = \sec^2\theta$$

$$1 + \cot^2\theta = \csc^2\theta$$

Reciprocal Relations: $\csc\theta = \dfrac{1}{\sin\theta}$ $\qquad \sec\theta = \dfrac{1}{\cos\theta}$ $\qquad \cot\theta = \dfrac{1}{\tan\theta}$

tan and cot in terms of sin and cos: $\tan\theta = \dfrac{\sin\theta}{\cos\theta}$ $\qquad \cot\theta = \dfrac{\cos\theta}{\sin\theta}$

Addition Formulas: $\sin(\alpha \pm \beta) = \sin\alpha\cos\beta \pm \cos\alpha\sin\beta$

$$\cos(\alpha \pm \beta) = \cos\alpha\cos\beta \mp \sin\alpha\sin\beta$$

$$\tan(\alpha \pm \beta) = \frac{\tan\alpha \pm \tan\beta}{1 \mp \tan\alpha\tan\beta}$$

Double-Angle Formulas: $\sin 2\theta = 2\sin\theta\cos\theta$

$$\cos 2\theta = \cos^2\theta - \sin^2\theta$$

$$= 2\cos^2\theta - 1$$

$$= 1 - 2\sin^2\theta$$

$$\tan 2\theta = \frac{2\tan\theta}{1 - \tan^2\theta}$$

$\sin^2$ and $\cos^2$ in terms of cos: $\sin^2\theta = \dfrac{1 - \cos 2\theta}{2}$ $\qquad \cos^2\theta = \dfrac{1 + \cos 2\theta}{2}$

Half-Angle Formulas: $\sin\frac{1}{2}\theta = \pm\sqrt{\dfrac{1 - \cos\theta}{2}}$

$$\cos\tfrac{1}{2}\theta = \pm\sqrt{\frac{1 + \cos\theta}{2}}$$

$$\tan\tfrac{1}{2}\theta = \frac{1 - \cos\theta}{\sin\theta} = \frac{\sin\theta}{1 + \cos\theta}$$

Sum Formulas: $\sin A + \sin B = 2\sin\dfrac{A + B}{2}\cos\dfrac{A - B}{2}$

$$\sin A - \sin B = 2\cos\frac{A + B}{2}\sin\frac{A - B}{2}$$

$$\cos A + \cos B = 2\cos\frac{A + B}{2}\cos\frac{A - B}{2}$$

$$\cos A - \cos B = -2\sin\frac{A + B}{2}\sin\frac{A - B}{2}$$

Product Formulas: $\sin\alpha\cos\beta = \frac{1}{2}[\sin(\alpha + \beta) + \sin(\alpha - \beta)]$

$$\sin\alpha\sin\beta = \tfrac{1}{2}[\cos(\alpha - \beta) - \cos(\alpha + \beta)]$$

$$\cos\alpha\cos\beta = \tfrac{1}{2}[\cos(\alpha + \beta) + \cos(\alpha - \beta)]$$

Reduction Formulas: If f is any trigonometric function and cf is its cofunction,

$$f(n\pi \pm \theta) = \pm f(\theta) \qquad f\left(\frac{(2n + 1)\pi}{2} \pm \theta\right) = \pm cf(\theta)$$

The sign on the right can be determined by considering θ to be acute.

Trigonometry

Trigonometry

LEONARD I. HOLDER
Gettysburg College

Wadsworth Publishing Company
Belmont, California
A Division of Wadsworth, Inc.

ISBN 0-534-01014-8

Mathematics Editor: Richard Jones
Production: Greg Hubit Bookworks
Technical Illustrator: Carl Brown
Cover art: "AL 3" by Laszlo Moholy-Nagy. Blue Four–Galka Scheyer Collection, Norton Simon Museum of Art at Pasadena

Printed in the United States of America

2 3 4 5 6 7 8 9 10—86 85 84 83

Library of Congress Cataloging in Publication Data

Holder Leonard Irvin,
Trigonometry.

Includes index.
1. Trigonometry. I. Title.
QA531.H666 516.2′4 81-14696
ISBN 0-534-01014-8 AACR2

Contents

Preface

Trigonometry is written as a college-level text to provide a basis for future work in mathematics and its applications. For students planning to take calculus, chemistry, engineering, or physics the material covered is essential.

I have avoided a formal "theorem-proof" style of writing and have instead used a more conversational approach because I believe this makes the book more readable for students. Nevertheless, the book is mathematically sound, and certain proofs appropriate to the level of mathematical maturity of the users are given. Many worked-out examples are included, illustrating the main types of problems in the exercise sets. The exercise sets are divided into A and B problems, with the more challenging problems to be found in the B category. There are numerous drill-type problems in the A exercises designed to fix concepts and provide practice in the technique being studied.

Each chapter is introduced by an example taken from a textbook in a field which employs trigonometry. The example illustrates an application of one or more of the concepts to be studied in the chapter. Students are not expected to follow the details of these introductory examples but simply to observe the application of trigonometry. The examples thus serve as motivational devices, showing the student that what they will be studying has immediate application in an area they may well be pursuing.

I have introduced numerical trigonometry before analytical trigonometry because I believe this provides greater motivation for students and is pedagogically the sounder approach. The trigonometric functions are introduced first in terms of right triangles, then for general angles, and finally as functions of real numbers. The solution of right triangles is taken up in Chapter 2, after which the analytical aspects of the subject are studied in Chapters 3 and 4. Chapters 5 and 6 on logarithms and the solution of oblique triangles could follow Chapter 2 if the instructor prefers this order. I have used the order given so as to avoid too great a delay in introducing analytical trigonometry. The concluding chapter on complex numbers brings out the beautiful relationships this subject provides between algebra, geometry, and trigonometry.

The advent of hand calculators and their widespread availability at moderate prices have had a dramatic impact on the computational aspects of many courses, and this is particularly true in trigonometry. Lengthy computations involving trigonometric functions and logarithms can now be done in a matter of seconds on a scientific calculator. Nevertheless, there is merit in having students learn to use tables, and I have included a section on tables of trigonometric functions in Chapter 2 and a brief treatment of logarithms in Chapter 5. Instructors who prefer to omit either or both of these subjects and rely solely on calculators may do so without loss of continuity. Throughout the text I have included problems and examples which can best be done with the aid of a hand calculator. I have discussed the mechanics of how to operate a calculator only briefly because of the differences in the various models and the availability of instructions for their use.

There are more than 1,900 problems in the book. These include many applications to a wide variety of fields. Answers to odd-numbered problems are given in the back of the book. A solutions manual with completely worked solutions to all even-numbered problems is also available. At the end of each chapter a review exercise set is given.

I wish to thank Richard Jones, Mathematics Editor of Wadsworth, for his help and encouragement, Donna Cullison for the excellent job she did in typing the manuscript, and John Spellman for his fine contribution to this work.

I also wish to thank the following persons who reviewed the manuscript and who offered many constructive criticisms:

Richard A. Alo′
Lamar University

Bill D. Anderson
East Texas State University

Robert A. Chaffer
Central Michigan University

Pat C. Cook
Weatherford College

K. Joseph Davis
East Carolina University

Richard Hill
Michigan State University

Russell Floyd
Texas Wesleyan College

Allen Hansen
Riverside City College

Virgil Kowalik
Texas A&M University

Maurice L. Monahan
South Dakota State University

Donna Supra
Community College of Allegheny

Arnold R. Vobach
University of Houston

1 Introduction to the Trigonometric Functions

The trigonometric functions which we will introduce in this chapter play an essential role in calculus as well as other mathematics, science, and engineering courses. The following excerpt from a calculus textbook illustrates one of these functions evaluated at certain angles.*

> . . . Consider the equation
>
> $$r = \sin\theta$$
>
> By substituting values for θ at increments of $\pi/3$ (30°), and calculating r we can construct the following table:
>
> | θ (radians) | 0 | $\frac{\pi}{6}$ | $\frac{\pi}{3}$ | $\frac{\pi}{2}$ | $\frac{2\pi}{3}$ | $\frac{5\pi}{6}$ | π | $\frac{7\pi}{6}$ | $\frac{4\pi}{3}$ | $\frac{3\pi}{2}$ | $\frac{5\pi}{3}$ | $\frac{11\pi}{6}$ | 2π |
> |---|---|---|---|---|---|---|---|---|---|---|---|---|---|
> | $r = \sin\theta$ | 0 | $\frac{1}{2}$ | $\frac{\sqrt{3}}{2}$ | 1 | $\frac{\sqrt{3}}{2}$ | $\frac{1}{2}$ | 0 | $-\frac{1}{2}$ | $-\frac{\sqrt{3}}{2}$ | -1 | $-\frac{\sqrt{3}}{2}$ | $-\frac{1}{2}$ | 0 |
>
> The pairs of values listed in this table may be plotted . . . .
>
> The points . . . appear to lie on a circle. That this is indeed the case may be seen by expressing the equation in terms of x and y . . . .

Sin θ is one of the six trigonometric functions we will consider in this chapter. Also, we will learn the meaning of the *radian* measure of an angle and will see how to calculate all of the values of sin θ given in the table.

1.1 Introduction

The introduction of trigonometry as a separate area of study is generally believed to have been due to the Greek Hipparchus in the second century B.C., but his works are lost. The first exposition of the subject still in existence comes from the work

---

*Howard Anton, *Calculus with Analytic Geometry* (New York: John Wiley & Sons, 1980), pp. 713–714. Reprinted by permission.

of Ptolemy, the Greek astronomer, who lived in the second century A.D. Rudiments of the subject are much older, however, and can be traced to Babylon and ancient Egypt. The principal use of trigonometry up until the fifteenth century was in astronomy, because it afforded a means of making indirect measurements. The name *trigonometry* was first used by Pitiscus, a German, in 1595 as the title of his book. The name is a combination of three Greek words meaning "three-angle measurement," or triangle measurement.

As the name implies, trigonometry originally meant a study of triangles, or more generally of angles of triangles, and its primary utility lay in the fact that by means of the trigonometric functions inaccessible quantities could be measured. In more modern times, while trigonometry still is of value for this purpose, it is the analytical aspects of the subject that make it indispensable in the study of many physical phenomena. We will consider both the so-called "numerical trigonometry" and "analytical trigonometry." The present chapter lays the groundwork for the numerical aspect of the subject.

1.2 The Measurement of Angles

We all know what an angle is, yet it really is not very easy to define. When two **rays** emanate from the same point, we say they form an **angle**. We might define an angle,

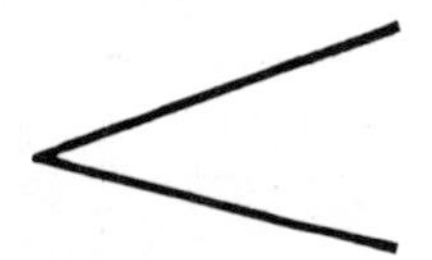

in fact, to be the set of points comprising two such rays. The common initial point of the two rays is called the **vertex** of the angle. There are some difficulties in this, since we want to distinguish between the interior angle and the exterior angle. We

usually get around this by the device shown of drawing a small arc to indicate the angle in question.

It is sometimes useful to think of an angle as having been formed by two rays that were initially coincident, with one ray then rotated while the other remains fixed. This enables us to speak of **directed angles**, and they are indicated by adding an appropriate arrow to the arc, as in Figure 1.1. The fixed ray is called the **initial side** of the angle, and the rotated ray the **terminal side**. An angle formed by a counterclockwise rotation is called **positive** and one formed by a clockwise rotation is called **negative**. We sometimes wish to indicate that the terminal side has been rotated more than one entire revolution, as in Figure 1.2. Every such angle is **coterminal**

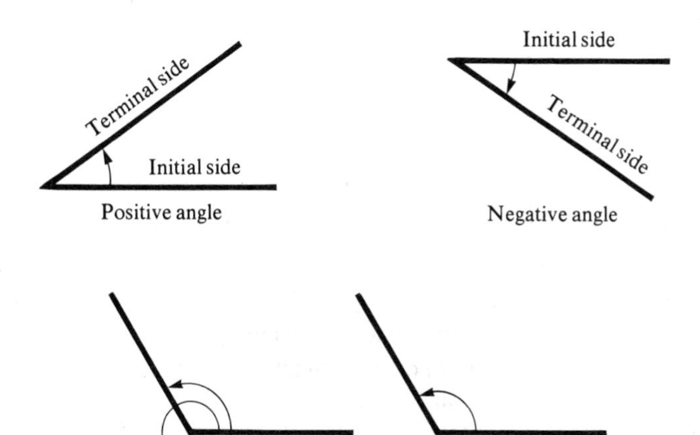

Figure 1.1

Figure 1.2

with an angle of less than one full revolution, that is, their terminal sides coincide if their vertices and initial sides are made to coincide.

The most familiar way of measuring the size of angles is by means of the degree.* If a circle is divided into 360 equal parts, then an angle formed by two rays from the center passing through two consecutive points on the circle is said to have a measure of one **degree**, indicated 1°. Degrees are divided into 60 equal parts called **minutes**, and minutes are further divided into **seconds**, with 60 seconds equalling one minute. One minute is designated $1'$ and one second $1''$. If we write, for example, $A = 25° \, 32' \, 15''$, we mean that A is an angle whose measure is 25 degrees, 32 minutes, and 15 seconds. In our work we will seldom use seconds. An alternate way of indicating fractions of a degree is to use decimal notation. For example, we might have $B = 47.3°$. This form is especially useful with scientific hand calculators, since most of these require angles to be written in decimal form. It is important to be able to go from one form to the other. For example, since $1° = 60'$, we see that $0.3° = (0.3)(60') = 18'$. So $47.3° = 47° \, 18'$. Also, to change minutes to decimal parts of degrees, divide the number of minutes by 60. This is illustrated by

$$35° \, 15' = \left(35 \, \frac{15}{60}\right)^{\circ} = 35.25°$$

While the measurement of angles by degrees has the weight of history and the advantage of familiarity on its side, there is another method of measurement that for many purposes is more useful, especially in the study of calculus and its applications. This is the **radian**. To define it we consider a given angle and construct any circle with its center at the vertex of the angle, as in Figure 1.3. Designate the angle

---

*No one knows for sure the origin of degree measurement, but it can be traced back to the ancient Sumerians (4000–2000 B.C.). One theory is that 360 was used because of their erroneous calculation of 360 days in a year. They and the Babylonians (2000–600 B.C.) used 60 as a base in their arithmetic.

by θ and let s be the length of the arc on the circle subtended by θ. If r is the length of the radius, then we define the **radian measure** of θ to be the ratio s/r. It might appear that this definition is dependent on which circle we use, but this is not the case. For if we take another circle of radius r' and subtended arc s', say, then the two circular sectors are similar (just as if they were triangles), and so $s/r = s'/r'$. Thus we get the same number for the radian measure of θ regardless of which circle is used. As with degree measure, we often use the same symbol for the name of an angle and for its measure. So we can write

$$\theta = \frac{s}{r} \tag{1.1}$$

As a practical matter, it is useful to write this in the equivalent form

$$s = r\theta \tag{1.2}$$

since we can measure angles and straight lines more easily than arcs of circles. Equation (1.2) says that the length of arc subtended on a circle of radius r by an angle of θ *radians* (not degrees) equals the product $r \cdot \theta$. This is an important result and should be remembered.

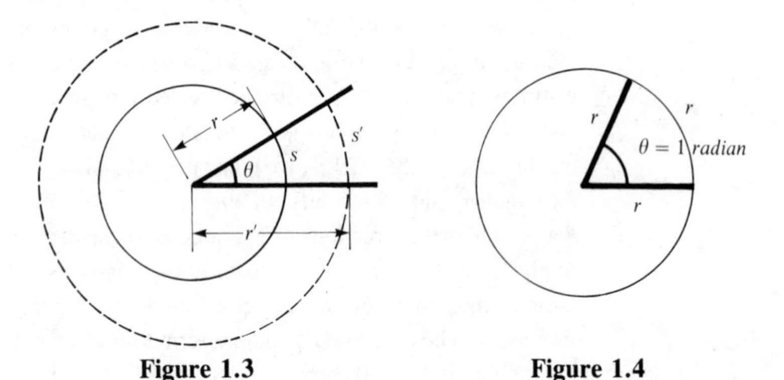

Figure 1.3 **Figure 1.4**

Another observation is that according to Equation (1.1) the radian measure of θ is the ratio of two lengths, and hence the radian is dimensionless. In analyzing the dimensional properties of an expression, then, if an angle in radians appears in the expression, the dimensionality is not affected.

According to Equation (1.1), an angle of one radian subtends, on any circle with its center at the vertex, an arc equal in length to the radius (Figure 1.4). This helps us to visualize the relative size of a radian. We know that the circumference of a circle is $2\pi r$. That is, we could mark off 2π arcs on the circle each equal in length to the radius (since $\pi \approx 3.1416$, we could mark off 6 full arcs equal to the radius and have a little more than a quarter of a radius left over). But for each arc on the circle of length r, there is an angle of one radian at the center. It follows then that there are 2π radians in one revolution. Since there are also 360° in one revolution, we have the relationships:

$$2\pi \text{ radians} = 360°$$

$$\pi \text{ radians} = 180°$$

$$1 \text{ radian} = \frac{180°}{\pi} \approx 57.3° \tag{1.3}$$

$$1 \text{ degree} = \frac{\pi}{180} \text{ radians}$$

Probably the easiest of these to remember is the second

$$\pi \text{ radians} = 180°$$

and the others can be obtained from it. In the future, unless the degree symbol is used in speaking of the measure of an angle, we will automatically mean radian measure. Thus, if we write $\theta = \pi$, $\pi/6$, or $\theta = 2$, we will mean the measure of θ is π radians, $\pi/6$ radians, or 2 radians, respectively.

The relationships (1.3) enable us to find the radian measure of an angle whose degree measure is known, and vice versa. The following example illustrates this.

EXAMPLE 1.1 Convert the degree measures in parts **a** and **b** to radians, and convert the radian measures in parts **c** and **d** to degrees.

a. 120° **b.** 540° **c.** $\frac{7\pi}{6}$ **d.** $\frac{2}{3}$

Solution **a.** Since $1° = \pi/180$ radians, we multiply 120 by $\pi/180$.

$$120 \cdot \frac{\pi}{180} = \frac{2\pi}{3} \text{ radians}$$

b.

$$540 \cdot \frac{\pi}{180} = 3\pi \text{ radians}$$

c. Since 1 radian $= 180/\pi$ degrees, we multiply $7\pi/6$ by $180°/\pi$.

$$\frac{7\pi}{6} \cdot \frac{180°}{\pi} = 210°$$

d.

$$\frac{2}{3} \cdot \frac{180°}{\pi} = \left(\frac{120}{\pi}\right)^{\circ} \approx 38.2°$$

Note. We will often leave the radian measure of an angle as a multiple of π, as in parts **a** and **b**, rather than give the decimal equivalent.

The next two examples illustrate uses of Equations (1.1) and (1.2).

EXAMPLE 1.2 A central angle in a circle of diameter 8 centimeters subtends an arc of 3 centimeters. Find the degree measure of the angle.

Solution Since the diameter is 8, the radius is 4. Since $s = 3$, we have by (1.1)

$$\theta = \frac{s}{r} = \frac{3}{4} \text{ radians}$$

To convert to degrees, we multiply by $180°/\pi$.

$$\theta = \frac{3}{4} \cdot \frac{180°}{\pi} = \left(\frac{135}{\pi}\right)^{\circ} \approx 42.97°$$

EXAMPLE 1.3 Find the arc length on a circle of radius 7 feet subtended by a central angle of 60°.

Solution We use $s = r\theta$, but first we must find θ in radians.

$$\theta = 60° \cdot \frac{\pi}{180°} = \frac{\pi}{3}$$

So

$$s = 7 \cdot \frac{\pi}{3} = \frac{7\pi}{3} \text{ feet} \approx 7.33 \text{ feet}$$

Suppose now that an object is traveling in a circular path at a constant rate. Its **linear velocity** v is the distance along the circular arc covered per unit of time. Thus, if in time t it covers a distance of s units of arc, then

$$v = \frac{s}{t}$$

There is also an **angular velocity**, which we designate by ω, associated with the moving object. This is the number of radians turned through per unit of time. Thus, if in time t a central angle of θ radians has been turned through, then

$$\omega = \frac{\theta}{t}$$

Since $s = r\theta$, we have

$$\frac{s}{t} = r \cdot \frac{\theta}{t}$$

or

$$v = r\omega \tag{1.4}$$

EXAMPLE 1.4 A point on a flywheel of radius 2 feet is turning at an angular velocity of 300 radians per second. What is the linear velocity of the point?

Solution

$$v = r\omega = 2 \cdot 300 = 600 \text{ feet per second}$$

EXAMPLE 1.5 A wheel of diameter 4 feet is rotating at the rate of 1,200 revolutions per minute. Find the linear velocity of a point on the rim in feet per second.

Solution In order to find the angular velocity ω in radians per second, we note that each revolution contains 2π radians, and since there are 1,200 revolutions each minute, which is equivalent to 20 revolutions each second, we have

$$\omega = 2\pi(20) = 40\pi \text{ radians per second}$$

Thus,

$$v = r\omega = 2(40\pi) = 80\pi \text{ feet per second}$$
$$\approx 251.33 \text{ feet per second}$$

EXERCISE SET 1.2

A In Problems 1–10 find the radian measure of the angles whose degree measures are given. Leave answers in terms of π.

1. **a.** $30°$ **b.** $45°$
2. **a.** $60°$ **b.** $90°$
3. **a.** $120°$ **b.** $135°$
4. **a.** $150°$ **b.** $225°$
5. **a.** $240°$ **b.** $270°$
6. **a.** $300°$ **b.** $315°$
7. **a.** $330°$ **b.** $15°$
8. **a.** $-50°$ **b.** $72°$
9. **a.** $600°$ **b.** $-144°$
10. **a.** $36°$ **b.** $-80°$

In Problems 11–20 give the degree measure of the angles having the given radian measures.

11. **a.** $\frac{3\pi}{4}$ **b.** $\frac{5\pi}{3}$
12. **a.** 4π **b.** $\frac{5\pi}{2}$
13. **a.** $\frac{5\pi}{6}$ **b.** $\frac{-4\pi}{3}$
14. **a.** $\frac{-7\pi}{2}$ **b.** $\frac{3\pi}{10}$
15. **a.** $\frac{5\pi}{9}$ **b.** 2
16. **a.** $\frac{3\pi}{2}$ **b.** $\frac{5\pi}{4}$
17. **a.** 3 **b.** $\frac{\pi}{12}$
18. **a.** $\frac{7\pi}{12}$ **b.** $\frac{\pi}{5}$

19. **a.** -4 **b.** $\frac{11\pi}{6}$ **20.** **a.** $\frac{9\pi}{4}$ **b.** $\frac{13\pi}{3}$

In Problems 21–23 use a calculator to convert to decimal degrees. Express answer to nearest hundredth of a degree.

21. **a.** $32° 51'$ **b.** $102° 35'$

22. **a.** $18° 03'$ **b.** $321° 48'$

23. **a.** $27° 13' 24''$ **b.** $-56° 19' 12''$

In Problems 24–27 use a calculator to convert to degrees and minutes, to the nearest minute.

24. **a.** $86.32°$ **b.** $29.78°$

25. **a.** $-13.05°$ **b.** $153.41°$

26. **a.** $\frac{\pi}{13}$ radians **b.** 2.563 radians

27. **a.** 3 radians **b.** 7.092 radians

In Problems 28–30 convert to radians, to the nearest thousandth of a radian.

28. **a.** $25.426°$ **b.** $115.750°$

29. **a.** $13° 22' 31''$ **b.** $290° 12' 24''$

30. **a.** $30°$ **b.** $-46.023°$

In Problems 31–34, θ is a central angle on a circle of radius r, and s is the arc on the circle subtended by θ. Find the value (s, r, or θ) which is not given.

31. **a.** $r = 3,\ \theta = 2$ **b.** $s = 4,\ \theta = 3$

32. **a.** $r = 5,\ s = 12$ **b.** $r = 2,\ \theta = 30°$

33. **a.** $s = 8,\ \theta = 45°$ **b.** $s = 4,\ r = 6$

34. **a.** $\theta = \frac{\pi}{3},\ s = 6$ **b.** $\theta = \frac{3\pi}{4},\ r = 4$

35. Find the arc length on the equator (in miles) subtended by an angle of 1° at the center of the earth. Assume the radius of the earth is 3,960 miles.

36. What is the length of the arc swept out by the tip of the minute hand of a clock during the time interval from 6:00 to 6:20 if the minute hand is 4 inches long?

37. The diameter of the steering wheel of a car is 15 inches, and it has 3 equally spaced spokes. Find the distance on the steering wheel between consecutive spokes.

38. A wheel is revolving at an angular velocity of $5\pi/3$ radians per second. Find the number of revolutions per minute (rpm) through which the wheel is turning.

39. A flywheel is turning at 1,800 rpm. Through how many radians per second is it turning?

B **40.** The earth completes one revolution about its axis in 24 hours. Find the velocity in miles per hour of a point on the equator (take $r = 3{,}960$ miles). What is the velocity in feet per second?

41. A train is traveling on a circular curve of $\frac{1}{2}$ mile radius at the rate of 30 miles per hour. Through what angle (in radians) will the train turn in 45 seconds? What is the angle in degrees?

42. A flywheel 4 feet in diameter is revolving at the rate of 50 rpm. Find the speed of a point on the rim in feet per second.
43. Assume the earth moves around the sun in a circular orbit with radius 93,000,000 miles. A complete revolution takes approximately 365 days. Find the approximate speed of the earth in its orbit in miles per hour.
44. An automobile tire is 28 inches in diameter. If the car is traveling 30 miles per hour, through how many revolutions per minute is the wheel turning?

The following two problems are taken from an engineering textbook.*

45. Two points B and C lie on a radial line of a rotating disk. The points are 2 in. apart. $v_B = 700$ fpm [feet per minute] and $v_C = 880$ fpm. Find the radius of rotation for each of these points.
46. The tire of an automobile has an outside diameter of 27 in. If the rpm of the wheel is 700, determine the speed of the automobile (a) in miles per hour, (b) in feet per second, and (c) the angular speed of the wheel in radians per second.

1.3 Some Special Angles

An angle of 90° is called a **right angle**, and an angle of 180° is called a **straight angle**. An **acute angle** is an angle between 0° and 90°, and an **obtuse angle** is one between 90° and 180°. A triangle in which one angle is a right angle is called a **right triangle**, and the side opposite the 90° angle is called the **hypotenuse**. The other two sides are often referred to as the **legs** of the triangle. A triangle which is not a right triangle is said to be an **oblique** triangle. The following fundamental result about right triangles is proved in plane geometry (a proof is outlined in Problem 34 at the end of this section, and an alternate proof is suggested in Problem 38 of Section 1.6).

THE PYTHAGOREAN THEOREM

The sum of the squares of the lengths of the legs of a right triangle equals the square of the length of the hypotenuse.

In Figure 1.5 we have labeled a right triangle with angles named by capital letters and sides opposite the angles named with the corresponding lowercase letters. This

Figure 1.5

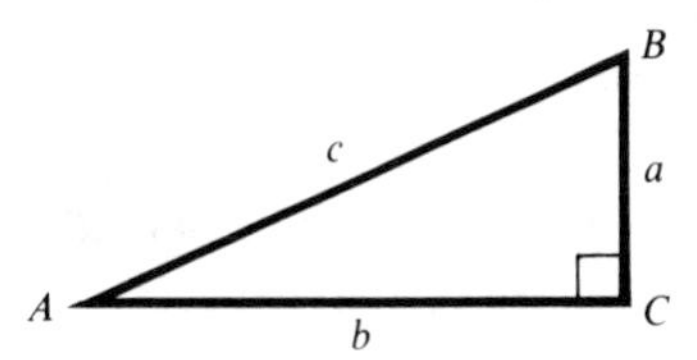

*George H. Martin, *Kinematics and Dynamics of Machines*, rev. printing (New York: McGraw-Hill Book Company, 1969), p. 43. Reprinted by permission.

is a standard notation, and we will understand all right triangles to be designated this way unless specific instructions are given otherwise. The right angle is at C. We will follow the common practice of using these capital letters both to name the angles and to designate their measure. Similarly, we will use the lowercase letters as names of the sides as well as to designate their lengths. For example, we might say $A = 30°$, rather than "the degree measure of angle A is 30," and we might write $a = 2$ to mean "the length of side a is 2." This dual interpretation should not cause any difficulty. With this notation the Pythagorean theorem states that

$$a^2 + b^2 = c^2$$

The converse of this theorem is also true, and we call for a proof of this in Problem 35 at the end of this section. That is, if for a triangle with sides a, b, and c it is true that $a^2 + b^2 = c^2$, then the triangle is a right triangle, with the 90° angle opposite side c.

Another fundamental property of triangles proved in plane geometry is that the sum of the three angles in *any* triangle (not just a right triangle) is 180°. In particular, if the triangle is a right triangle, so that one of the angles is 90°, the sum of the two acute angles is 90°. Such angles are said to be **complementary**, and each is said to be the **complement** of the other. In Figure 1.5, angles A and B are complementary.

There are two right triangles that we single out for special attention. The first of these is referred to as a **30°–60° right triangle** because the acute angles are 30° and 60°, respectively. Let ABC be a 30°–60° right triangle with $A = 30°$, and $B = 60°$. If we flip the triangle over with side AC as an axis, as in Figure 1.6, we obtain a triangle ABB' in which all angles are 60°. Thus, this triangle is **equiangular** (all

Figure 1.6

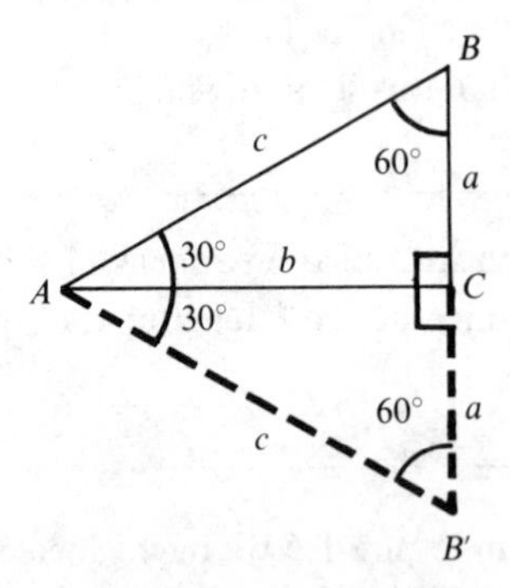

angles equal), and hence also **equilateral** (all sides equal). So $2a = c$, or $a = c/2$. This is the result we wish to emphasize.

> In a **30°–60° right triangle**, the side opposite the 30° angle is one-half the hypotenuse.

As a special case, consider a right triangle ABC with $A = 30°$, $B = 60°$, and $c = 2$. Then by what we have just shown, we know that $a = 1$. We use the Pythagorean theorem to find b.

$$b^2 = c^2 - a^2 = 4 - 1 = 3$$

$$b = \sqrt{3}$$

Figure 1.7 illustrates the result, and it is useful to commit this to memory. All other 30°–60° right triangles are similar to this one, and so the sides are proportional to the corresponding sides of this triangle.

The second special type of right triangle we want to consider is the **isosceles right triangle**, that is, one in which the two legs are equal. It follows that the two acute angles are also equal, and hence each is 45°. If we let each leg be 1 unit, we find by the Pythagorean theorem that the hypotenuse is $\sqrt{1^2 + 1^2} = \sqrt{2}$. The situation is shown in Figure 1.8, which again should be committed to memory.

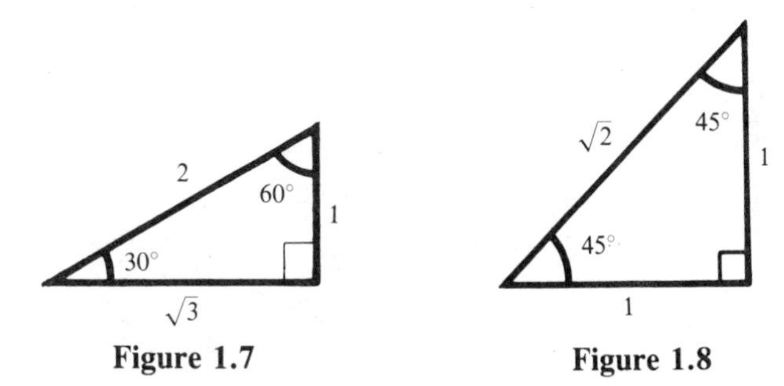

Figure 1.7 **Figure 1.8**

Example 1.6 illustrates how the concepts discussed in this section can be used to find unknown sides of certain right triangles.

EXAMPLE 1.6 In a right triangle ABC, with right angle at C, find the unknown sides from the given information.

a. $a = 3$, $c = 5$ **b.** $A = 30°$, $c = 10$

c. $B = 45°$, $a = 8$ **d.** $A = 60°$, $b = 12$

Solution **a.** In Figure 1.9, by the Pythagorean theorem

$$b^2 = c^2 - a^2 = 25 - 9 = 16$$

$$b = 4$$

Note. This is known as a **3–4–5 right triangle**, and it is useful to remember that a triangle with sides 3, 4, and 5, or with sides respectively proportional to 3, 4, and 5, is a right triangle.

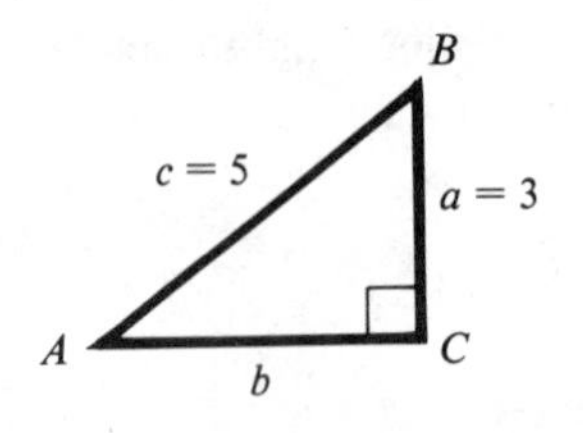

Figure 1.9

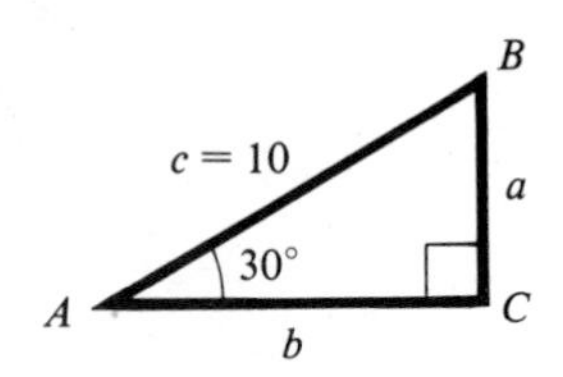

Figure 1.10

b. Since $c = 10$, we know that $a = 5$. Using proportionality and Figure 1.7, we see that in Figure 1.10 $b = 5\sqrt{3}$. Alternatively, we could use the Pythagorean theorem to find b:

$$b = \sqrt{c^2 - a^2} = \sqrt{100 - 25} = \sqrt{75} = 5\sqrt{3}$$

c. Since $B = 45°$, the triangle is isosceles. So $b = 8$, and from Figure 1.8 and proportionality we determine in Figure 1.11 that $c = 8\sqrt{2}$. Again, we could also find c by the Pythagorean theorem:

$$c = \sqrt{a^2 + b^2} = \sqrt{64 + 64} = \sqrt{(64) \cdot 2} = 8\sqrt{2}$$

d. In Figure 1.12, angle $B = 90° - A = 30°$, and since the side opposite the 30° angle is half the hypotenuse, it follows that $c = 24$. Comparison with the basic 30°–60° right triangle of Figure 1.7, or use of the Pythagorean theorem, gives $a = 12\sqrt{3}$.

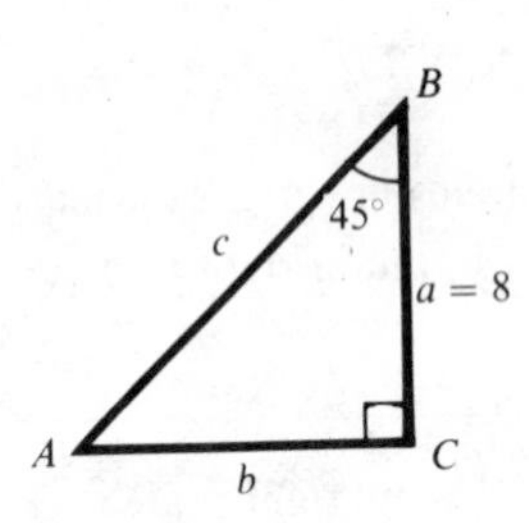

Figure 1.11

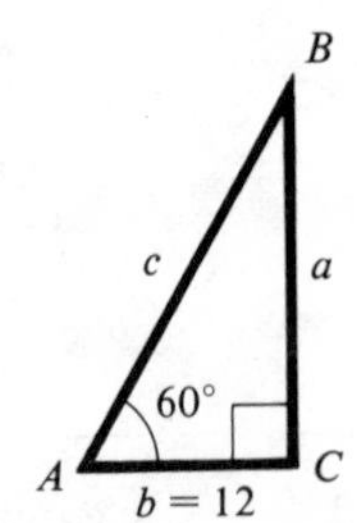

Figure 1.12

EXERCISE SET 1.3

A In Problems 1–20 ABC is a right triangle with right angle at C. In each problem find the unknown sides.

1. $A = 30°$, $c = 5$

2. $B = 60°$, $a = 3$

3. $A = 60°$, $b = 10$

4. $A = 30°$, $b = 4\sqrt{3}$

5. $A = 30°$, $b = 6$

6. $B = 30°$, $a = 12$

7. $B = 45°$, $c = 4\sqrt{2}$

8. $A = 45°$, $c = 4$

9. $A = 60°$, $c = 5$

10. $B = 60°$, $b = 3$

11. $B = 45°$, $a = 3$

12. $A = 60°$, $a = 8$

13. $a = 2$, $b = 3$

14. $a = 5$, $c = 7$

15. $a = 7$, $c = 25$

16. $b = 12$, $c = 13$

17. $a = 14$, $c = 50$

18. $a = 3$, $b = \sqrt{3}$
Also find angles A and B.

19. $a = 1$, $b = 2\sqrt{2}$

20. $a = 8$, $c = 17$

In Problems 21–26 two angles of a triangle are given. Find the third angle.

21. $23° 17'$, $75° 34'$

22. $43.21°$, $68.35°$

23. $52.76°$, $103.42°$

24. $92° 47'$, $21° 54'$

25. $\frac{\pi}{4}$ radians, $\frac{\pi}{3}$ radians

26. 1.25 radians, 0.78 radians

27. If a, b, and c are positive integers for which $a^2 + b^2 = c^2$, the numbers are said to form a **Pythagorean triple**, and we know that a triangle having sides of lengths a, b, and c, respectively, is a right triangle. Show that the following are Pythagorean triples.

a. 6, 8, 10
b. 5, 12, 13
c. 8, 15, 17
d. 7, 24, 25
e. 9, 40, 41

B **28.** It is proved in more advanced mathematics that all Pythagorean triples (see Problem 27) are of the form $a = m^2 - n^2$, $b = 2mn$, and $c = m^2 + n^2$, where m and n are positive integers, with $m > n$.

a. Verify that all such numbers do form a Pythagorean triple.
b. Find the values of m and n which produce each triple in Problem 27.
c. Determine two additional Pythagorean triples not proportional to any of those in Problem 27.

29. In the accompanying figure the 45° right triangle is constructed so that one of its legs is the hypotenuse of the 30°–60° right triangle. Find

a. the hypotenuse of the 45° triangle
b. the combined areas of the two triangles.

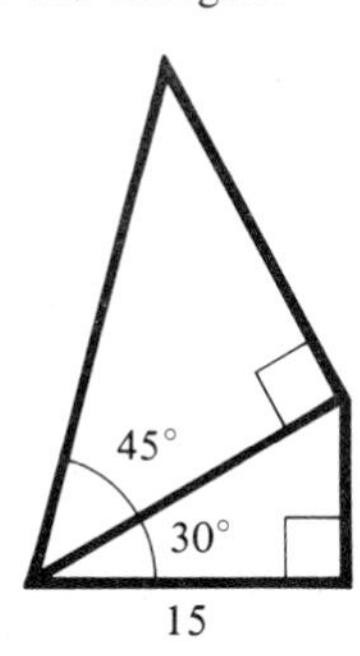

30. Find a formula for the area of an equilateral triangle with side of length a.

31. Prove that in a triangle ABC with $A = 30°$ and $C = 90°$, the median drawn from C has length a. (A **median** is the line segment from a vertex to the midpoint of the opposite side.)

32. Find a formula for the area of the inscribed circle in an equilateral triangle with side of length a.

Hint. The angle bisectors, which in an equilateral triangle coincide with the medians and the altitudes, intersect at the center of the inscribed circle.

33. Find the area of the circumscribed circle to an equilateral triangle with side of length a. (See hint to Problem 32. The inscribed circle and circumscribed circle are concentric.)

34. By the following steps prove the Pythagorean theorem.

a. Consider a right triangle with legs of lengths a and b and hypotenuse of length c. Construct a square of side $a + b$ as shown, and draw four replicas (I, II, III, and IV) of the given triangle inside the square as shown.

b. Show that the quadrilateral V is a square. (Work with angles.)

c. Write the equation that puts into mathematical form the fact that the sum of the areas I, II, III, IV, and V equals the area of the large square.

d. By simplifying the equation in **c**, draw the desired conclusion.

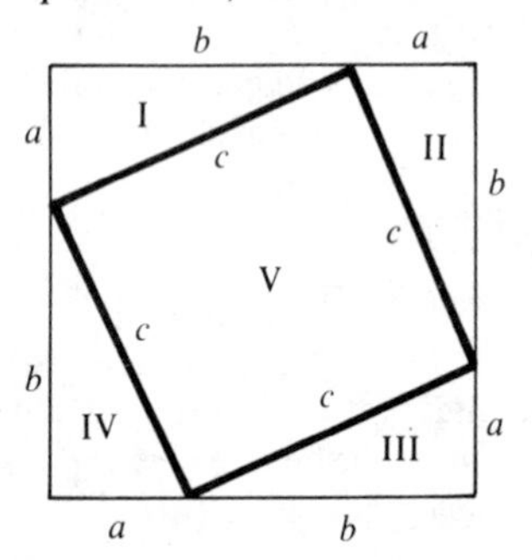

35. Follow the steps given to prove the converse of the Pythagorean theorem.

a. Let ABC be a triangle in which $c^2 = a^2 + b^2$.

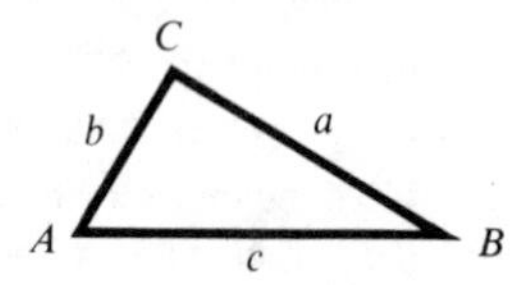

b. Construct a right triangle $A'B'C'$ having legs a and b. Let the hypotenuse be of length c'. Use the Pythagorean theorem to express c' in terms of a and b.

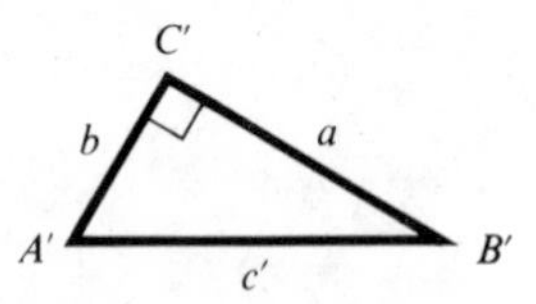

c. How can you conclude that $c' = c$?

d. What does this say about triangle ABC compared with triangle $A'B'C'$?

e. What do you conclude about angle C?

1.4 Definitions of Trigonometric Functions Using Right Triangles

In this section we introduce the concept of the trigonometric functions for acute angles. In Section 1.5 the definition will be extended to arbitrary angles, and in Chapter 3 the trigonometric functions will be defined for arbitrary real numbers. We first review the meaning of a function.

In mathematics, a **function** is a rule which assigns to each element of a given set one and only one element of another set. The first set is called the **domain** of the function, and the elements of the second set which correspond under the rule to the elements of the domain comprise the **range** of the function. We often designate a function by a letter, such as f, and if x is an element of the domain, the corresponding element of the range is written $f(x)$ (read "f of x"). For example, we might have $f(x) = \sqrt{25 - x^2}$. Then $f(0) = 5$, $f(4) = 3$, $f(-3) = 4$, and so on. The domain in this case consists of all real numbers x such that $-5 \leq x \leq 5$, since the function yields real values for these, and only these, values of x. In general, when the domain is not explicitly stated, we understand it to be the maximum set for which the values of the function are real.

For our present purposes we are going to define six functions, called the **trigonometric functions**, and these are given special names as indicated in Definition 1.1 below. The domain of each consists of the set of all acute angles, and the range in each case is a subset of the set $\mathbb{R}$ of real numbers.

Let θ designate any acute angle, and consider a right triangle having θ as one of its angles. In the definition that follows, we refer to the sides of the triangle in relation to θ, namely, the side **opposite** to θ, the side **adjacent** to θ, and the **hypotenuse** (see Figure 1.13). The definition assigns names to the six ratios possible, using the lengths of the three sides. When we write "opposite," "adjacent," and "hypotenuse," we will understand that we mean the *lengths* of these sides.

Figure 1.13

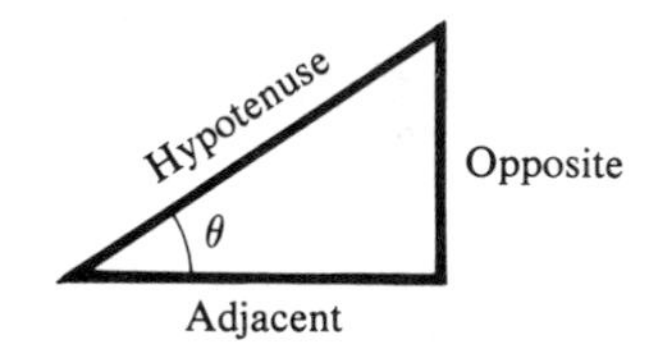

DEFINITION 1.1

The Trigonometric Functions

Name	*Abbreviation*	*Value*
sine of θ	$\sin\theta$	$\dfrac{\text{opposite}}{\text{hypotenuse}}$
cosine of θ	$\cos\theta$	$\dfrac{\text{adjacent}}{\text{hypotenuse}}$

Name	*Abbreviation*	*Value*
tangent of θ	$\tan \theta$	$\dfrac{\text{opposite}}{\text{adjacent}}$
cotangent of θ	$\cot \theta$	$\dfrac{\text{adjacent}}{\text{opposite}}$
secant of θ	$\sec \theta$	$\dfrac{\text{hypotenuse}}{\text{adjacent}}$
cosecant of θ	$\csc \theta$	$\dfrac{\text{hypotenuse}}{\text{opposite}}$

Although in this definition we refer to a right triangle, it should be emphasized that, except for the requirement that one of its acute angles be θ, it is not important which right triangle is used. The values of the trigonometric functions will be the same regardless of which such triangle is used. To see this, consider any two right triangles having θ as an acute angle, as in Figure 1.14. Since $A = A' = \theta$, it follows that $B = B'$, and so the triangles are similar. Thus, corresponding sides are proportional. In particular,

$$\frac{a}{c} = \frac{a'}{c'}, \qquad \frac{b}{c} = \frac{b'}{c'}, \qquad \frac{a}{b} = \frac{a'}{b'}$$

and so on. Therefore, whether we use triangle ABC or $A'B'C'$ to compute the values of the functions of θ, we get the same result. It follows that the values of the trigonometric functions of θ depend only on the size of θ and not on which triangle is used.

Figure 1.14

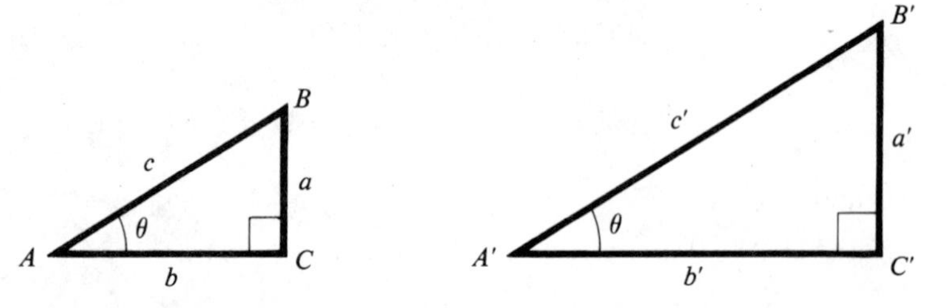

If we designate the angles and sides of a right triangle ABC in the usual way, as in Figure 1.15, then by Definition 1.1 we have

$$\sin A = \frac{a}{c} \qquad \csc A = \frac{c}{a}$$

$$\cos A = \frac{b}{c} \qquad \sec A = \frac{c}{b}$$

$$\tan A = \frac{a}{b} \qquad \cot A = \frac{b}{a}$$

The side opposite angle B is b, and the adjacent side to B is a, so the functions of angle B are:

$$\sin B = \frac{b}{c} \qquad \csc B = \frac{c}{b}$$

$$\cos B = \frac{a}{c} \qquad \sec B = \frac{c}{a}$$

$$\tan B = \frac{b}{a} \qquad \cot B = \frac{a}{b}$$

Figure 1.15

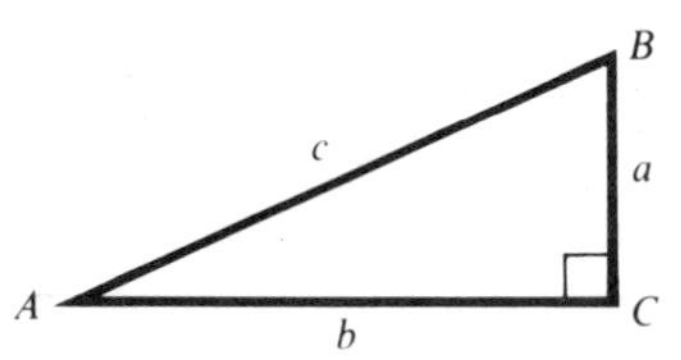

We observe that $\sin A = \cos B$ and $\cos B = \sin A$, and similar relationships hold for the other functions. The sine and cosine are said to be **cofunctions**, each being the cofunction of the other. Similarly, the tangent and the cotangent are cofunctions, as are the secant and cosecant. Cofunctions are easily remembered since their names differ only in the prefix *co* for one of them. Since A and B are complementary angles, we conclude the following:

> Any trigonometric function of an acute angle equals the corresponding cofunction of its complement.

It is important that the trigonometric functions of acute angles be learned in terms of the words *opposite*, *adjacent*, and *hypotenuse*, rather than any particular set of letters, for while we usually use the standard notation for the angles and sides of a right triangle, this will not always be the case. For example, for the angle R in the triangle of Figure 1.16 the values of the six functions are:

$$\sin R = \frac{n}{s} \qquad \csc R = \frac{s}{n}$$

$$\cos R = \frac{m}{s} \qquad \sec R = \frac{s}{m}$$

$$\tan R = \frac{n}{m} \qquad \cot R = \frac{m}{n}$$

Figure 1.16

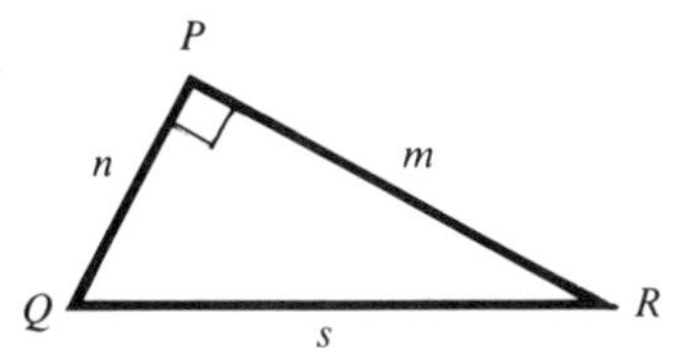

Extensive tables of the values of the six trigonometric functions of angles between 0° and 90° have been calculated. Two such tables (Table I and Table II) are given in the back of this book, one using degree measure and the other radian measure. As you might suspect, the values in the tables were not calculated by drawing right triangles and measuring sides. Much more sophisticated and accurate techniques, not relying on geometry, were used. Ways of doing this are studied in calculus.

The modern scientific hand calculator provides a more efficient and accurate way of obtaining values of the trigonometric functions. The widespread availability of these calculators at moderate prices has made tables of functions less important than they once were. However, even if you have a calculator, you should also learn how to use tables, so that you will not be dependent on just one means of evaluating trigonometric functions. In Chapter 2 we will be dealing with problems requiring either a calculator or tables, and we will discuss their use further at that time. For the present we limit discussion to problems in which the values of the functions can be obtained from our knowledge of special triangles.

EXAMPLE 1.7 Find the six trigonometric functions of 30°.

Solution We first draw a 30°–60° right triangle with sides as shown in Figure 1.17. (Remember, we can use *any* right triangle having a 30° angle.) Then by Definition 1.1, we have,

$$\sin 30° = \frac{1}{2} \qquad \csc 30° = 2$$

$$\cos 30° = \frac{\sqrt{3}}{2} \qquad \sec 30° = \frac{2}{\sqrt{3}}$$

$$\tan 30° = \frac{1}{\sqrt{3}} \qquad \cot 30° = \sqrt{3}$$

Figure 1.17

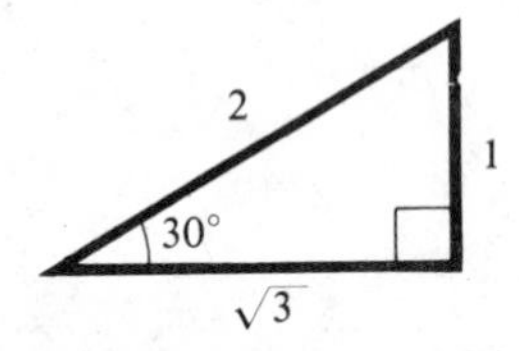

We could get decimal approximations to these, but for many purposes it is best to leave the answers in the form given. We say these answers are in **exact form**.

EXAMPLE 1.8 In the right triangle ABC shown in Figure 1.18, $a = 6$ and $b = 8$. Find the six trigonometric functions of angle B.

Solution We could use the Pythagorean theorem to find the hypotenuse c; however, we can also observe that the triangle is similar to a 3–4–5 right triangle, so that c must be 10.

Figure 1.18

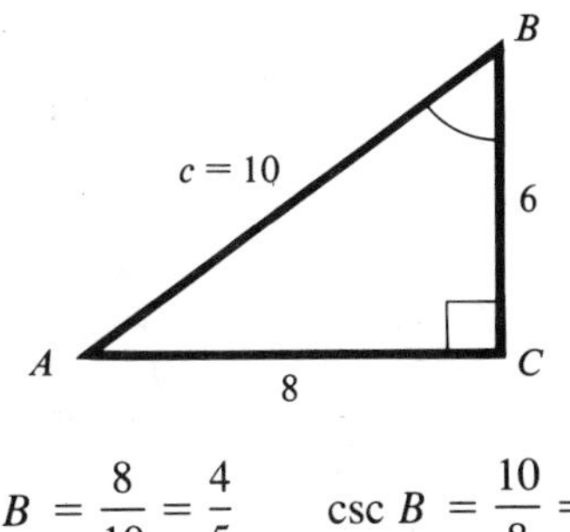

$$\sin B = \frac{8}{10} = \frac{4}{5} \qquad \csc B = \frac{10}{8} = \frac{5}{4}$$

$$\cos B = \frac{6}{10} = \frac{3}{5} \qquad \sec B = \frac{10}{6} = \frac{5}{3}$$

$$\tan B = \frac{8}{6} = \frac{4}{3} \qquad \cot B = \frac{6}{8} = \frac{3}{4}$$

Remark on notation. When powers of trigonometric functions are employed (as in Problems 18, 19, 25–27 of the next exercise set), it is customary to use notations such as $\sin^2 \theta$ to mean $(\sin \theta)^2$, and in general $\sin^n \theta = (\sin \theta)^n$.* Similar notations are used for powers of the other trigonometric functions.

EXERCISE SET 1.4

Leave all answers in exact form unless otherwise specified.

A

1. Find all six trigonometric functions of

a. 60° **b.** 45°

Problems 2–7 refer to a right triangle ABC with standard notation. Find all six trigonometric functions of the specified angle.

2. $a = 5$, $c = 13$; A **3.** $b = 8$, $c = 17$; B

4. $a = 7$, $b = 24$; B **5.** $a = 1$, $b = 2$; A

6. $b = 2$, $c = 3$; A **7.** $a = 8$, $c = 12$; B

In Problems 8–11 write the six trigonometric functions of the specified angle for the right triangle given.

8. Angle M

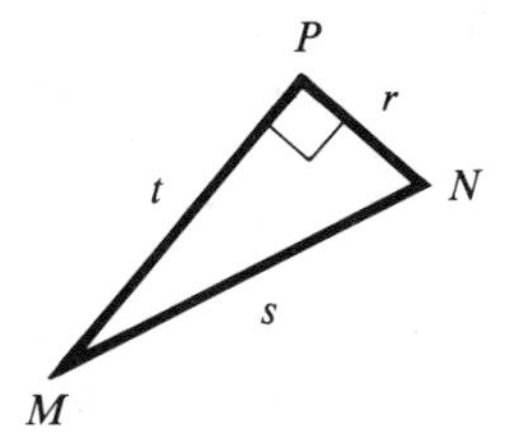

9. Angle U

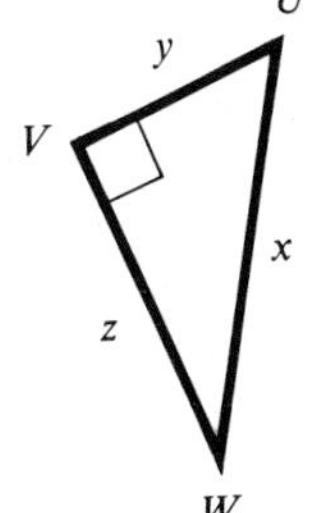

---

*The single exception is when $n = -1$. This exception is discussed in Chapter 4.

10. Angle D

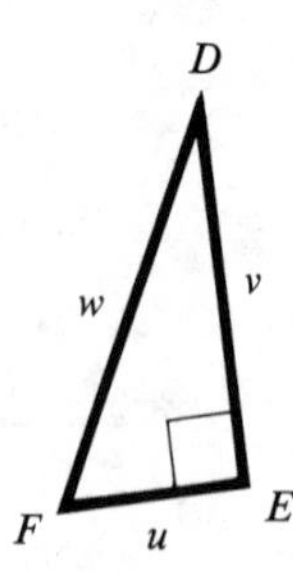

11. Angle M

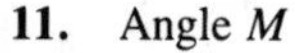

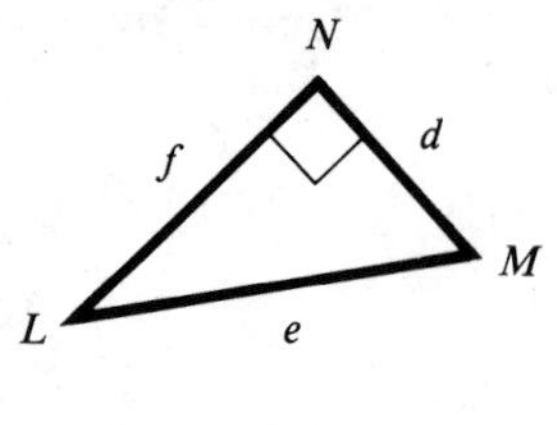

Evaluate the expressions in Problems 12–21, leaving answers in exact form.

12. $3 \sin 30° - 5 \cos 60°$

13. $\cos \dfrac{\pi}{3} + 2 \sin \dfrac{\pi}{4}$

14. $\csc \dfrac{\pi}{6} - \cot \dfrac{\pi}{4}$

15. $3 \tan 60° - 2 \cos 30°$

16. $\dfrac{1 - \cos \dfrac{\pi}{3}}{2}$

17. $\sec \dfrac{\pi}{6} \left(\sin \dfrac{\pi}{3} + \tan \dfrac{\pi}{3} \right)$

18. $\cos^2 \dfrac{\pi}{6} - \sin^2 \dfrac{\pi}{6}$

19. $\sec^2 \dfrac{\pi}{4} - \tan^2 \dfrac{\pi}{4}$

20. $\sin 45° \cos 30° - \cos 45° \sin 30°$

21. $\cos 60° \cos 45° + \sin 60° \sin 45°$

22. From tables the following approximate values of the trigonometric functions of 32° are found. Use these to write the six trigonometric functions of 58°.

$$\begin{array}{ll} \sin 32° = 0.5299 & \csc 32° = 1.887 \\ \cos 32° = 0.8480 & \sec 32° = 1.179 \\ \tan 32° = 0.6249 & \cot 32° = 1.600 \end{array}$$

23. Verify the following relationships among the trigonometric functions of any acute angle θ.

$$\csc \theta = \frac{1}{\sin \theta}, \qquad \sec \theta = \frac{1}{\cos \theta}, \qquad \cot \theta = \frac{1}{\tan \theta},$$
$$\tan \theta = \frac{\sin \theta}{\cos \theta}, \qquad \cot \theta = \frac{\cos \theta}{\sin \theta}$$

24. Use the relationships given in Problem 23 to find the other four trigonometric functions of an acute angle θ for which $\sin \theta = 2/\sqrt{13}$ and $\cos \theta = 3/\sqrt{13}$.

In Problems 25–30 verify each formula for each of the angles $\theta = \pi/6$, $\theta = \pi/4$, and $\theta = \pi/3$.

25. $\sin^2 \theta + \cos^2 \theta = 1$

26. $1 + \tan^2 \theta = \sec^2 \theta$

27. $1 + \cot^2 \theta = \csc^2 \theta$

28. $\dfrac{1 + \sin \theta}{\tan \theta} = \cos \theta + \cot \theta$

29. $\dfrac{\sin \theta}{1 - \cos \theta} = \csc \theta + \cot \theta$

30. $\sin \theta \tan \theta = \sec \theta - \cos \theta$

B In Problems 31–39 draw a right triangle having θ as one of its acute angles. Find the remaining five trigonometric functions of θ. Leave answers in exact form.

31. $\cos\theta = \dfrac{1}{3}$

32. $\sin\theta = \dfrac{2}{7}$

33. $\tan\theta = 2$

34. $\cot\theta = \dfrac{1}{2}$

35. $\sec\theta = \dfrac{3}{2}$

36. $\csc\theta = \dfrac{3\sqrt{2}}{4}$

37. $\sin\theta = x$, where $0 < x < 1$

38. $\sec\theta = x$, where $x > 1$

39. $\tan\theta = x$, where $x > 0$

40. It can be shown that

$$\cos 15° = \frac{\sqrt{6} + \sqrt{2}}{4}$$

Find the other five trigonometric functions of 15°. Simplify your results, but leave answers in exact form. Also give the six trigonometric functions of 75°.

1.5 Trigonometric Functions of General Angles

We will show in this section how to extend the definitions of the trigonometric functions to angles of any size. In order to do this we will need to make use of a rectangular (or Cartesian) coordinate system, and we begin by reviewing this concept.

Consider two mutually perpendicular lines, one horizontal and one vertical. Their point of intersection is called the **origin**, labeled 0. We call the horizontal line the ***x* axis** and the vertical line the ***y* axis**. A unit of distance is selected, and we use this to mark off integer points, both positive and negative, on each axis. Positive numbers are to the right, negative numbers to the left on the x axis, and positive numbers are upward, negative numbers downward on the y axis. If P is a point in the plane, then the horizontal directed distance from the y axis to P is called the ***x* coordinate**, or **abscissa**, of P, and the vertical directed distance from the x axis to P is called the ***y* coordinate**, or **ordinate** of P. The x coordinate and y coordinate together are called the **coordinates** of P, and we write them in the form (x, y). This is called an **ordered pair** of numbers since the order is important. For example, $(3, -4)$ is a different ordered pair from $(-4, 3)$. In Figure 1.19 we have shown points with coordinates $(3, 4)$, $(-3, 2)$, $(-4, -2)$, and $(2, -3)$. In the manner described we see that every point in the plane determines a unique ordered pair of coordinates.

Conversely, if we are given any ordered pair (x, y) of numbers, we can locate the point in the plane having these coordinates. So there is a one-to-one correspondence between points in the plane and ordered pairs of numbers. Because of this

Figure 1.19

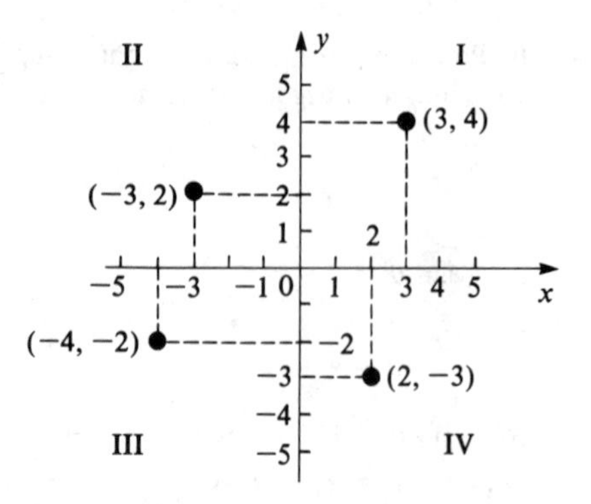

identification we frequently say "the point (x, y)" rather than the more precise "the point with coordinates (x, y)."

The x and y axes divide the plane into four regions, called **quadrants**, which are numbered I, II, III, and IV, in a counterclockwise direction as shown in Figure 1.19. The signs of the coordinates in the various quadrants are seen to be as follows:

Quadrant	Signs of coordinates
I	x and y both positive
II	x negative, y positive
III	x and y both negative
IV	x positive, y negative

The identification of ordered pairs of numbers with points in the plane seems to us now like a simple and natural concept, but it was a step forward in mathematics of great importance when introduced by the French philosopher and mathematician René Descartes (hence the name Cartesian coordinate system) in 1637. For the first time a relationship between equations, which are algebraic in nature, and graphs, which are geometric in nature, could be established. This provided a necessary foundation for the invention of calculus which followed soon afterward.

Before proceeding to our main purpose of this section, we derive a formula for the distance between two points in the plane. We will have need for this in Chapter 3, and it has many uses outside of trigonometry. Let P_1 and P_2 be any two points in the plane, and suppose the coordinates of P_1 are (x_1, y_1) and those of P_2 are (x_2, y_2). Let d denote the distance between P_1 and P_2. We draw a horizontal line through P_1 and a vertical line through P_2 and label their point of intersection Q, as shown in Figure 1.20. The point Q therefore has coordinates (x_2, y_1). Now the points P_1, P_2, and Q are the vertices of a right triangle, with right angle at Q, and so we can apply the Pythagorean theorem to find d. Since P_1Q is a horizontal segment, its length is the difference in the x coordinate of the rightmost point and the x coordinate of the leftmost point. As the figure is drawn, this distance is $x_2 - x_1$. If,

Figure 1.20

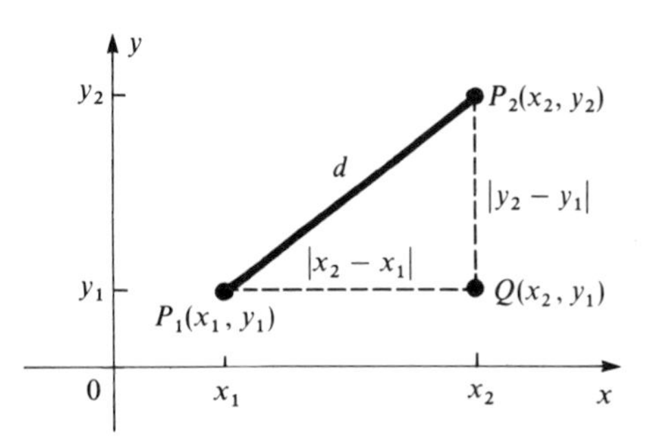

however, Q were to the left of P_1, the distance would be $x_1 - x_2$. In either case the distance is the absolute value of the difference in the x coordinates, $|x_2 - x_1|$. Similarly, the length of the segment QP_2 is $|y_2 - y_1|$. When we square the absolute value of a number, we get the same thing as the square of the number itself. So, by the Pythagorean theorem we have $d^2 = (x_2 - x_1)^2 + (y_2 - y_1)^2$, from which we obtain

The Distance Formula

$$d = \sqrt{(x_2 - x_1)^2 + (y_2 - y_1)^2}$$

EXAMPLE 1.9 Find the distance between the points $(-3, 4)$ and $(5, -2)$.

Solution It does not matter which point we call (x_1, y_1) and which we call (x_2, y_2) in the distance formula. The distance is

$$d = \sqrt{[5 - (-3)]^2 + [(-2) - 4]^2} = \sqrt{8^2 + 6^2} = \sqrt{100} = 10$$

We return now to the problem of defining the trigonometric functions for arbitrary angles. For this purpose we consider an angle θ, and we orient it with respect to a rectangular coordinate system so that its vertex is at the origin and its initial side coincides with the positive x axis, as in Figure 1.21. When an angle is placed in this way, we say it is in **standard position**. We select any point (x, y), other than the origin, on the terminal side of θ, and we designate the distance from $(0, 0)$ to (x, y) by r. The six trigonometric functions of θ are defined as follows.

Figure 1.21

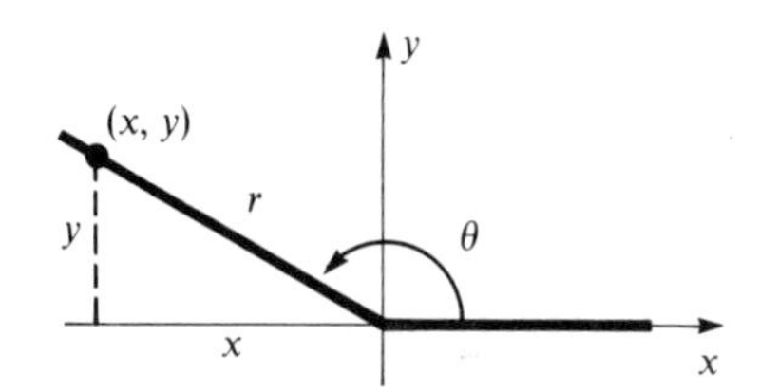

DEFINITION 1.2

$$\sin\theta = \frac{y}{r} \qquad \cos\theta = \frac{x}{r} \qquad \tan\theta = \frac{y}{x} \quad (x \neq 0)$$

$$\csc\theta = \frac{r}{y} \quad (y \neq 0) \qquad \sec\theta = \frac{r}{x} \quad (x \neq 0) \qquad \cot\theta = \frac{x}{y} \quad (y \neq 0)$$

The choice of the particular point (x, y) on the terminal side of θ is immaterial since, by similar triangles, the ratios would be the same if any other point were used (except the origin, which we exclude).

We note that if θ is acute, this definition is consistent with Definition 1.1, as can be seen by considering the right triangle having hypotenuse r, adjacent side x, and opposite side y.

Since r is always positive, the sine and cosine are defined for all angles θ. But there are certain values of θ excluded from the domain of each of the other functions, namely those angles which cause the denominator in the fraction defining the function to be zero. In particular, $\tan\theta$ and $\sec\theta$ are undefined when $x = 0$, and $\cot\theta$ and $\csc\theta$ are undefined when $y = 0$. Thus $\theta = 90°, 270°$, and all angles coterminal with these are excluded from the domains of the tangent and secant. Similarly, $\theta = 0°, 180°$, and all angles coterminal with these are excluded from the domains of the cotangent and cosecant.

We name the positive acute angle between the terminal side of θ and the x axis the **reference angle** for θ (Figure 1.22). For example, the reference angle of 150° is 30°, and the reference angle of 315° is 45°. There is a close relationship between the functions of an angle θ and of its reference angle. When θ is in standard position and we have chosen a point (x, y) on its terminal side, if we superimpose a right triangle with hypotenuse r so that the reference angle for θ is one of the angles of the triangle, then the legs of the triangle have lengths equal in absolute value to x and y respectively (Figure 1.21). But whereas x and y may be either positive or negative, the lengths of the sides of a triangle are positive. We conclude that *any trigonometric function of θ is equal in absolute value to the same function of its reference angle*. The correct sign must be affixed according to the function and the signs of x and y in the quadrant in which the terminal side lies. We can illustrate this with several examples.

Figure 1.22

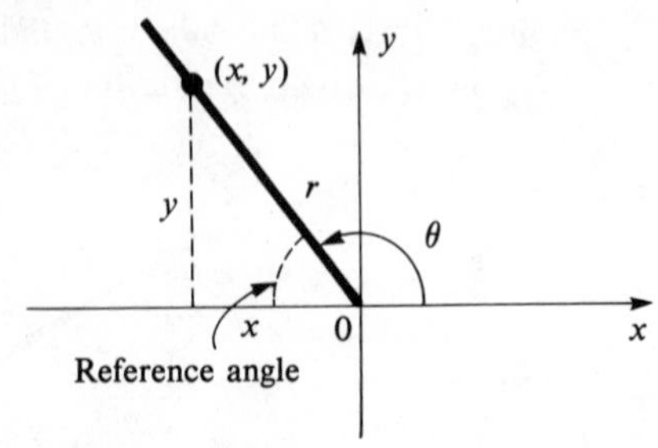

EXAMPLE 1.10 Find the six trigonometric functions of 150°.

Solution As we have already noted, the reference angle for 150° is 30°. If we choose that point on the terminal side of θ for which $r = 2$, as shown in Figure 1.23, our knowledge

Figure 1.23

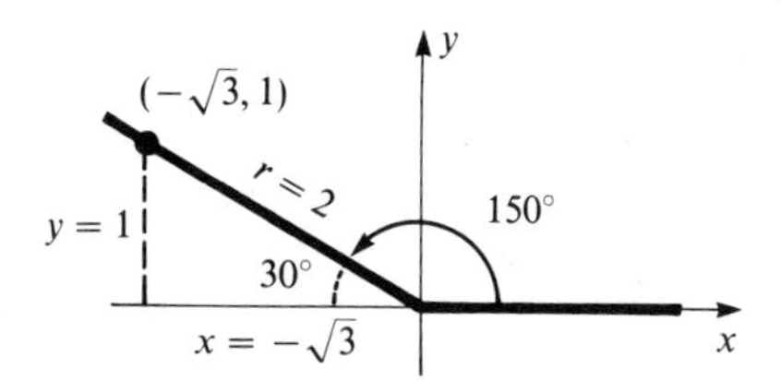

of a 30°–60° right triangle tells us that $x = -\sqrt{3}$ and $y = 1$. Notice that since the terminal side is in quadrant II, x must be negative and y positive. Now we use Definition 1.2 to obtain

$$\sin 150° = \frac{1}{2} \qquad \cos 150° = -\frac{\sqrt{3}}{2} \qquad \tan 150° = -\frac{1}{\sqrt{3}}$$

$$\csc 150° = 2 \qquad \sec 150° = -\frac{2}{\sqrt{3}} \qquad \cot 150° = -\sqrt{3}$$

EXAMPLE 1.11 Find all six trigonometric functions of 225°.

Solution The terminal side lies in quadrant III, and the reference angle is 45°, as shown in Figure 1.24. A convenient point to choose on the terminal side is $(-1, -1)$, and then $r = \sqrt{2}$. So we have

$$\sin 225° = -\frac{1}{\sqrt{2}} \qquad \cos 225° = -\frac{1}{\sqrt{2}} \qquad \tan 225° = 1$$

$$\csc 225° = -\sqrt{2} \qquad \sec 225° = -\sqrt{2} \qquad \cot 225° = 1$$

Figure 1.24

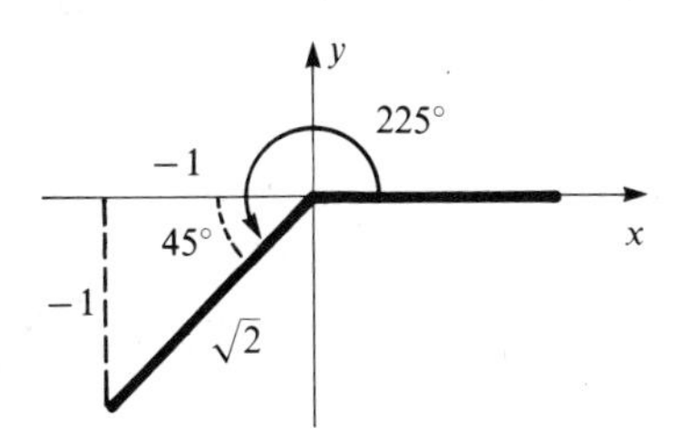

EXAMPLE 1.12 Find all six trigonometric functions of $5\pi/3$.

Solution This time the angle is given in radians. With practice you will find it as convenient to work with radians as with degrees, but at this stage you may prefer to convert to degrees.

$$\frac{5\pi}{3} \text{ radians} = \frac{5\pi}{3} \cdot \frac{180°}{\pi} = 300°$$

The terminal side is in the fourth quadrant, and the reference angle is 60° (Figure 1.25). So we have a 30°–60° right triangle as shown and choose $r = 2$, $x = 1$, $y = -\sqrt{3}$. Thus we have,

$$\sin\frac{5\pi}{3} = -\frac{\sqrt{3}}{2} \qquad \csc\frac{5\pi}{3} = -\frac{2}{\sqrt{3}}$$

$$\cos\frac{5\pi}{3} = \frac{1}{2} \qquad \sec\frac{5\pi}{3} = 2$$

$$\tan\frac{5\pi}{3} = -\sqrt{3} \qquad \cot\frac{5\pi}{3} = -\frac{1}{\sqrt{3}}$$

Figure 1.25

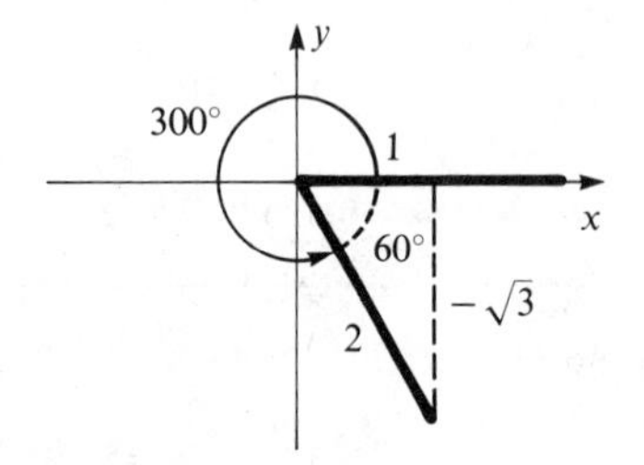

EXAMPLE 1.13 Find all six trigonometric functions of 510°.

Solution The first thing to do is observe that 510° is coterminal with 150° (Figure 1.26). Thus, the values of the trigonometric functions of 510° are the same as those of 150°. The angle 150° lies in the second quadrant and has a reference angle of 30°. Placing our standard 30°–60° triangle as shown in the figure, we read the following function values.

Figure 1.26

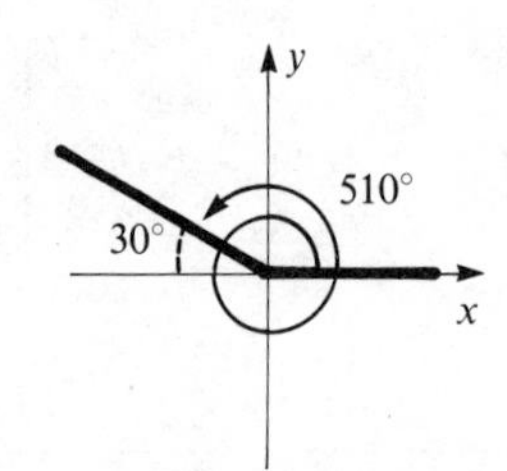

$$\sin 510° = \frac{1}{2} \qquad \csc 510° = 2$$

$$\cos 510° = -\frac{\sqrt{3}}{2} \qquad \sec 510° = -\frac{2}{\sqrt{3}}$$

$$\tan 510° = \frac{1}{\sqrt{3}} \qquad \cot 510° = -\sqrt{3}$$

These examples illustrate how we can find exact values of all angles having as reference angles 30°, 45°, or 60°. In particular, we can find the functions of 30°, 45°, 60°, 120°, 135°, 150°, 210°, 225°, 240° 300°, 315°, and 330°, as well as angles coterminal with any of these.

The angles 0°, 90°, 180°, 270°, and all angles coterminal with these are called **quadrantal angles**. Using Definition 1.2 we can find the values of the trigonometric functions of all such angles. We illustrate how to do this for 0° and 90°.

EXAMPLE 1.14 Find the six trigonometric functions of 0°.

Solution In applying Definiton 1.2 we may choose any point on the terminal side of the angle. For simplicity we choose (1, 0) in this case (Figure 1.27). Thus $x = 1$, $y = 0$, and $r = 1$. So we have

$$\sin 0° = \frac{0}{1} = 0 \qquad \csc 0° = \frac{1}{0} \quad \text{undefined}$$

$$\cos 0° = \frac{1}{1} = 1 \qquad \sec 0° = \frac{1}{1} = 1$$

$$\tan 0° = \frac{0}{1} = 0 \qquad \cot 0° = \frac{1}{0} \quad \text{undefined}$$

Remark. In working with quadrantal angles it will be typical that two of the functions will be undefined. The thing to remember is that *division by 0 is not defined*.

Figure 1.27

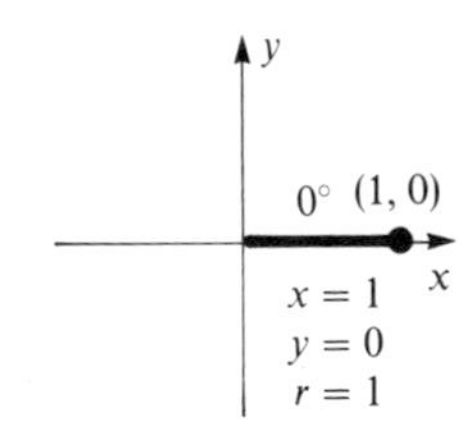

EXAMPLE 1.15 Find the six trigonometric functions of 90°.

Solution We select the point (0, 1) on the terminal side, so that $x = 0$, $y = 1$, and $r = 1$ (Figure 1.28). Definition 1.2 gives the following.

$$\sin 90° = \frac{1}{1} = 1 \qquad \csc 90° = \frac{1}{1} = 1$$

$$\cos 90° = \frac{0}{1} = 0 \qquad \sec 90° = \frac{1}{0} \text{ undefined}$$

$$\tan 90° = \frac{1}{0} \text{ undefined} \qquad \cot 90° = \frac{0}{1} = 0$$

Figure 1.28

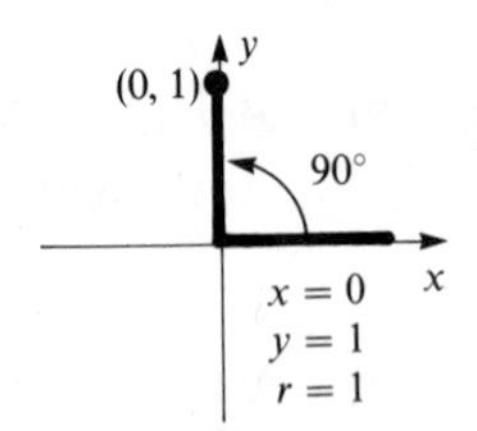

It is important to learn the special angles we have considered in terms of radians as well as degrees. We summarize these in the following table.

Degrees	Radians	Degrees	Radians
0°	0	180°	π
30°	$\pi/6$	210°	$7\pi/6$
45°	$\pi/4$	225°	$5\pi/4$
60°	$\pi/3$	240°	$4\pi/3$
90°	$\pi/2$	270°	$3\pi/2$
120°	$2\pi/3$	300°	$5\pi/3$
135°	$3\pi/4$	315°	$7\pi/4$
150°	$5\pi/6$	330°	$11\pi/6$

EXERCISE SET 1.5

Do all problems without using tables or a calculator.

A **1.** Find the distance between each pair of points.

a. $(2, -1), (-3, 4)$ **b.** $(-3, -5), (6, -2)$

c. $(-5, 2), (7, -3)$

In Problems 2–15 find exact values of the six trigonometric functions of the given angle.

2. $135°$ **3.** $330°$ **4.** $240°$ **5.** $120°$ **6.** $180°$ **7.** $270°$

8. $\frac{5\pi}{6}$ **9.** $\frac{7\pi}{4}$ **10.** $\frac{7\pi}{6}$ **11.** $-\frac{\pi}{6}$ **12.** $-\frac{2\pi}{3}$ **13.** $480°$

14. $600°$ **15.** 3π

In Problems 16–21 evaluate each function.

16. **a.** $\sin 120°$ **b.** $\cos 315°$ **c.** $\tan 210°$ **d.** $\sec 240°$ **e.** $\csc 330°$

17. **a.** $\cos \frac{3\pi}{4}$ **b.** $\tan \frac{2\pi}{3}$ **c.** $\csc \frac{5\pi}{6}$ **d.** $\sin \frac{5\pi}{4}$ **e.** $\cot \frac{7\pi}{6}$

18. **a.** $\tan 480°$ **b.** $\sec 900°$ **c.** $\sin(-270°)$ **d.** $\cos \frac{10\pi}{3}$ **e.** $\cot \frac{3\pi}{2}$

19. **a.** $\sin \frac{7\pi}{6}$ **b.** $\cos 330°$ **c.** $\tan \frac{7\pi}{4}$ **d.** $\sec 135°$ **e.** $\cot \frac{3\pi}{4}$

20. **a.** $\csc \frac{11\pi}{6}$ **b.** $\tan 600°$ **c.** $\sin \frac{4\pi}{3}$ **d.** $\cos 225°$ **e.** $\sec \frac{7\pi}{6}$

21. **a.** $\cot(-300°)$ **b.** $\sec \frac{9\pi}{4}$ **c.** $\tan\left(-\frac{17\pi}{6}\right)$ **d.** $\cos 660°$ **e.** $\csc(-240°)$

In Problems 22–25 give an equivalent expression as a function of an acute angle.

22. **a.** $\sin 250°$ **b.** $\cos 132°$ **c.** $\tan 97°$ **d.** $\sec 280°$ **e.** $\csc 310°$

23. **a.** $\cot 212°$ **b.** $\sin 562°$ **c.** $\tan 620°$ **d.** $\cos(-430°)$ **e.** $\csc(-491°)$

24. **a.** $\cos 123° 40'$ **b.** $\tan 312° 15'$ **c.** $\csc(-240° 38')$
d. $\cot 263° 24'$ **e.** $\sin(-158° 13')$

25. **a.** $\tan \frac{7\pi}{12}$ **b.** $\sec \frac{7\pi}{8}$ **c.** $\sin \frac{9\pi}{5}$ **d.** $\csc \frac{8\pi}{9}$ **e.** $\cos \frac{13\pi}{12}$

In Problems 26–31 a point on the terminal side of an angle θ in standard position is given. Find all six trigonometric functions of θ.

26. $(-3, 4)$ **27.** $(5, -12)$

28. $(-15, -8)$ **29.** $(2, -3)$

30. $(-1, 2)$ **31.** $(-7, -24)$

32. Evaluate the expression

$$\frac{\sin \theta + \cos 2\theta}{\tan \theta \cot 2\theta}$$

when $\theta = 5\pi/6$.

33. Evaluate the expression

$$\left(\sec \theta - \csc \frac{\theta}{4}\right)(1 + \cos 3\theta)$$

when $\theta = 2\pi/3$.

B In Problems 34–39 draw the angle θ in standard position for which $0 \leq \theta < 2\pi$ and find the other five trigonometric functions of θ.

34. $\sin\theta = -\frac{4}{5}$, θ in quadrant III

35. $\cos\theta = \frac{5}{13}$, θ in quadrant IV

36. $\sec\theta = -\frac{3}{2}$, $\tan\theta > 0$ **37.** $\csc\theta = 3$, $\cos\theta < 0$

38. $\tan\theta = \frac{1}{2}$, $\csc\theta < 0$ **39.** $\cot\theta = -\frac{1}{3}$, $\sin\theta > 0$

40. If $\sin\theta = y > 0$, find expressions for the other five trigonometric functions in terms of y. (Two solutions.)

41. If $\tan\theta = t > 0$, find expressions for the other five trigonometric functions of θ in terms of t. (Two solutions.)

42. If $\sec\theta = r > 0$, find expressions for the other five trigonometric functions in terms of r. (Two solutions.)

43. For $0 < \theta < \pi/2$, express all six trigonometric functions of each of the following in terms of θ alone.

a. $\pi - \theta$ **b.** $\pi + \theta$ **c.** $\frac{\pi}{2} + \theta$ **d.** $\frac{3\pi}{2} - \theta$ **e.** $2\pi - \theta$

Hint. Find the reference angle in each case.

44. Prove that $\cos n\pi = (-1)^n$, where n is any integer.

45. Prove that $\sin\frac{(2n+1)\pi}{2} = (-1)^n$, where n is any integer.

46. Prove that for all integers k,

$$\cos\frac{\pi}{3}(1 + 3k) = \frac{(-1)^k}{2}$$

47. Prove that for all integers k,

$$\tan\frac{\pi}{4}(1 + 2k) = (-1)^k$$

48. Using distances, prove that the points $A(8, 3)$, $B(7, -5)$, and $C(4, -3)$ are vertices of a right triangle. Find the six trigonometric functions of angle A.

1.6 Review Exercise Set

A

1. Give the radian measure of each of the following.

a. 240° **b.** 315° **c.** 15° **d.** 540° **e.** 160°

2. Give the degree measure of each of the angles whose radian measures are given.

a. $\frac{5\pi}{6}$ **b.** $\frac{2\pi}{9}$ **c.** 3π **d.** 5 **e.** $\frac{7\pi}{4}$

3. Use a calculator to convert to decimal degrees, to the nearest hundredth of a degree.

a. 59° 37′ **b.** 205° 23′ **c.** −452° 46′ **d.** 14° 34′ 21″ **e.** 256° 51′ 43″

4. Use a calculator to convert to degrees and minutes, to the nearest minute.

a. 23.17° **b.** 152.13° **c.** −46.08° **d.** 3.25 radians **e.** $\frac{\pi}{7}$ radians

5. Use a calculator to convert to radians, to the nearest thousandth of a radian.

a. 34° 27′ **b.** 133.172° **c.** 15° 17′ 34″ **d.** −24° 52′ 13″ **e.** 135°

6. What is the length of an arc on a circle of radius 20 inches subtended by an angle of 36°?

7. An arc of 8 feet is subtended by a central angle θ on a circle of radius 6 feet. Find θ in degrees.

8. A ceiling fan 4 feet in diameter is rotating at the rate of 300 revolutions per minute. Through how many radians will a point on the tip of the fan turn in 1 second? What is the linear velocity of the point in feet per second?

9. A wheel 18 inches in diameter is rotating at a constant rate. The linear velocity of a point on the periphery is 20 feet per second. What is the angular velocity in radians per second? Through how many revolutions per minute is the wheel turning?

In Problems 10–17, ABC is a right triangle with right angle at C. Find all unknown sides.

10. $A = 60°,\ b = 12$

11. $B = 45°,\ c = 4$

12. $B = 30°,\ c = 3$

13. $A = 30°,\ b = 5$

14. $A = 45°,\ b = 6$

15. $B = 60°,\ c = 5$

16. $a = 5,\ b = 7$

17. $c = 9,\ a = 4$

18. Write the six trigonometric functions of angle D for the given triangle.

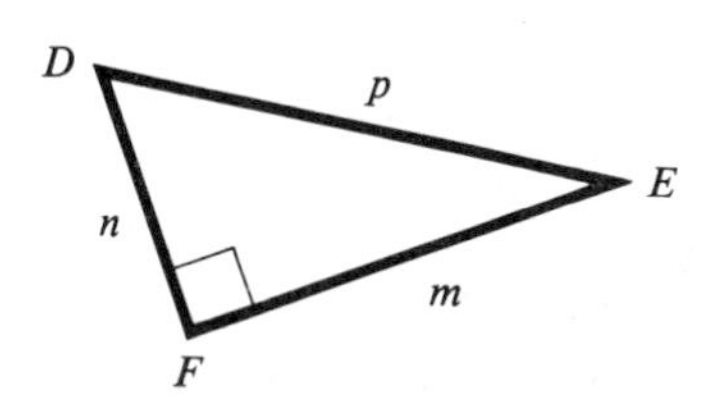

Evaluate the expressions in Problems 19 and 20.

19. **a.** $\dfrac{\tan 60° - \cot 45°}{\cos 30° + \sin 60°}$

b. $\sec 45° (\sin 30° \tan 60° - \csc 60°)$

20. **a.** $\dfrac{1 + \tan^2 \dfrac{\pi}{3}}{\cot^2 \dfrac{\pi}{6}} + \dfrac{\sin^2 \dfrac{\pi}{4}}{\csc \dfrac{\pi}{6}}$

b. $\sqrt{\dfrac{1 + \cos \dfrac{\pi}{3}}{2}}$

21. Complete the following table.

	0°	30°	45°	60°	90°	120°	135°	150°	180°	210°	225°	240°	270°	330°	315°	330°
sin																
cos																
tan																
cot																
sec																
csc																

22. Write each of the following as a function of an acute angle.

a. $\sin 237°$ **b.** $\tan 124° 32'$ **c.** $\sec(-310°)$ **d.** $\cos \dfrac{11\pi}{12}$ **e.** $\csc 750° 54'$

23. A point on the terminal side of an angle θ in standard position is given. Find all six trigonometric functions of θ.

a. $(-3, -4)$ **b.** $(5, -12)$ **c.** $(-2, 3)$ **d.** $(2\sqrt{2}, -1)$

24. Find the distance between the following pairs of points.

a. $(3, -1), (-2, -4)$ **b.** $(-4, -5), (-6, 2)$

25. Show that each of the formulas

a. $\sin 2\theta = 2 \sin \theta \cos \theta$ **b.** $\cos 2\theta = 2 \cos^2 \theta - 1$

is true for each of the values $\theta = \pi/6, \pi/4, \pi/3$, and $\pi/2$.

26. Verify the formulas in Problem 25 for $\theta = 2\pi/3, 3\pi/4, 5\pi/6$, and π.

27. Give all values of θ in the interval $0 \le \theta < 2\pi$ for which

a. $\sin \theta = -\dfrac{1}{2}$ **b.** $\cos \theta = -1$ **c.** $\tan \theta = -\sqrt{3}$ **d.** $\sec \theta = 2$

e. $\cot \theta = 0$

28. Evaluate and simplify

a. $\dfrac{\tan 135° \csc 210°}{\sin 270° + \sec 660°}$ **b.** $\left(\sec^2 \dfrac{5\pi}{6} - \cot^2 \dfrac{2\pi}{3}\right)\left(\csc \dfrac{\pi}{2} - \cos 3\pi\right)$

29. A window is in the shape of a square 3 feet on a side surmounted by an isosceles triangle having base angles of 30° each (see sketch). Find

a. the area of the window **b.** the overall perimeter.

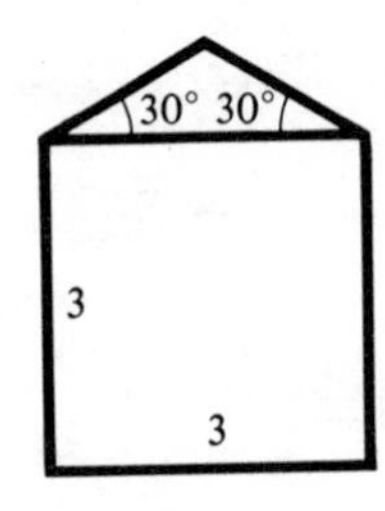

30. Find the arc length s in the given figure.

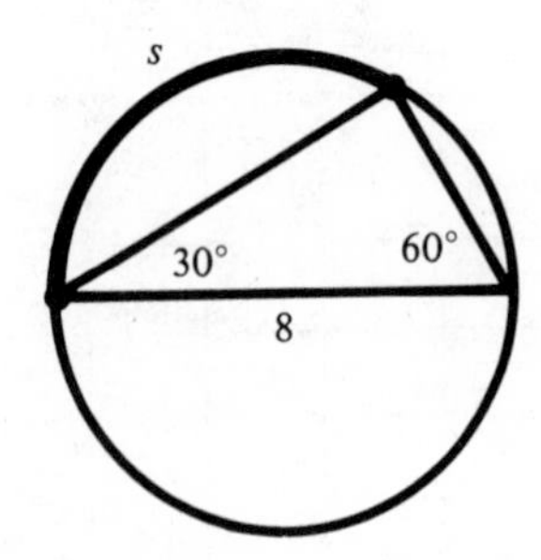

In Problems 31 and 32 draw a right triangle ABC for which the given equation is valid, and find the other five trigonometric functions of the specified angle.

31. **a.** $\cos A = \dfrac{3}{5}$ **b.** $\tan B = \dfrac{12}{5}$

32. **a.** $\sec B = \dfrac{17}{8}$ **b.** $\sin A = \dfrac{7}{25}$

In Problems 33 and 34 draw the angle θ in standard position for which $0 \le \theta < 2\pi$ and the given information is satisfied. Find the other five trigonometric functions of θ.

33. **a.** $\tan \theta = \dfrac{2}{3}$, $\csc \theta < 0$ **b.** $\sin \theta = -\dfrac{1}{4}$, $\cos \theta > 0$

34. $\cos \theta = -\dfrac{4}{5}$, $\cot \theta < 0$ **b.** $\sec \theta = \dfrac{3}{2\sqrt{2}}$, $\sin \theta < 0$

B **35.** The tires of a bicycle are 26 inches in diameter. When the bicycle is moving at 15 miles per hour, find the angular velocity of a point on the tire in radians per second. Through how many revolutions per minute is the tire turning?

36. The linear velocity of a point on the spoke of a wheel is 20 feet per second, and the linear velocity of a point 3 inches farther out on the same spoke is 30 feet per second. Find:

a. the angular velocity of the wheel

b. the distance of each point from the center of the wheel.

37. Find the area of the triangle shown.

Hint. Find the altitude to the base.

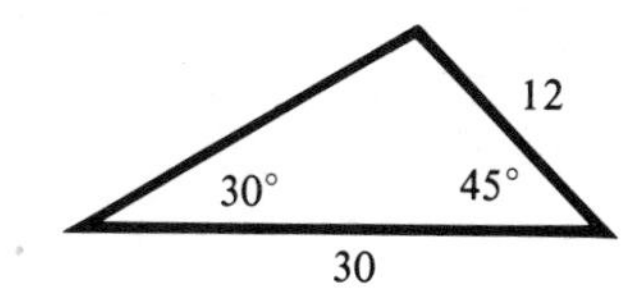

38. Complete the following proof of the Pythagorean theorem. Let ABC be a right triangle with right angle at C. Draw CD perpendicular to AB. Show that angle β = angle B and angle α = angle A. Conclude that the triangles ABC, CBD, and ACD are all similar. Show that

$$\frac{x}{a} = \frac{a}{c} \quad \text{and} \quad \frac{y}{b} = \frac{b}{c}$$

Solve for x and y and use the fact that $x + y = c$.

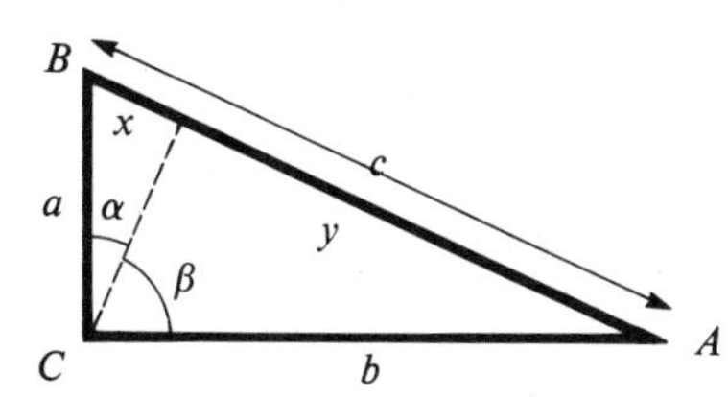

39. Using Definiton 1.2, prove the following for any angle θ.

a. $\sin^2 \theta + \cos^2 \theta = 1$ **b.** $1 + \tan^2 \theta = \sec^2 \theta$

40. If $\cos \theta = x$ and $x < 0$, find the other five trigonometric functions of θ. (Two solutions.)

41. If $\cot \theta = m$ and $m > 0$, find the other five trigonometric functions of θ. (Two solutions.)

2 The Solution of Right Triangles

The following excerpt from a physics textbook shows one way in which the solution of a right triangle arises in an applied problem.*

Example A ladder in equilibrium leans against a vertical frictionless wall. The forces on the ladder are (1) its weight $\boldsymbol{w}$, (2) the force $\boldsymbol{F}_1$, exerted on the ladder by the vertical wall . . . , and (3) the force $\boldsymbol{F}_2$ exerted by the ground on the base of the ladder . . . . In part (b) the forces have been transferred to the point of intersection of their lines of action, and applying the equations for the equilibrium of a particle at this point, we get

$$\Sigma \boldsymbol{F}_x = \boldsymbol{F}_2 \cos \theta - \boldsymbol{F}_1 = 0$$
$$\Sigma \boldsymbol{F}_y = \boldsymbol{F}_2 \sin \theta - \boldsymbol{w} = 0.$$

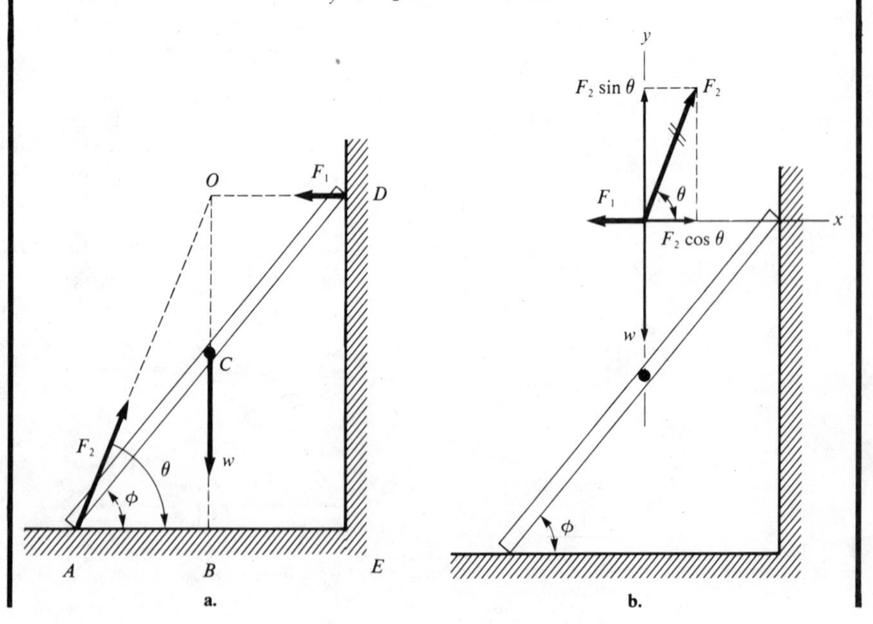

*Sears/Zemansky/Young, *University Physics*, © 1976, 5/e, Addison-Wesley Publishing Company, Inc., Chapter 2, figure 2–8, a, b, page 24 and page 25. Reprinted with permission.

As a numerical example, suppose the ladder weighs $400N$, is $10\ m$ long, has its center of gravity at its center, and makes an angle of $\phi = 53°$ with the ground. We wish to find the angle θ and the forces $\boldsymbol{F}_1$ and $\boldsymbol{F}_2$ . . . . From the right triangle ABC we have

$$\overline{AB} = \overline{AC}\ \cos\ \phi = (5\ m)(0.60) = 3.0\ m,$$

and from right triangle AED,

$$\overline{DE} = \overline{AD}\ \sin\ \phi = (10\ m)(0.80) = 8\ m.$$

Then, from the right triangle AOB, since $\overline{OB} = \overline{DE}$, we have

$$\tan\ \theta = \frac{\overline{OB}}{\overline{AB}} = \frac{8\ m}{3\ m} = 2.67.$$

Then

$$\theta = 69.5°, \quad \sin\ \theta = 0.937, \quad \cos\ \theta = 0.350.$$

$$\boldsymbol{F}_2 = \frac{w}{\sin\ \theta} = \frac{400N}{0.937} = 427N$$

and . . . ,

$$\boldsymbol{F}_1 = \boldsymbol{F}_2\ \cos\ \theta = (427N)(0.350) = 150N.$$

In addition to illustrating the solution of a right triangle, this example employs **vectors**, which we will also study in this chapter. The letter N in the example stands for a *newton* which is the unit of force (or of weight) in the meter-kilogram-second system, and the letter m stands for one meter.

2.1 Introduction

In this chapter we will study what has historically been one of the most important aspects of trigonometry, that of finding the unknown sides and angles in right triangles when certain information is given. This is called **solving the triangle**. The solution of right triangles, as well as the solution of oblique triangles, continues to be an indispensable tool in applications, even though what is known as analytical trigonometry has assumed even greater importance, especially in the study of calculus and more advanced mathematics. In Chapter 3 we will begin a systematic treatment of the analytical aspects of the subject. In Chapter 6 we will return to the solution of triangles, dealing there with oblique triangles.

The groundwork was laid in Chapter 1 for solving right triangles, and for specially chosen angles the complete solution can be obtained with the techniques learned there. For example, suppose we are given a right triangle ABC, with $A = 30°$ and $b = 12$, as shown in Figure 2.1. We know first of all that $B = 60°$, since A and B are complementary. While we could find sides a and c using proportionality and our

Figure 2.1

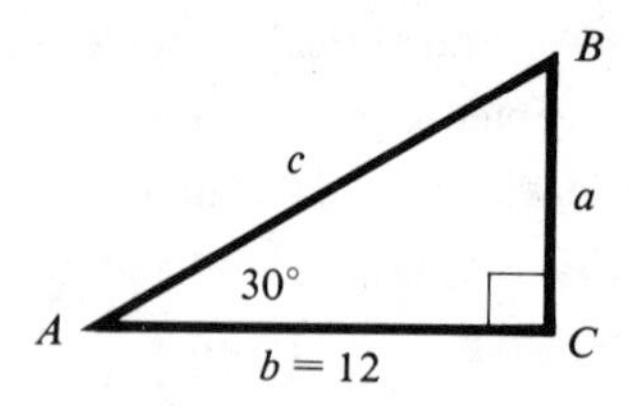

knowledge of the sides of the special 30°–60° right triangle studied in Chapter 1, we choose to proceed in another way in order to indicate a general method applicable to other angles. We select a trigonometric function of angle A which involves the known side b and one of the unknown sides. For example, let us choose the tangent function,

$$\tan A = \frac{a}{b}$$

Substituting known values,

$$\tan 30° = \frac{a}{12}$$

and then solving for a gives

$$a = 12 \tan 30°$$

Now we know from Chapter 1 that $\tan 30° = 1/\sqrt{3}$, so

$$a = 12 \cdot \frac{1}{\sqrt{3}} = 12\frac{\sqrt{3}}{3} = 4\sqrt{3}$$

Knowing a and b, we can find side c by the Pythagorean theorem, or alternatively we could choose one of the trigonometric functions of angle A which involves side c. Since it is best in general to work insofar as is possible with the given data rather than computed results, we could choose the secant function to get

$$\sec A = \frac{c}{b}$$

or

$$\sec 30° = \frac{c}{12}$$

from which

$$c = 12 \sec 30° = 12 \cdot \frac{2}{\sqrt{3}} = 24\frac{\sqrt{3}}{3} = 8\sqrt{3}$$

The triangle is now completely solved.

The problem is that for most angles we do not know the values of the trigonometric functions. This is where a hand calculator or tables become a necessity. We discuss their use in the next section.

2.2 Values of Trigonometric Functions from Calculators and Tables

The scientific hand calculator provides a rapid and highly accurate way of obtaining values of trigonometric functions and of performing calculations with them. When a calculator is available, there is no reason to use trigonometric tables. There is some merit, however, in learning to use tables in the event they are needed, and so we do devote some time in this section to their use. *Throughout the remainder of the text you are encouraged to use a hand calculator for all numerical calculations unless specific instructions to the contrary are given.*

There are many makes and models of calculators and, while they all share common features, the differences are sufficient to make it impractical to include here specific instructions on their operation. You should refer to the instructions for the particular calculator you are using. We will indicate general instructions, where appropriate, which are applicable to most calculators which use so-called algebraic logic. If your calculator uses reverse Polish logic, you will have to make appropriate modifications.

For purposes of finding values of the trigonometric functions, it is necessary first to indicate whether the angle is to be in degrees or in radians. We say that we are working in the degree mode or in the radian mode. This is accomplished on the calculator by pressing the appropriate key (or keys), or switch. When the angle is in degrees, fractions of degrees in most cases must be expressed in decimal form, rather than in minutes and seconds. (There are some calculators, however, which accept minutes and seconds.) We use the fact that $1' = (\frac{1}{60})^\circ$ and $1'' = (\frac{1}{3600})^\circ$. For example, $24^\circ\ 39' = (24\frac{39}{60})^\circ = 24.65^\circ$. To find the sine, cosine, or tangent, we enter the angle (in degrees or radians, as appropriate) and then press the function key. For the other three functions we make use of the following reciprocal relations, which are seen from the definition to hold true for all admissible values of θ:

$$\csc\theta = \frac{1}{\sin\theta}, \qquad \sec\theta = \frac{1}{\cos\theta}, \qquad \cot\theta = \frac{1}{\tan\theta}$$

To find the cosecant, for example, we first find the sine and then press the reciprocal key, marked $\boxed{1/x}$ on most calculators.

With the calculator we do not need to use reference angles. The calculator is designed to give the correct value regardless of the size, or the sign, of the angle.* For example, we get immediately

$$\sin 256.54^\circ = -0.972532659$$

and

$$\cos(-1{,}023^\circ) = 0.544639035$$

---

*The size, of course, is subject to the limitation of the largest number which can be entered on the calculator.

Another advantage of the calculator is that when using a trigonometric function as a part of a calculation, we need not read the value of the function, but simply have the calculator perform the operations called for. Thus, if we want to calculate $c = 2.54 \sin 32.7°$ we proceed as follows (with algebraic logic): First put the calculator in degree mode and then carry out the operations

2.54 [×] 32.7 [SIN] [=]

The answer, rounded, is 1.37.

To use the calculator to find the angle when we know the value of a trigonometric function of the angle, we enter the number and then press the keys which give the *inverse* of the function. On some calculators this is done by pressing a key labeled [INV], followed by pressing the appropriate function key. On others, instead of an [INV] key, there is a key labeled [ARC] which serves the same purpose. For example, if we know that $\tan\theta = 1.234$ and want to find θ in degrees, we enter the number 1.234 and press the [INV] (or [ARC]) key followed by pressing the [TAN] key. The answer is displayed as

50.97963309

We would probably want to round this to 50.98°.

In finding angles in this way on the calculator, the answer given will be a positive acute angle if the function value is positive. If the function value is negative, the answer will be given as a negative angle between 0° and −90° for the sine and tangent, and as a positive angle between 90° and 180° for the cosine. The reason for these choices will be made clear when we study inverse trigonometric functions in Chapter 4.

Working in the radian mode on a calculator presents no special difficulties. Problems are handled just as with degrees.

EXAMPLE 2.1 Use a calculator to find the following, correct to four decimal places.

a. $\sin 102.6°$ **b.** $\tan(-37° 42')$ **c.** $\sec 2.675$

Solution **a.** With the calculator in degree mode we enter 102.6, press [SIN] and [=] to get, after rounding,

$$\sin 102.6° \approx 0.9759$$

b. First change 42′ to $(\frac{42}{60})° = 0.7°$. Put the calculator in degree mode and carry out the following sequence:

37.7 [+/−] [TAN]

The answer is approximately −0.7729. The [+/−] key changes 37.7 to −37.7.

c. Since a degree symbol is not written, we understand the angle to be in radians and so place the calculator in radian mode. We will find the cosine of the angle and then take the reciprocal, as shown in the sequence

$$2.675 \ \boxed{\text{COS}} \quad \boxed{1/x}$$

This gives

$$\sec 2.675 \approx -1.1197$$

EXAMPLE 2.2 Use a calculator to find the angle θ having the degree of accuracy as indicated.

a. $\cos \theta = 0.2134, \quad 0 \leq \theta \leq 90°$ (nearest hundredth of a degree)
b. $\cot \theta = -2.765, \quad -\pi/2 < \theta < 0°$ (nearest thousandth of a radian)
c. $\sin \theta = 0.6281, \quad 90° \leq \theta \leq 180°$ (nearest minute)

Solution **a.** Put the calculator in degree mode, and proceed as follows:

$$.2134 \ \boxed{\text{INV}} \quad \boxed{\text{COS}}$$

The answer shown is rounded to 77.68. In this and the other parts of this example, replace $\boxed{\text{INV}}$ by $\boxed{\text{ARC}}$ if appropriate on your calculator.

b. We use the fact that $\tan \theta = 1/\cot \theta$. First place the calculator in radian mode. Then carry out the following sequence of operations.

$$2.765 \ \boxed{+/-} \quad \boxed{1/x} \quad \boxed{\text{INV}} \quad \boxed{\text{TAN}}$$

The answer to the desired degree of accuracy is $\theta = -0.347$.

c. Place the calculator in degree mode. Since $\sin \theta$ is given as a positive number, the inverse will be given by the calculator as a positive acute angle. Thus, to find θ in the second quadrant, we treat the angle shown by the calculator as the reference angle for θ and subtract it from 180° to find θ. The sequence of operations is

$$.6281 \ \boxed{\text{INV}} \quad \boxed{\text{SIN}} \quad \boxed{+/-} \quad \boxed{+} \ 180 \ \boxed{=}$$

This gives $\theta = 141.08992°$. To change the decimal part of the answer to minutes we multiply by 60. This gives

$$\theta \approx 141° \ 05'$$

Tables giving approximate values of the trigonometric functions for angles between 0° and 90° are available with varying degrees of accuracy. Four-place tables are sufficient for most purposes, and these are included in the back of this book. Scientific hand calculators typically give eight- to ten-figure accuracy. In Table I, angles are given to the nearest tenth of a degree and equivalently to the nearest 6′ (since $0.1° = 6'$). This dual representation in terms of decimals as well as degrees and minutes makes the use of this table compatible with hand calculators, where angles typically must be in decimal form. Table II gives angles in radians in intervals of 0.01 radian.

To use Table I, if the angle is between 0° and 45°, we read the angle in the left-hand column and find the desired function column by reading from the top down. For example, we read sin 25° 30′ = 0.4147.* If the angle is between 45° and 90°, we read the angle in the right-hand column and read the function column from the bottom up. For example, sin 65° 30′ = 0.9100. It is the fact that cofunctions of complementary angles are equal that allows the table to be constructed in this way. For angles not in the range from 0° to 90°, we first express the function in terms of the reference angle.

When we want to find a function of an angle that lies between tabular values, we use what is known as **linear interpolation**. Suppose, for example, that we want to find sin 35° 22′. From Table I we find sin 35° 18′ = 0.5779 and sin 35° 24′ = 0.5793. Now 35° 22′ is $\frac{4}{6}$, or $\frac{2}{3}$, of the way from 35° 18′ to 35° 24′. So it seems reasonable that sin 35° 22′ would be approximately $\frac{2}{3}$ of the way between 0.5779 and 0.5793. The difference in the values is 0.0014, and $\frac{2}{3}$ (0.0014) = 0.0009, approximately. Since the sine *increases* in value as the angle increases, we *add* this amount to the smaller value to get the desired estimate, sin 35° 22′ = 0.5788. These calculations can be arranged in the following compact form.

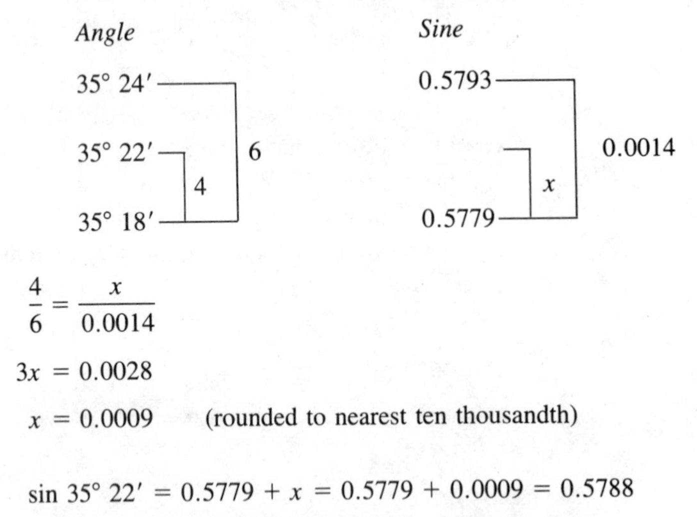

$$\frac{4}{6} = \frac{x}{0.0014}$$

$$3x = 0.0028$$

$$x = 0.0009 \quad \text{(rounded to nearest ten thousandth)}$$

$$\sin 35^\circ\, 22' = 0.5779 + x = 0.5779 + 0.0009 = 0.5788$$

Now suppose we want to find cos 35° 22′. Up to the final step the work proceeds in a similar way.

---

*Throughout the remainder of this chapter we will be dealing with approximations, but rather than to write, for example, sin 25° 30′ ≈ 0.4147 to mean the value is approximate, we will use equal signs and understand that the numbers normally are not exact.

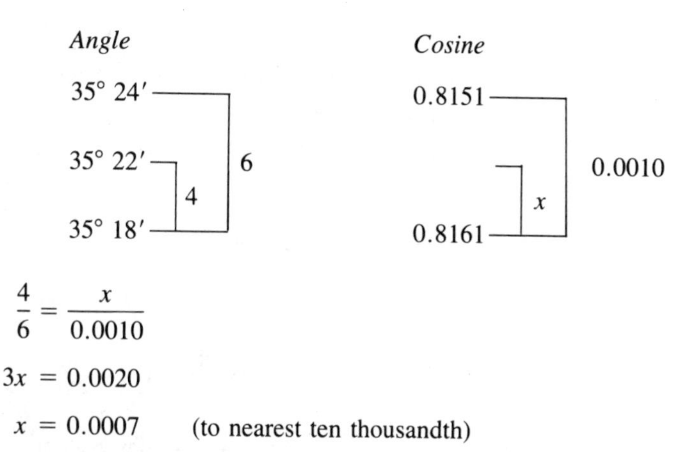

$$\frac{4}{6} = \frac{x}{0.0010}$$

$$3x = 0.0020$$

$$x = 0.0007 \qquad \text{(to nearest ten thousandth)}$$

We observe in this case that as the angle increases the value of the cosine *decreases*, so that we must *subtract* x from 0.8161. Thus,

$$\cos 35°\ 22' = 0.8161 - x = 0.8161 - 0.0007 = 0.8154$$

Remark. For angles in the range from 0° to 90°, the sine, tangent, and secant *increase* as the angle increases, whereas the cosine, cotangent, and cosecant *decrease* as the angle increases. Thus, when interpolating, the number x is added to the smaller value when working with the sine, tangent, or secant, and is subtracted from the larger value in the case of the cosine, cotangent, and cosecant. As a check, you should always obtain an interpolated value which lies between the tabular values.

In Chapter 4 we will study graphs of the trigonometric functions, and we will see that a magnified section of the graph of the sine and cosine curves in the vicinity of 35° 22′ would resemble what we have shown in Figure 2.2. When we use linear interpolation, we are effectively replacing the curve in each case by a straight line, which explains the word "linear." There is therefore an error associated with this procedure, but it is quite small.

Now suppose we are given the value of one of the trigonometric functions of an angle and want to determine the angle. If the value does not coincide with one in the table, we will need to interpolate. For example, if we have found that for a

Figure 2.2

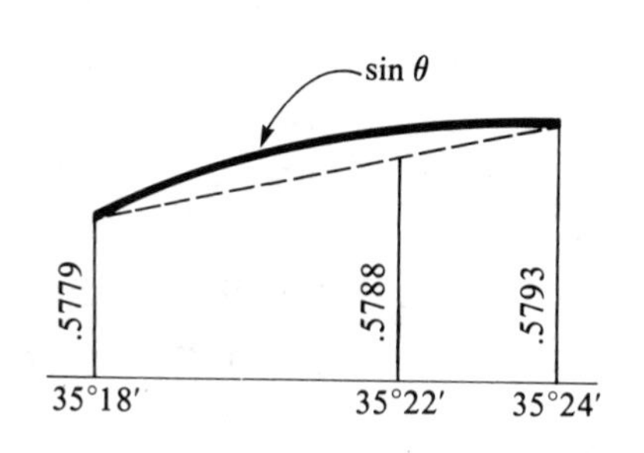

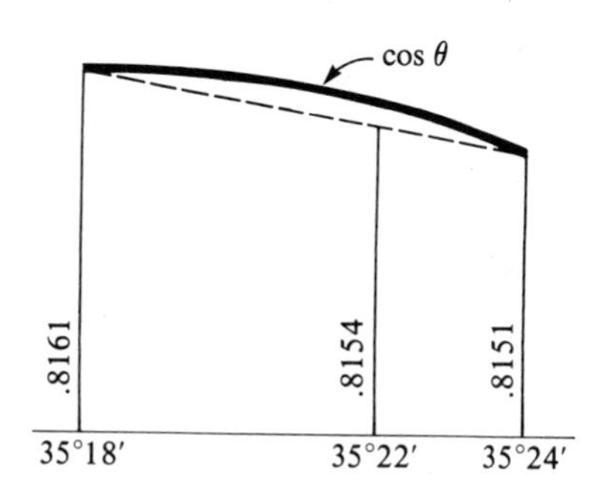

certain acute angle A, $\tan A = 1.644$, and we want to find A to the nearest minute, we proceed as follows. We search the tangent column in Table I until we find the closest values above and below 1.644 and then read the corresponding angles. The display below indicates how to proceed.

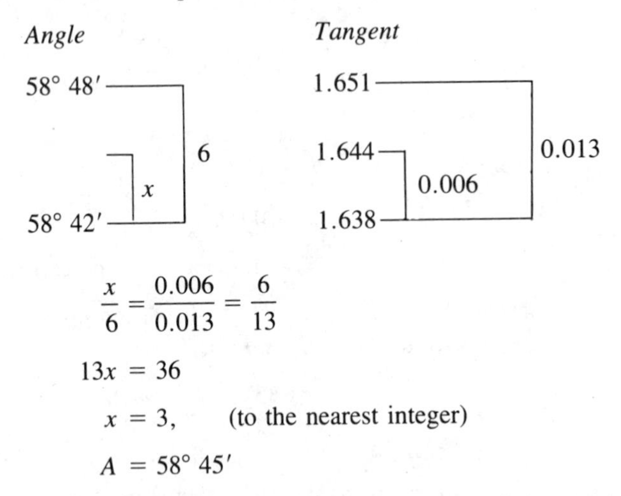

$$\frac{x}{6} = \frac{0.006}{0.013} = \frac{6}{13}$$

$$13x = 36$$

$$x = 3, \qquad \text{(to the nearest integer)}$$

$$A = 58^\circ\ 45'$$

When an angle is expressed in radians and is not one of the special angles discussed in Chapter 1, we can use Table II to find the trigonometric functions. Interpolation is handled in a similar way to that for angles in degrees.

The following examples further illustrate linear interpolation, with angles expressed in various ways.

EXAMPLE 2.3 Use linear interpolation to approximate each of the following:

a. $\tan 18^\circ\ 23'$ **b.** $\cos 0.254$
c. $\sin 37.23^\circ$ **d.** $\sec 135^\circ\ 14'$
e. $\csc(-2.167)$

Solution **a.**

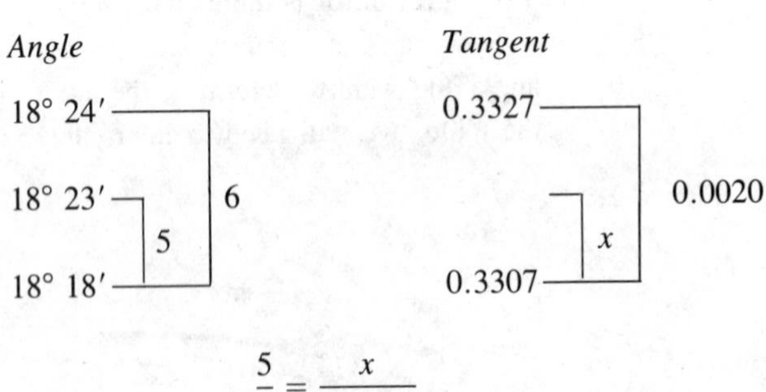

$$\frac{5}{6} = \frac{x}{0.0020}$$

$$6x = 0.0100$$

$$x = 0.0017$$

$$\tan 18^\circ\ 23' = 0.3307 + x = 0.3307 + 0.0017 = 0.3324$$

b. Since there is no degree symbol, we understand the angle to be 0.254 radians. From Table II we get,

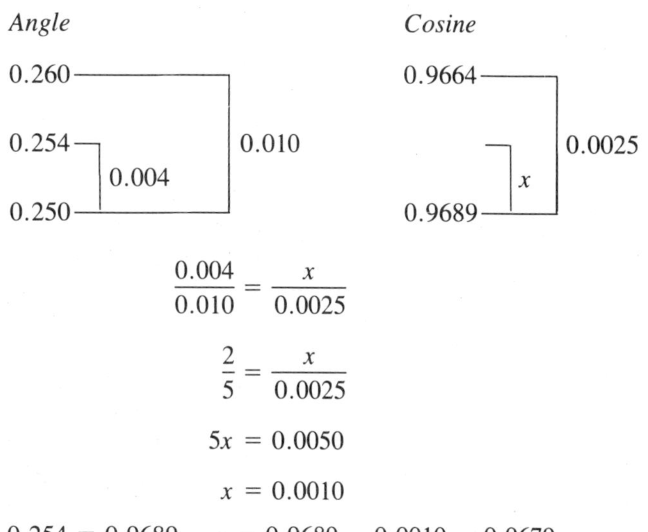

$$\frac{0.004}{0.010} = \frac{x}{0.0025}$$

$$\frac{2}{5} = \frac{x}{0.0025}$$

$$5x = 0.0050$$

$$x = 0.0010$$

$$\cos 0.254 = 0.9689 - x = 0.9689 - 0.0010 = 0.9679$$

c.

Angle | *Sine*

37.30° — 0.6060

37.23° (0.03 from 37.20°; 0.10 between 37.30° and 37.20°) — x (0.0014 between 0.6060 and 0.6046)

37.20° — 0.6046

$$\frac{0.03}{0.10} = \frac{x}{0.0014}$$

$$\frac{3}{10} = \frac{x}{0.0014}$$

$$10x = 0.0042$$

$$x = 0.0004$$

$$\sin 37.23° = 0.6046 + x = 0.6046 + 0.0004 = 0.6050$$

d. The reference angle is $180° - 135°\,14' = 179°\,60' - 135°\,14' = 44°\,46'$. Since the secant is negative in quadrant II, $\sec 135°\,14' = -\sec 44°\,46'$.

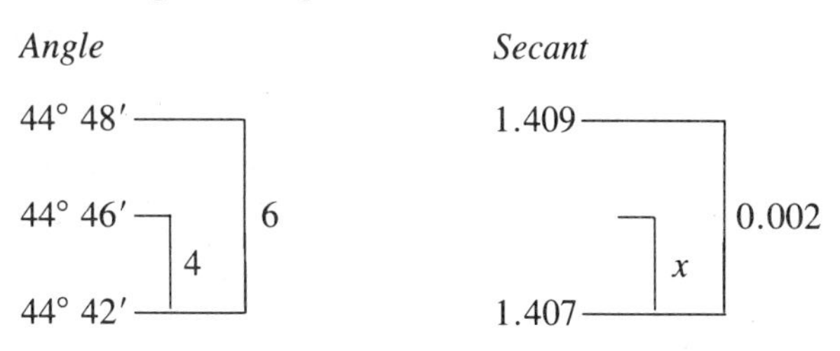

$$\frac{\not{4}^{2}}{\not{6}_{3}} = \frac{x}{0.002}$$

$$3x = 0.004$$

$$x = 0.001 \qquad \text{(to nearest thousandth)}$$

$$\sec 44^\circ\, 46' = 1.407 + x = 1.408$$

So

$$\sec 135^\circ\, 14' = -1.408$$

e. To find the reference angle of -2.167 radians, we recall that $\pi \approx 3.142$, and hence $\pi/2 \approx 1.571$. So -2.167 is in quadrant III, as shown in Figure 2.3. The

Figure 2.3

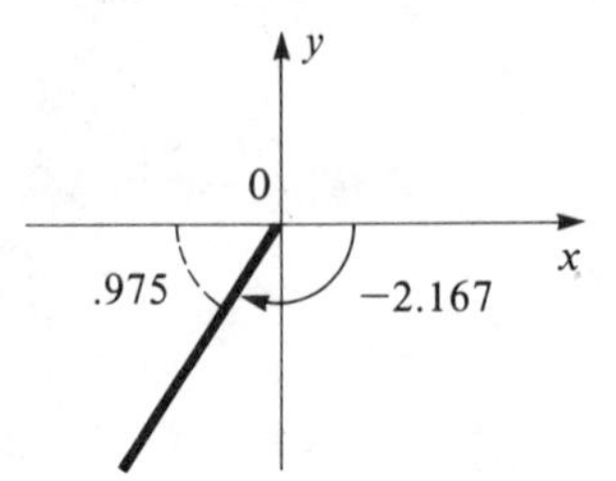

reference angle is $3.142 - 2.167 = 0.975$. Since the cosecant is negative in quadrant III, we have

$$\csc(-2.167) = -\csc 0.975$$

Using Table II, we obtain,

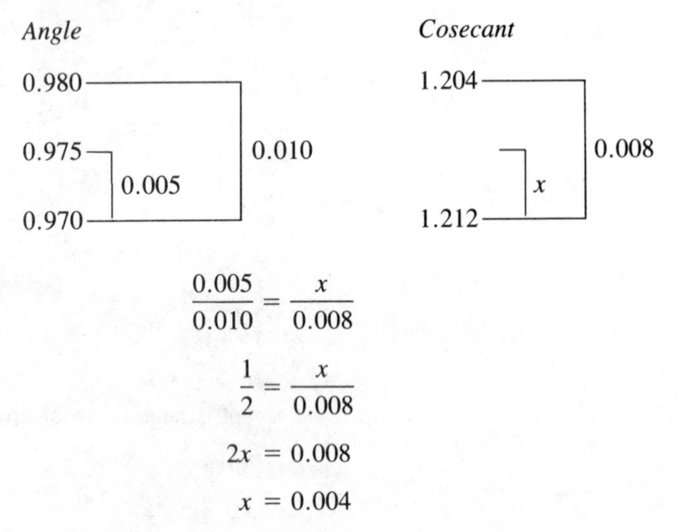

$$\frac{0.005}{0.010} = \frac{x}{0.008}$$

$$\frac{1}{2} = \frac{x}{0.008}$$

$$2x = 0.008$$

$$x = 0.004$$

The cosecant is a decreasing function for angles between 0 and $\pi/2$, so

$$\csc 0.975 = 1.212 - x = 1.212 - 0.004 = 1.208$$

and therefore

$$\csc(-2.167) = -1.208$$

EXAMPLE 2.4

a. Find θ to the nearest thousandth of a radian if $\sin \theta = 0.5244$ and θ is acute.

b. Find θ to the nearest minute if $90^\circ \leq \theta \leq 180^\circ$ and $\cos \theta = -0.2342$.

Solution **a.** Using Table II we get

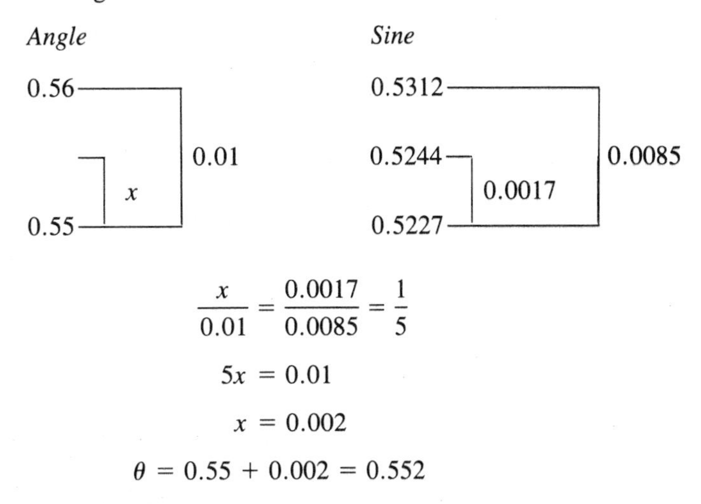

$$\frac{x}{0.01} = \frac{0.0017}{0.0085} = \frac{1}{5}$$

$$5x = 0.01$$

$$x = 0.002$$

$$\theta = 0.55 + 0.002 = 0.552$$

b. We first find the reference angle and then subtract this from 180°.

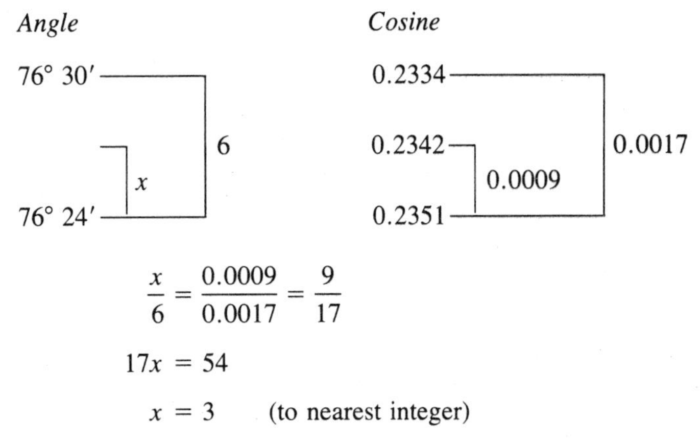

$$\frac{x}{6} = \frac{0.0009}{0.0017} = \frac{9}{17}$$

$$17x = 54$$

$$x = 3 \quad \text{(to nearest integer)}$$

Reference angle $= 76^\circ\ 27'$.

$$\theta = 180^\circ - 76^\circ\ 27' = 103^\circ\ 33'$$

EXERCISE SET 2.2

A In Problems 1–5 use a hand calculator to find the function values. Round off answers to four decimal places.

1. **a.** sin 38.35° **b.** cos 52.78° **c.** tan 12.17° **d.** cot 43.94°

2. **a.** sec 21.52° **b.** csc 82.06° **c.** sin 52° 13′ **d.** cos 113° 25′

3. **a.** tan 250° 51′ **b.** sec 27° 32′ **c.** sin(−433.92°) **d.** cos(−121° 52′)

4. **a.** sin 0.234 **b.** cos 0.568 **c.** tan 1.053 **d.** cot 0.862

5. **a.** sec 0.487 **b.** csc 0.218 **c.** sin(−2.345) **d.** tan(−5.201)

In Problems 6–14 use a hand calculator to find the angle correct to the nearest hundredth of a degree. Assume θ is a positive acute angle unless otherwise specified.

6. **a.** $\sin \theta = 0.4213$ **b.** $\sec \theta = 1.2015$

7. **a.** $\tan \theta = 0.5678$ **b.** $\cos \theta = 0.9625$

8. **a.** $\csc \theta = 2.5348$ **b.** $\cot \theta = 1.0537$

9. **a.** $\tan \theta = \dfrac{27}{64}$ **b.** $\sin \theta = \dfrac{7}{13}$

10. **a.** $\cos \theta = \dfrac{2\sqrt{2}}{3}$ **b.** $\cot \theta = \dfrac{4}{\sqrt{5}}$

11. **a.** $\csc \theta = \dfrac{\sqrt{5}}{2}$ **b.** $\sec \theta = \dfrac{65}{63}$

12. **a.** $\sin \theta = -0.4237, \quad -90° \le \theta \le 0°$
b. $\cos \theta = -0.8732, \quad 90° \le \theta \le 180°$

13. **a.** $\tan \theta = -1.238, \quad -90° < \theta \le 0°$
b. $\sec \theta = -2.435, \quad 90° \le \theta \le 180°$

14. **a.** $\csc \theta = -1.037, \quad -90° \le \theta \le 0°$
b. $\cot \theta = -0.7862, \quad -90° < \theta < 0°$

In Problems 15–20 use a hand calculator to find the angle θ correct to the nearest thousandth of a radian. Assume θ is a positive acute angle unless otherwise specified.

15. **a.** $\tan \theta = 0.2786$ **b.** $\cos \theta = 0.7258$

16. **a.** $\sin \theta = 0.3972$ **b.** $\cot \theta = 1.239$

17. **a.** $\csc \theta = 2.483$ **b.** $\sec \theta = 1.864$

18. **a.** $\sin \theta = -0.5684, \quad -\pi/2 \le \theta \le 0$
b. $\cos \theta = -0.3946, \quad \pi/2 \le \theta \le \pi$

19. **a.** $\tan \theta = -0.4967, \quad -\pi/2 < \theta \le 0$
b. $\sec \theta = -1.387, \quad \pi/2 < \theta \le \pi$

20. **a.** $\csc \theta = -3.425, \quad -\pi/2 \le \theta < 0$
b. $\cot \theta = -1.823, \quad -\pi/2 < \theta < 0$

In Problems 21–24 use Table I to find the function value.

21. **a.** sin 32° 12′ **b.** cos 18° 48′ **c.** tan 26° 30′ **d.** csc 41° 54′

22. **a.** $\cot 72^\circ 36'$ **b.** $\sec 46^\circ 18'$ **c.** $\cos 52.3^\circ$ **d.** $\sin 63.8^\circ$

23. **a.** $\tan 73.5^\circ$ **b.** $\sec 31.4^\circ$ **c.** $\sin 123^\circ 24'$ **d.** $\cot 212^\circ 42'$

24. **a.** $\cos(-32.7^\circ)$ **b.** $\tan(-254.6^\circ)$ **c.** $\sec 488^\circ$ **d.** $\csc 610^\circ 30'$

In Problems 25–28 use Table II to find the function value.

25. **a.** $\tan 0.21$ **b.** $\sin 1.05$ **c.** $\sec 0.48$ **d.** $\cos 0.13$

26. **a.** $\csc 1.12$ **b.** $\cot 0.98$ **c.** $\cos 0.83$ **d.** $\tan 1.22$

27. **a.** $\sin 0.57$ **b.** $\sec 0.09$ **c.** $\cot 2.03$ **d.** $\csc 1.94$

28. **a.** $\sin 2.54$ **b.** $\cos 4.15$ **c.** $\tan(-0.25)$ **d.** $\sec(-5.32)$

In Problems 29 and 30 find the acute angle θ in degrees having the given function value, using Table I.

29. **a.** $\sin \theta = 0.5402$ **b.** $\cos \theta = 0.4415$ **c.** $\tan \theta = 0.3959$ **d.** $\sec \theta = 1.3563$

30. **a.** $\csc \theta = 1.1618$ **b.** $\cot \theta = 2.0872$ **c.** $\cos \theta = 0.1323$ **d.** $\sin \theta = 0.9505$

In Problems 31–39 use linear interpolation and the appropriate table to find the approximate value of the specified function.

31. **a.** $\sin 36^\circ 28'$ **b.** $\cos 25^\circ 14'$

32. **a.** $\tan 12^\circ 45'$ **b.** $\sec 43^\circ 16'$

33. **a.** $\csc 59^\circ 34'$ **b.** $\cot 72^\circ 47'$

34. **a.** $\tan 46.32^\circ$ **b.** $\sec 63.58^\circ$

35. **a.** $\sin 81.67^\circ$ **b.** $\cos 53.07^\circ$

36. **a.** $\csc 21.92^\circ$ **b.** $\cot 40.52^\circ$

37. **a.** $\sin 0.134$ **b.** $\cos 1.237$

38. **a.** $\tan 1.305$ **b.** $\cot 1.368$

39. **a.** $\sec 0.832$ **b.** $\csc 0.543$

In Problems 40–42 use Table I and linear interpolation to find the acute angle A to the nearest minute.

40. **a.** $\sin A = 0.4765$ **b.** $\sec A = 1.2934$

41. **a.** $\tan A = 1.2762$ **b.** $\cos A = 0.9416$

42. **a.** $\cot A = 0.1521$ **b.** $\csc A = 1.2992$

In Problems 43–45 use Table II and linear interpolation to find the acute angle B to the nearest thousandth of a radian.

43. **a.** $\cos B = 0.9330$ **b.** $\sin B = 0.6233$

44. **a.** $\sec B = 2.057$ **b.** $\tan B = 5.696$

45. **a.** $\csc B = 1.143$ **b.** $\cot B = 0.6993$

B In Problems 46–51 approximate the value to four places using the appropriate table and linear interpolation.

46. **a.** $\sin 132^\circ 45'$ **b.** $\cos 257^\circ 13'$

47. a. $\tan 318° 22'$ b. $\cot(-159° 47')$
48. a. $\sec(-485° 51')$ b. $\csc 953° 52'$
49. a. $\sec 13$ b. $\sin 4.031$
50. a. $\tan(-9.872)$ b. $\cos 3.245$
51. a. $\cot 5$ b. $\csc(-2.154)$

In Problems 52–54 find the angle θ (a) to the nearest minute, and (b) to the nearest thousandth of a radian, using tables and linear interpolation.

52. $\sin \theta = 0.7839, \quad \pi/2 \leq \theta \leq \pi$
53. $\cos \theta = -0.2226, \quad \pi \leq \theta \leq 3\pi/2$
54. $\tan \theta = -0.7231, \quad -\pi/2 \leq \theta \leq 0$

Use a hand calculator for Problems 55–60.

55. Find the value of each of the following to five decimal places.
    a. $\cot \dfrac{5\pi}{8}$ b. $\csc \dfrac{13\pi}{12}$
56. Find θ in degrees and minutes if
    a. $\sin \theta = 0.3127$ and $90° \leq \theta \leq 180°$
    b. $\sec \theta = 1.3482$ and $270° \leq \theta \leq 360°$
57. Find θ to the nearest hundredth of a degree if
    a. $\cos \theta = -0.4278$ and $180° \leq \theta \leq 270°$
    b. $\cot \theta = -1.396$ and $90° \leq \theta \leq 180°$
58. Find θ in radians correct to five decimal places if
    a. $\tan \theta = 1.2345$ and $\pi \leq \theta \leq 3\pi/2$
    b. $\csc \theta = -2.1342$ and $3\pi/2 \leq \theta < 2\pi$
59. Find angle A in degrees and minutes if A is acute and
$$\sin A = \frac{12.35 \sin 118° 39'}{15.72}$$
60. Find b correct to three decimal places if
$$b^2 = (2.014)^2 + (4.372)^2 - 2(2.014)(4.372) \cos 143° 28'$$

2.3 Accuracy of Computed Results

As previously stated, most of the values of trigonometric functions found in tables are only approximations to the true values. Even though calculators give greater accuracy than most tables, the values are still approximations. Most of the values of trigonometric functions, in fact, are irrational and so can never be written down as exact decimal quantities. So whenever we use tables or a calculator for trigonometric functions and then use the values to compute some quantity, the question of the accuracy of our result naturally arises. The same is true whenever we use an approximation to any quantity in a calculation. For example, when measured quantities, such as distances, are employed, there is inevitably a question of degree of accuracy of the result.

To illustrate the type of question we are concerned with, suppose we are able to measure the radius of a circle to the nearest hundredth of an inch, and we find it to be 2.35 inches. Suppose we then use a calculator to compute the area by the formula $A = \pi r^2$ and get

$$A = 17.34944543 \text{ square inches}$$

Would we be justified in claiming that we have found the area to this degree of accuracy? Clearly, the answer is no. But to what degree of accuracy can we reasonably write the result? The best we can say is that the area is approximately 17.3, and even then there is some doubt about the last digit. To see why this is so, we need to understand the concepts of *rounding off* and of *significant digits*, which we discuss below.

When we **round off** a number, the last digit retained is either left unchanged or increased by one unit in that place, according to whether the amount dropped is less than or greater than one-half unit in the last place retained. When it is exactly one-half, we follow the convention of rounding so that the last digit retained is *even*. Here, then, are some examples of numbers rounded to two decimal places:

Number	Rounded number	
32.576	32.58	
5.2349	5.23	
163.045	163.04	we rounded to the *even* number
77.875	77.88	
2.045001	2.05	

To find the number of **significant digits** in a number, we count all non-zero digits, and we count zero except when it is used solely to place the decimal. For example, 0 is significant in the numbers 201 and 1.350, but it is not significant in 0.002. When a whole number ends in 0, such as 1,200, the situation is ambiguous. Without more information we would not know if this is to the nearest 100, the nearest 10, or to the nearest unit, and so we would not know whether there are 2, 3, or 4 significant digits. We can remove the ambiguity by employing **scientific notation**, in which we express the number as the product of a number between 1 and 10 and an appropriate power of 10. Here are some examples:

Number	Scientific notation
2,135	2.135×10^3
102.5	1.025×10^2
1,500,000	1.5×10^6
0.0023	2.3×10^{-3}
0.000000587	5.87×10^{-7}

To determine the correct power of 10, place the decimal to the right of the first non-zero digit and count the number of places it would have to be moved from there to give the number. If it has to be moved to the right, the power of 10 is positive, and if to the left, it is negative. Check the above examples to verify this rule.

Consider again the number 1,200. If we mean only two significant digits (accuracy to the nearest 100), we can write 1.2×10^3. If we mean three significant digits (accuracy to the nearest 10), we write 1.20×10^3, and if we mean all four digits are significant (accuracy to the nearest unit), we write 1.200×10^3. Here are some further examples illustrating significant digits:

Number	Number of significant digits
2.37	3
1.02	3
0.045	2
0.0450	3
2.001	4
1.234×10^{12}	4
3.00×10^{-5}	3

One further concept is needed before we state rules for computational accuracy, namely that of the relative *precision* of two numbers. A number A is said to be **more precise** than B if the last significant digit of A occurs in a position to the right of the last significant digit of B. Roughly speaking, A has more decimal places of accuracy than B. For example, 0.0032 is more precise than 2.753, even though the latter number has more significant digits. Note that precision is affected by the position of the decimal, but the number of significant digits is not.

We now state without proof two rules for calculating with approximate numbers. Both are based on the general principle that the accuracy of the result of a calculation is limited by the least accurate of the data involved.

Rule for Addition and Subtraction of Approximate Data

Carry out the computations with the given data, and round off the answer so that its precision is that of the least precise of the data.

Note. The more precise data may first be rounded to *one place greater* than will be retained in the final answer, but they should not be rounded all the way to that of the least precise data. (An exception is when only two numbers are added or subtracted.)

EXAMPLE 2.3 Find the following sum of approximate numbers, and express the answer to the appropriate precision.

$$32.155 + 17.2 + 69.1048$$

Solution The least precise number is 17.2, so we initially round off the other numbers to two decimal places and then add

$$\begin{array}{r} 32.16 \\ 17.2 \\ \underline{69.10} \\ 118.46 \end{array}$$

Finally, we round off to 118.5.

Rule for Multiplication and Division of Approximate Data

Carry out the computations with the given data and round off the answer so that it has the same number of significant digits as the number in the data with the least number of significant digits.

Note. Again, it is permissible to round the more accurate data initially to *one more* significant digit than will ultimately be retained.

In computations involving powers and roots, the rule for multiplication should be used.

EXAMPLE 2.4 Find the product of the following approximate numbers, and express the answer with the appropriate number of significant digits.

$$(3.2568)(0.0204)(729.1)$$

Solution With the aid of a calculator we obtain 48.440471. Since the least accurate of the data is 0.0204, which has three significant digits, we are justified in expressing the answer with three significant digits only, namely 48.4.

The subject of errors in computed results requires far more analysis for a complete understanding than we can go into here, but the rules stated provide reasonable results. It should be noted, however, that even when these rules are followed, there can be some doubt about the reliability of the answer, especially in the last digit. This uncertainty can be illustrated by returning to the example of the area of the circle where the radius was measured to be 2.35 inches. This measurement means that the true radius is between 2.345 inches and 2.355 inches. Using these two extreme values, we find from a calculator that the area lies between 17.276 and 17.423 square inches. If we apply our rule for multiplication, we get the answer 17.3. So by the above calculations we see there is uncertainty in the final digit.

When working with triangles, the approximate relationships between accuracy of angular measure and the number of significant digits in the measured sides is given by the following table:

Angle to nearest	Significant digits
1 degree	2
10′ or 0.1 degree	3
1′ or 0.01 degree	4
10″ or 0.001 degree	5

EXERCISE SET 2.3

A

1. Round off each number to the nearest hundredth.
a. 321.547 **b.** 2.1051 **c.** 0.2049 **d.** 1.035 **e.** 25.645

2. State the number of significant digits in each of the following.
a. 2.32 **b.** 0.025 **c.** 2,003 **d.** 1.0030 **e.** 0.0200

3. Write each of the following in scientific notation.
a. 24,000 (3 significant digits) **b.** 24,000 (4 significant digits)
c. 0.00052 **d.** 3,201.5 **e.** 85,000,000 (4 significant digits)

4. Round off each of the following to three significant digits.
a. 23,815 **b.** 0.4035 **c.** 576.5 **d.** 3.1552 **e.** 9.2349×10^{17}

In Problems 5–18 all of the numbers are approximate. Perform the indicated operations and express the answer to the appropriate accuracy.

5. $234.15 + 1.0236 - 15.7$

6. $5.0267 - 13.34 + 17.326 - 0.02145$

7. $(2.13)(4.025)$

8. $\dfrac{0.784}{2.5}$

9. $\dfrac{(39.6)(0.054)}{6.6023}$

10. $(3.78215)(6.72) + (0.0346)(278.59)$

11. $32.35 \sin 25.4°$

12. $7.15 \tan 47° 14'$

13. $\dfrac{100.2}{\cos 37° 42'}$

14. $(1.3)^2 + (2.1)^2 - 2(1.3)(2.1) \cos 119° 32'$ (Treat 2 as exact)

15. $\dfrac{36.4 \sin 18.7°}{\sin 56.85°}$

B **16.** $\dfrac{\sqrt{3.04576}\,(18.3)^3}{(205.98)^2}$ **17.** $\dfrac{(3.14 \times 10^4)(0.020355)}{6.234 \times 10^{-2}}$

18. $\dfrac{72.45 \sin 134^\circ\, 43'}{\sqrt{(72.45)^2 + (59.27)^2 - 2(72.45)(59.27) \cos 134^\circ\, 43'}}$ (Treat 2 as exact)

19. If Q is the true value of some quantity and q is an approximation to Q, then the **absolute error**, e, is defined as $e = |Q - q|$, and the **relative error**, r, is defined as $r = e/Q$. Since in general Q is not known, we can usually obtain only an estimate for e and r. In each of the following approximate numbers, find the maximum absolute error and estimate the maximum relative error by using $r \approx e/q$.

a. 214.3 **b.** 0.025 **c.** 15.48 **d.** 5.723×10^{-6}

20. The precision of a number is related to its absolute error, and the accuracy is related to its relative error (see Problem 19). If the absolute error in A is less than the absolute error in B, then A is more precise than B. If the relative error in A is less than the relative error in B, then A is more accurate than B. Compare the precision and accuracy of the numbers in Problem 19.

21. The dimensions of a rectangular box are measured to be 24.1 inches by 18.3 inches by 12.5 inches. Find the volume of the box to the appropriate accuracy. By using the largest and smallest values possible for the dimensions as measured, find the largest and smallest possible values of the volume.

2.4 Applications of Right Triangle Trigonometry

We now return to the problem of solving right triangles. The examples below illustrate the techniques involved and show how the theory can be applied in calculating inaccessible distances.

EXAMPLE 2.5 Solve the right triangle ABC in which $A = 37^\circ\, 30'$ and $b = 17.3$ (Figure 2.4).

Figure 2.4

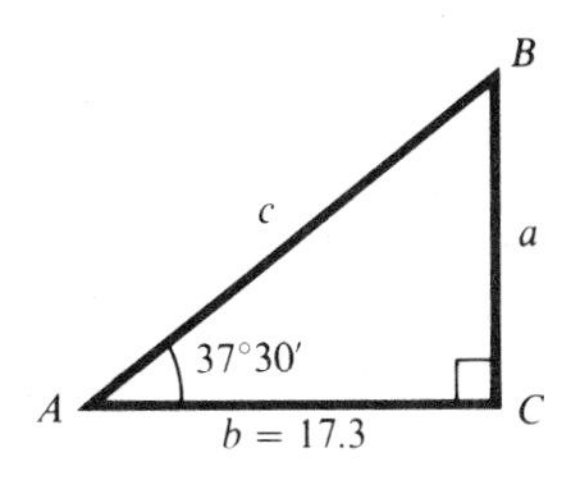

Solution Since A and B are complementary,

$$B = 90^\circ - 37^\circ\, 30' = 52^\circ\, 30'$$

To find a we use the tangent of angle A.

$$\tan A = \frac{a}{b}$$

$$a = b \tan A = 17.3 \tan 37^\circ\, 30' = 13.3$$

By calculator:

17.3 [×] 37.5 [TAN] [=] 13.274757

For c we use the cosine of A.

$$\cos A = \frac{b}{c}$$

$$c = \frac{b}{\cos A} = \frac{17.3}{\cos 37^\circ 30'} = 21.8$$

By calculator:

17.3 [÷] 37.5 [COS] [=] 21.806173

Remarks. In deciding which function to use when there is a choice, we are guided by the following considerations.

1. If possible, it should involve only one unknown quantity.
2. To the extent possible, it should involve original data rather than calculated quantities. This helps to reduce cumulative errors.
3. If tables are used, a function should be chosen, if possible, that results in multiplication rather than division. When a calculator is used, this is not an important consideration. (A calculator was used in Example 2.5. Otherwise, we would have used the secant function to calculate c.)

EXAMPLE 2.6 Figure 2.5 illustrates a river and two points P and Q on opposite sides. By means of a transit at P, a line of sight perpendicular to PQ is determined, and a distance of 100 feet along this line from P to R is measured. At R the transit is again set up and the angle from RP to RQ is measured and found to be 32.3°. It is desired to find the distance from P to Q.

Solution Let the distance from P to Q be denoted by x. Then

$$\tan 32.3^\circ = \frac{x}{100}$$

$$x = 100 \tan 32.3^\circ = 63.2 \text{ feet}$$

Figure 2.5

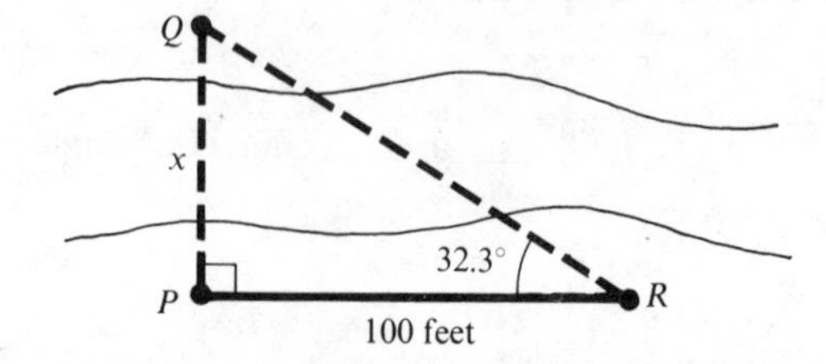

The following terminology will be employed in certain applications.

Angle of elevation. Angle from the horizontal upward to the line of sight of the observer.

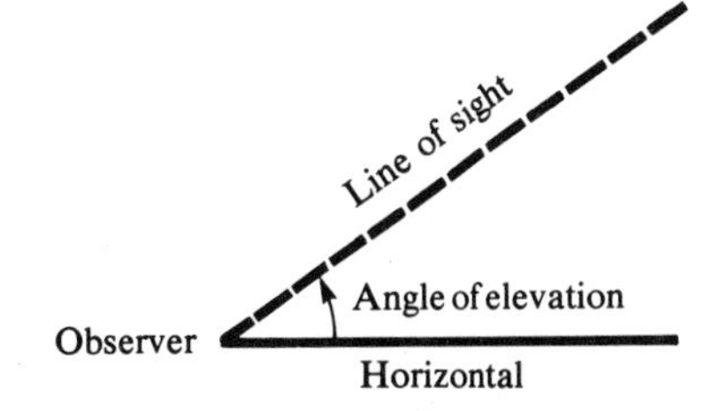

Angle of depression. Angle from the horizontal downward to the line of sight of the observer.

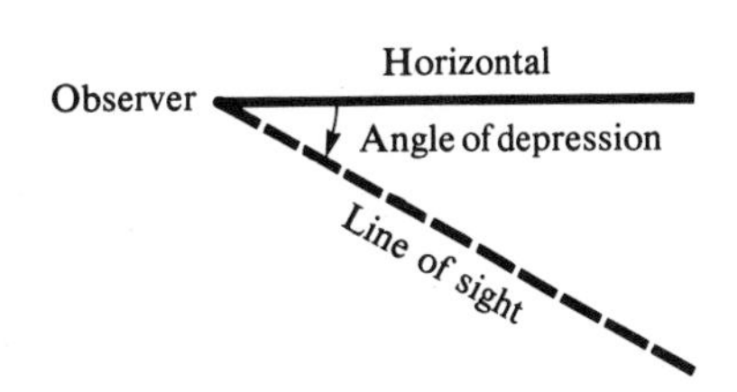

Bearing. A means of giving the direction of a path by using either North or South as a base line and then indicating an acute angle either East or West of the base line. For example, N 30° E means a path in a direction 30° to the East of North. Similarly, S 60° W means a direction which is 60° West of South.

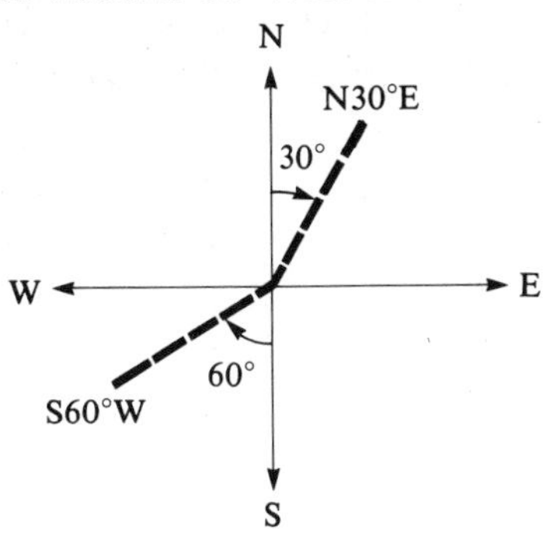

Heading. An alternate means of indicating direction in which an angle between 0° and 360° measured clockwise from North is given. The figure shows a heading of 225°. This means of indicating direction is widely employed, especially in air navigation.

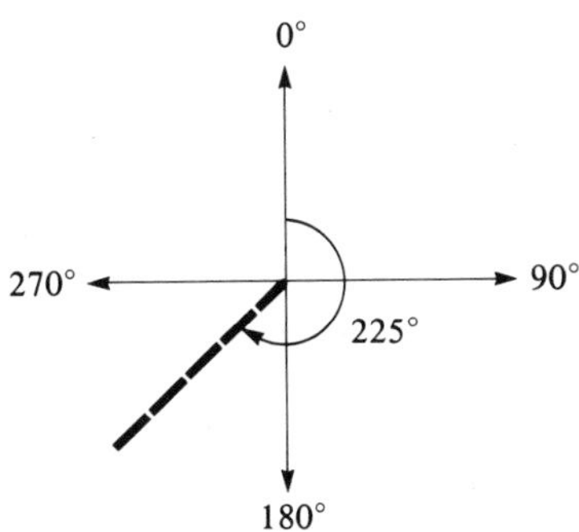

EXAMPLE 2.7 The angle of elevation of the top of a mountain from point A is 36.2°, and the angle of elevation of the top from point B, 1,000 feet nearer the base of the mountain and on the same level as A, is 41.7° (see Figure 2.6). Find the height of the mountain.

Figure 2.6

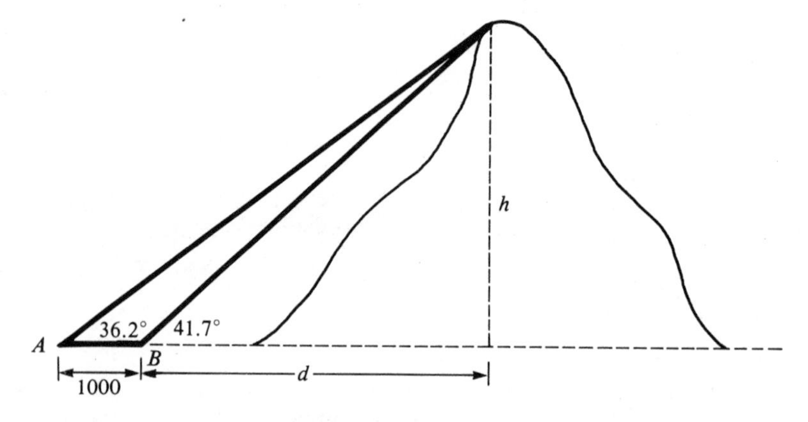

Solution We introduce the auxiliary unknown d in the figure, since it will be involved in the functions we use. There are two right triangles, each involving the unknown height h. In each triangle, h is the side opposite the known angle, and the adjacent side is the other about which we have at least some information. We could therefore use either the tangent or the cotangent. Either one would involve both h and d. By using both triangles we will get two equations in these two unknowns. Our object will be to eliminate d between the two equations, thereby obtaining an equation in h alone. We choose the cotangent function, since this facilitates the elimination of d.

$$\cot 36.2° = \frac{1{,}000 + d}{h}$$

$$1{,}000 + d = h \cot 36.2°$$

$$\cot 41.7° = \frac{d}{h}$$

$$d = h \cot 41.7°$$

Now we substitute for d from the last equation into the second, and then solve for h.

$$1{,}000 + h \cot 41.7° = h \cot 36.2°$$

$$h(\cot 36.2° - \cot 41.7°) = 1{,}000$$

$$h = \frac{1{,}000}{\cot 36.2° - \cot 41.7°}$$

$$= 4{,}099$$

By calculator:

1000 [÷] [(] 36.2 [TAN] [1/x] [−] 41.7 [TAN] [1/x] [)] [=] 4099.1782

So the mountain is approximately 4,100 feet above the surrounding countryside. If the elevation of points A and B above sea level is known, this can be added to 4,100 to give the height of the mountain above sea level.

EXAMPLE 2.8 Ships A and B leave port at the same time, ship A traveling in the direction N 39° 24′ E at an average speed of 25 miles per hour, and ship B traveling in the direction S 50° 36′ E at an average speed of 15 miles per hour. Find how far apart the ships are after two hours and the bearing of ship B from ship A.

Solution Since $39° 24' + 50° 36' = 90°$, we see that the paths of the two ships are at right angles to each other, as shown in Figure 2.7. After two hours ship A has gone 50 miles and ship B 30 miles. Let d denote the distance between the ships at that time. Then, by the Pythagorean theorem

$$d^2 = (50)^2 + (30)^2 = 2{,}500 + 900 = 3{,}400$$
$$d = 58.3 \text{ miles}$$

To find the bearing of B from A, first we find the angle θ at A in triangle AOB. We have

$$\tan \theta = \frac{30}{50} = 0.6000$$
$$\theta = 30° 58'$$

Now, if we draw a line from A in the south direction (see Figure 2.8), we see by geometry that the angle from this line to OA is 39° 24′. So the angle from this line to AB is

$$39° 24' - 30° 58' = 8° 26'$$

The bearing of B from A is therefore S 8° 26′ W.

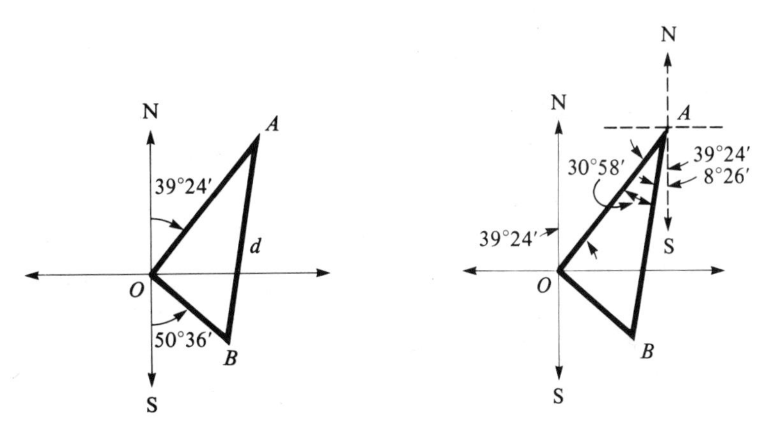

Figure 2.7 **Figure 2.8**

EXAMPLE 2.9 An airplane flies at a heading of 130° for 30 minutes and then changes to a heading of 220° and flies for 2 more hours. If the average speed of the plane is 250 miles per hour and the wind is negligible, find

a. The distance of the plane from its starting point.
b. The heading the plane would have taken to get directly to the final destination from the starting point.

Solution Let A be the starting point, B the point where a course change was made, and C the final destination (see Figure 2.9). Since $220° - 130° = 90°$, the angle at B in triangle ABC is a right angle. Let $d = \overline{AC}$. The distances $\overline{AB}$ and $\overline{BC}$ are determined from the given flight times to be 125 miles and 500 miles, respectively.

Figure 2.9

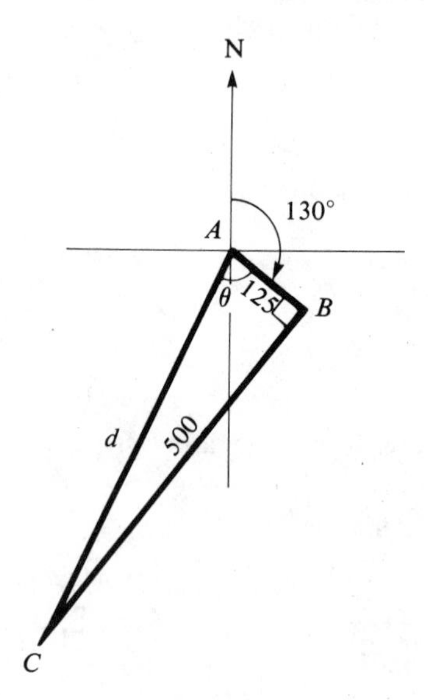

a. By the Pythagorean theorem,

$$d = \sqrt{(125)^2 + (500)^2} = 515.4$$

b. Let θ be the angle as shown in Figure 2.9. Then

$$\tan\theta = \frac{500}{125} = 4.0000$$

$$\theta = 75°\ 58'$$

The heading to go directly from A to C should therefore be

$$130° + 75°\ 58' = 205°\ 58'$$

EXERCISE SET 2.4

A In Problems 1–16 triangle ABC is a right triangle with right angle at C. Solve the triangle.

1. $A = 57.4°$, $b = 15.2$
2. $a = 32.5$, $b = 12.7$
3. $B = 67.3°$, $b = 13.5$
4. $A = 33°\ 20'$, $c = 120$
5. $A = 49.7°$, $c = 24.6$
6. $B = 37°\ 40'$, $c = 14.2$
7. $A = 32.5°$, $a = 24.7$
8. $B = 64.3°$, $a = 205$
9. $a = 15.35$, $c = 32.14$
10. $B = 62°\ 15'$, $c = 104.3$
11. $B = 22°\ 40'$, $b = 2.56$
12. $A = 43°\ 20'$, $b = 10.5$
13. $A = 35.57°$, $a = 3.462$
14. $a = 32.51$, $b = 21.28$
15. $b = 25.4$, $c = 38.2$
16. $a = 5.136$, $c = 8.423$

17. From a point on the ground 150 feet from the base of a building, the angle of elevation of the top of the building is 68.5°. How high is the building?

18. A rectangle is 14.3 inches long and 8.72 inches high. Find the angle between a diagonal and the base. What is the length of the diagonal?

19. From the top of a lighthouse the angle of depression of a boat is 32.4°. If the lighthouse is 80.5 feet high, how far is the boat from the base of the lighthouse?

20. A tower 102.6 feet high is on the bank of a river. From the top of the tower the angle of depression of a point on the opposite side of the river is 28° 14′. How long a cable would be required to reach from the top of the tower to the point on the opposite bank?

21. A guy wire for a telephone pole is anchored to the ground at a point 21.3 feet from the base of the pole, and it makes an angle of 64.7° with the horizontal. Find the height of the point where the guy wire is attached to the pole and the length of the guy wire.

22. A surveyor wants to find the distance between points A and B on opposite sides of a pond. With her transit at point A, she determines a line of sight at right angles to AB and establishes point C along this line 150 feet from A. Then she measures the angle at C from CA to CB and finds it to be 62° 10′. How far is it from A to B?

23. Town A is 12.0 miles due west of town B, and town C is 15.0 miles due south of town B. How far is it from town A to town C? Assuming straight roads connect the three towns, what is the angle at A between the roads leading to B and C?

24. A bar with a regular pentagonal (five-sided) cross-section is to be milled from round stock. What diameter should the stock be if the dimension of each flat section is to be 2.50 centimeters? (See figure.)

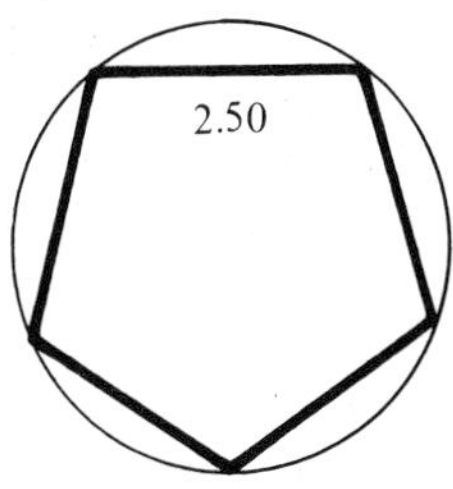

25. Points A and B are situated on the coast, with B 1,250 meters due south of A. From A the bearing of a lighthouse on an island is S 42.2° E, and from B the bearing of the lighthouse is N 47.8° E. Find the distance of the lighthouse from each of the points A and B.

26. An oil drilling platform is located in the Gulf of Mexico 3.25 miles from the nearest point, A, on shore. From a point B on the shore due east of A the bearing of the platform is S 51.2° W. How far is it from B to the platform?

27. Two forest ranger towers are 10.35 miles apart. Smoke is sighted from the first tower at a bearing of N 36° 24′ W, and simultaneously it is spotted from the second tower at a bearing of N 53° 36′ E. If the first tower is due east of the second, find the distance from each tower to the source of the smoke.

28. A jetliner travels at a heading of 312° for 45 minutes at an average speed of 480 miles per hour. Then the course is changed to a heading of 42°. After flying at this new heading for 25 minutes, maintaining the same average speed as before, what is the distance of the jetliner from the starting point? (Neglect the effect of the wind.)

29. An airplane flies in a straight path in a southwesterly direction. After one hour the plane is 205 miles west and 78 miles south of its starting point. If the velocity of the wind was negligible, determine the heading of the airplane. What was its average speed?

B 30. The accompanying figure shows the angles of elevation of a balloon at a certain instant from points A and B 1,200 meters apart. Find the height of the balloon at that instant.

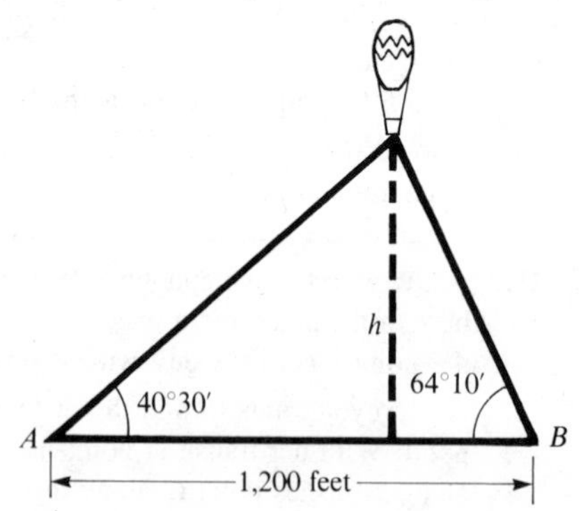

31. From a lighthouse 120 feet high, the angles of depression of two boats directly in line with and on the same side of the lighthouse are found to be 34° 30′ and 27° 40′. How far apart are the boats?

32. At a point 250 feet from the base of a building the angle of elevation of the top of the building is 27.5°, and the angle of elevation of the top of a statue on the top edge of the building is 30.2°. How high is the statue?

33. Find the length of 1° of longitude on the circle of latitude through Chicago, 41° 50′ north. (Assume the radius of the earth is 3,960 miles.)

34. Ship A leaves port at 2:00 P.M. and travels in the direction S 34° E at an average speed of 15 knots (nautical miles per hour). At 2:45 P.M. ship B leaves the same port and travels in the direction N 56° E at an average speed of 20 knots. Find:
    a. the distance between the ships at 4:00 P.M., and
    b. the bearing of ship B from ship A at 4:00 P.M.

35. From a ship, the bearings of two landmarks on shore are found to be S 37.2° W and S 21.7° W, respectively. If the landmarks are known to be 5 kilometers apart, find how far the ship is from the nearest point on shore.

36. A pilot leaves point A and flies for 35 minutes at a heading of 142° 20′ and arrives at point B. The pilot then flies at a heading of 232° 20′ for 1 hour and 12 minutes, arriving at point C. The average speed on each leg of the trip was 180 miles per hour, and wind

velocity was negligible. How far is it from A to C? If the pilot wanted to fly directly from A to C, what would be the heading?

2.5 Vectors

Certain physical quantities have both *magnitude* and *direction*. Examples are force, velocity, and acceleration. Such quantities can be represented geometrically by means of a line segment with an arrowhead indicating direction. The length of the segment represents the magnitude of the quantity, and the direction of the segment as measured by an angle from some base line (such as the x axis) represents the direction of the physical quantity. Such a directed line segment is called a **vector** (see Figure 2.10). The endpoint of the segment with the arrowhead is called the terminal point (or tip) of the vector, and the other endpoint is called the initial point (or tail).

There are various means of designating vectors. Frequently a single boldface letter, such as **V**, is used. Alternately, the symbol $\vec{V}$ is employed (this is especially useful in written work, where boldface letters are not easily made). If the endpoints of a vector are named by letters, the vector can also be indicated by using the two letters. For example, the vector from point A to point B could be designated either by **AB** or by $\overrightarrow{AB}$. The magnitude of a vector **V** is designated by |**V**|. Similarly, using the other notations, the magnitude is designated by $|\vec{V}|$, |**AB**|, or $|\overrightarrow{AB}|$.

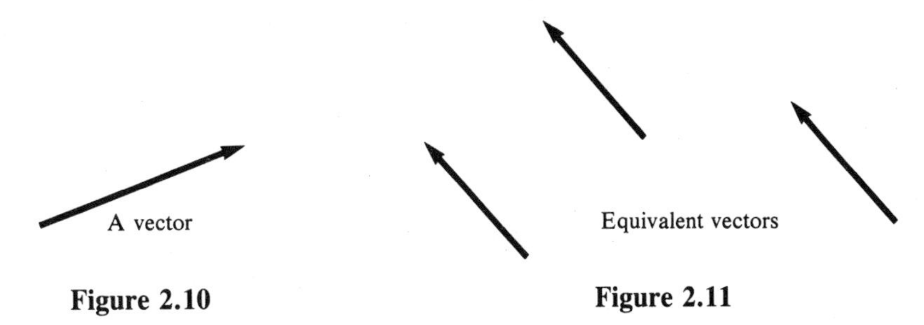

Figure 2.10 **Figure 2.11**

Two vectors are said to be **equivalent** if they have the same magnitude and direction. Figure 2.11 shows three equivalent vectors. If **V** and **W** are equivalent, we write **V** = **W**, and we may treat them as if they were the same vector. A consequence of this is that a vector can be moved from one place to another, without changing its effect, so long as it remains parallel to its original direction.

Two vectors having different directions are *added* as follows. Place the vectors so that their initial points coincide. Then construct a parallelogram with the two vectors as adjacent sides. The diagonal from the common initial point is then the sum of the two vectors. This is illustrated in Figure 2.12. It should be clear that **V** + **W** = **W** + **V**. That is, vector addition is commutative. When **V** and **W** are in the same direction, their sum is a vector in that direction with magnitude $|\mathbf{V}| + |\mathbf{W}|$. When **V** and **W** are in opposite directions, their sum is in the direction of the vector with the greater magnitude, and the magnitude of the sum vector is the magnitude of the larger minus the magnitude of the smaller vector. In case they are equal in magnitude and opposite in direction, their sum is the zero vector, denoted by **0**. No direction is assigned to the zero vector.

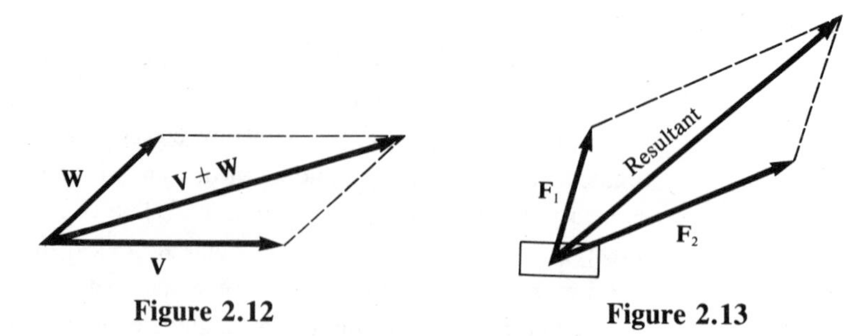

Figure 2.12 **Figure 2.13**

The sum of two vectors is often called their **resultant**. This name comes from the physical problem of the application of two forces to a body, as in Figure 2.13. The net result, or resultant, of the two forces is in the direction of the diagonal and has magnitude given by the length of the diagonal. That is, the resultant force is the sum $\mathbf{F}_1 + \mathbf{F}_2$. The two forces could be replaced by the single force represented by the resultant vector.

An alternate way of viewing the sum of two vectors **V** and **W** is to place the initial point of **W** at the terminal point of **V** and complete the triangle from the initial point of **V** to the terminal point of **W**, as in Figure 2.14. You can convince yourself that this definition of $\mathbf{V} + \mathbf{W}$ is equivalent to the one previously given.

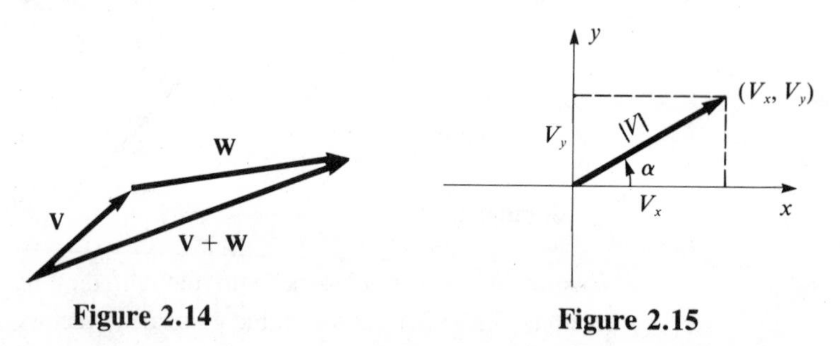

Figure 2.14 **Figure 2.15**

It is frequently convenient to introduce a rectangular coordinate system when working with vectors. Let the initial point of a non-zero vector **V** be placed at the origin, (we say the vector is then in **standard position**), and denote the coordinates of the terminal point of **V** by (V_x, V_y), as in Figure 2.15. These coordinates are called the **horizontal component** and the **vertical component**, respectively, of the vector **V**. If α denotes the angle in standard position from the positive x axis to **V**, we see that

$$\cos\alpha = \frac{V_x}{|\mathbf{V}|} \quad \text{and} \quad \sin\alpha = \frac{V_y}{|\mathbf{V}|}$$

or

$$V_x = |\mathbf{V}|\cos\alpha \quad \text{and} \quad V_y = |\mathbf{V}|\sin\alpha$$

Also, by the Pythagorean theorem

$$|\mathbf{V}|^2 = V_x{}^2 + V_y{}^2$$

Remark. As defined above, V_x and V_y are *numbers* and not vectors. However, it is also sometimes convenient to speak of **vector components** of $\mathbf{V}$. The horizontal *vector* component, $\mathbf{V}_x$, of $\mathbf{V}$ is the vector with initial point at the origin and terminal point at $(V_x, 0)$ on the x axis. The vertical *vector* component, $\mathbf{V}_y$, is the vector from the origin to the point $(0, V_y)$ on the y axis. Note that $|\mathbf{V}_x| = V_x$, $|\mathbf{V}_y| = V_y$, and $\mathbf{V}_x + \mathbf{V}_y = \mathbf{V}$. More generally, if $\mathbf{V} = \mathbf{U} + \mathbf{W}$, then $\mathbf{U}$ and $\mathbf{W}$ are vector components of $\mathbf{V}$. We sometimes wish to find particular vector components of a vector $\mathbf{V}$, and when this is done, we say we have *resolved* $\mathbf{V}$ *into its vector components*. Resolving a vector into horizontal and vertical vector components is particularly common.

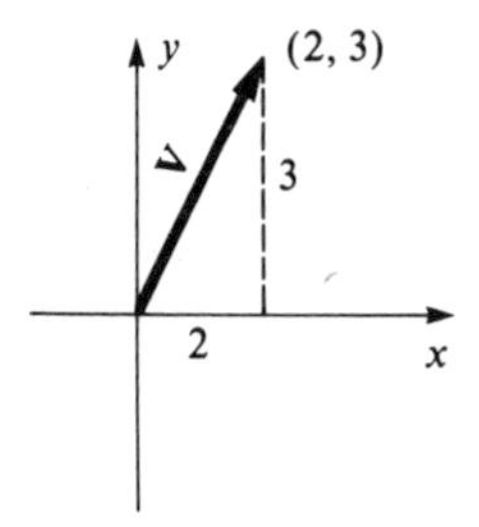

Figure 2.16

A vector $\mathbf{V}$ in standard position is completely determined by specifying its horizontal and vertical components V_x and V_y. Conversely, from $\mathbf{V}$, we can find V_x and V_y. For this reason we sometimes write

$$\mathbf{V} = (V_x, V_y)$$

For example, when we write $\mathbf{V} = (2, 3)$ we mean $\mathbf{V}$ is the vector whose horizontal component is 2 and whose vertical component is 3 (see Figure 2.16). With this understanding, consider two vectors $\mathbf{V} = (V_x, V_y)$ and $\mathbf{W} = (W_x, W_y)$. We define a certain kind of product of $\mathbf{V}$ and $\mathbf{W}$ as follows.

DEFINITION 2.1 The **dot product** of $\mathbf{V}$ and $\mathbf{W}$ is given by

$$\mathbf{V} \cdot \mathbf{W} = V_xW_x + V_yW_y$$

Note. It is essential to write the dot between $\mathbf{V}$ and $\mathbf{W}$. Thus $\mathbf{VW}$ and $\mathbf{V}(\mathbf{W})$ have no meaning. For example, if $\mathbf{V} = (2, -3)$ and $\mathbf{W} = (-5, -4)$, then

$$\mathbf{V} \cdot \mathbf{W} = (2)(-5) + (-3)(-4) = -10 + 12 = 2$$

Notice that the result of taking the dot product is a *number*, not a vector.*

*There is another type of vector product studied in more advanced mathematics courses called the **cross product**, written $\mathbf{V} \times \mathbf{W}$, which results in a vector.

An interesting relationship exists between the dot product of two non-zero vectors and the angle between them. If we designate this angle by θ (which is always between 0° and 180°) then we can show that

$$\mathbf{V} \cdot \mathbf{W} = |\mathbf{V}|\,|\mathbf{W}| \cos \theta, \qquad 0 \le \theta \le \pi \tag{2.1}$$

or equivalently, since neither $|\mathbf{V}|$ nor $|\mathbf{W}|$ is zero,

$$\cos \theta = \frac{\mathbf{V} \cdot \mathbf{W}}{|\mathbf{V}|\,|\mathbf{W}|} \tag{2.2}$$

You will be asked in Chapter 6 to prove Equation (2.1).

EXAMPLE 2.10 Let $\mathbf{V} = (2, 6)$ and $\mathbf{W} = (1, -2)$. Find

a. The dot product of **V** and **W**.
b. The angle between **V** and **W**.

Solution **a.** By Definiton 2.1,

$$\mathbf{V} \cdot \mathbf{W} = (2)(1) + (6)(-2) = 2 + (-12) = -10$$

b. By Equation (2.2),

$$\cos \theta = \frac{\mathbf{V} \cdot \mathbf{W}}{|\mathbf{V}|\,|\mathbf{W}|} = \frac{-10}{\sqrt{2^2 + 6^2}\,\sqrt{1^2 + 2^2}}$$

$$= \frac{-10}{\sqrt{40}\,\sqrt{5}} = \frac{-10}{10\sqrt{2}} = -\frac{1}{\sqrt{2}}$$

So

$$\theta = 135°$$

Since two non-zero vectors are perpendicular if, and only if, the angle between them is 90°, and since this is true if and only if $\cos \theta = 0$ (since $0° \le \theta \le 180°$), we obtain from Equation (2.2) the following criterion for perpendicular vectors.

> **Perpendicular Vectors**
> Two non-zero vectors are perpendicular if, and only if, their dot product is 0.

Remark. The zero vector is $\mathbf{0} = (0, 0)$, and if $\mathbf{V} = (V_x, V_y)$ is any vector, we have

$$\mathbf{V} \cdot \mathbf{0} = (V_x)(0) + (V_y)(0) = 0$$

So for consistency with the above criterion, we *define* the **0** vector to be perpendicular to every vector.

EXAMPLE 2.11 Prove that the vectors $\mathbf{A} = (3, 4)$ and $\mathbf{B} = (-8, 6)$ are perpendicular.

Solution

$$\mathbf{A} \cdot \mathbf{B} = (3)(-8) + (4)(6) = 0$$

So **A** and **B** are perpendicular.

EXAMPLE 2.12 A force of 100 pounds is acting horizontally to the right on an object, and a force of 65 pounds is acting vertically upward on the object. Find the direction and magnitude of the resultant force.

Solution We represent the two forces by vectors as shown in Figure 2.17, and let **R** be the resultant force. We have,

$$\tan\theta = \frac{65}{100} = 0.65$$

$$\theta \approx 33.0°$$

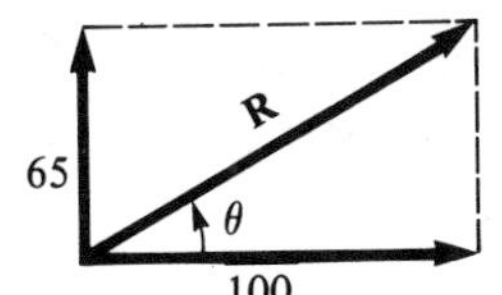

Figure 2.17

By calculator:

.65 [INV] [TAN] 33.023868

$$|\mathbf{R}| = \sqrt{(100)^2 + (65)^2} = \sqrt{14{,}225} \approx 119.3 \text{ pounds}$$

By calculator:

100 [x^2] [+] 65 [x^2] [=] [$\sqrt{x}$] 119.2686

EXAMPLE 2.13 An airplane is flying at a heading of 126° with an indicated airspeed of 225 miles per hour. The wind velocity is 90 miles per hour from 216°. Find the actual course of the airplane and its actual speed, relative to the ground.

Solution We represent the velocity of the airplane and the wind velocity by vectors, as shown in Figure 2.18. Since the wind is blowing from 216°, it is blowing in the direction 180° from this, or 36°. The angle between the vectors is therefore 90°. If we designate by θ the angle between the resultant and the velocity vector of the airplane, we have

$$\tan\theta = \frac{90}{225} = 0.4000$$

$$\theta = 21°\ 50'$$

Figure 2.18

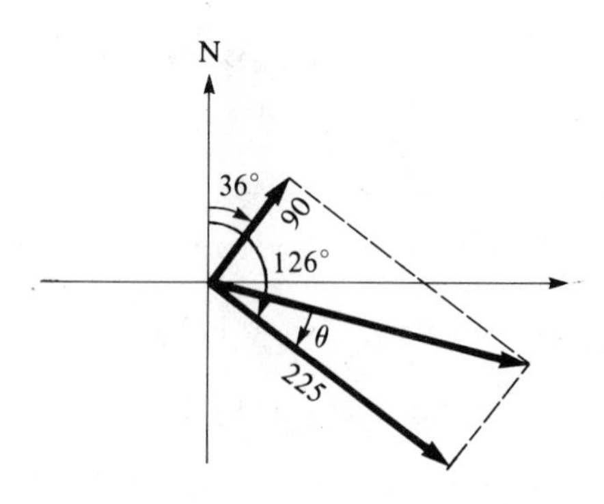

Thus, the true course of the airplane relative to the ground is 126° − 21° 48′ = 104° 10′. The true speed is the magnitude of the resultant,

$$|\mathbf{R}| = \sqrt{(225)^2 + (90)^2} = \sqrt{58{,}725} = 242 \text{ miles per hour}$$

EXAMPLE 2.14 In a rectangular coordinate system a vector is drawn from $A(3, -4)$ to $B(-5, 2)$, as shown in Figure 2.19. Find the horizontal and vertical components of $\overrightarrow{AB}$.

Figure 2.19

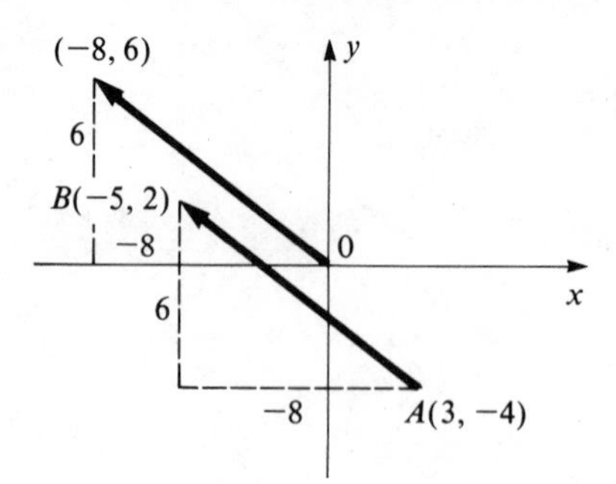

Solution The horizontal displacement from A to B is $-5 - 3 = -8$, and the vertical displacement is $2 - (-4) = 6$. So $\overrightarrow{AB}$ is equivalent to a vector drawn from the origin, with terminal point $(-8, 6)$. Therefore, the horizontal component of $\overrightarrow{AB}$ is -8, and the vertical component is 6.

EXAMPLE 2.15 A 1,000 pound weight is resting on an inclined plane which makes an angle of 30° with the horizontal. Find the components of the weight parallel to and perpendicular to the plane.

Figure 2.20

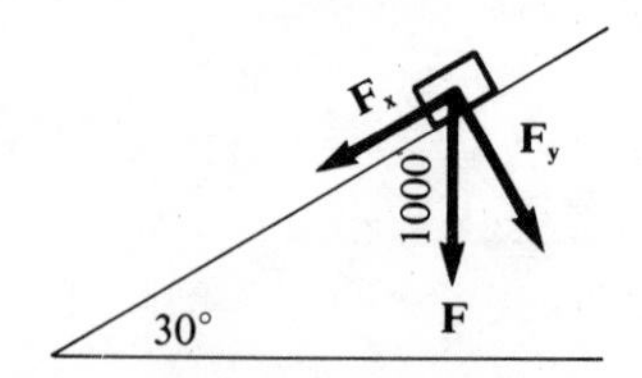

Solution We represent the weight by a vector $\mathbf{F}$, as shown in Figure 2.20. Its vector components $\mathbf{F}_x$ and $\mathbf{F}_y$ parallel and perpendicular to the inclined plane are also shown. By similar triangles we see that the angle between $\mathbf{F}$ and $\mathbf{F}_y$ is also 30°. The magnitudes of $\mathbf{F}_x$ and $\mathbf{F}_y$ are therefore,

$$|\mathbf{F}_x| = 1{,}000 \sin 30° = 500 \text{ pounds}$$

$$|\mathbf{F}_y| = 1{,}000 \cos 30° = 866 \text{ pounds}$$

EXAMPLE 2.16 A weight of 250 newtons* is supported by two cables, as shown in Figure 2.21. Find the tension in each cable.

Figure 2.21

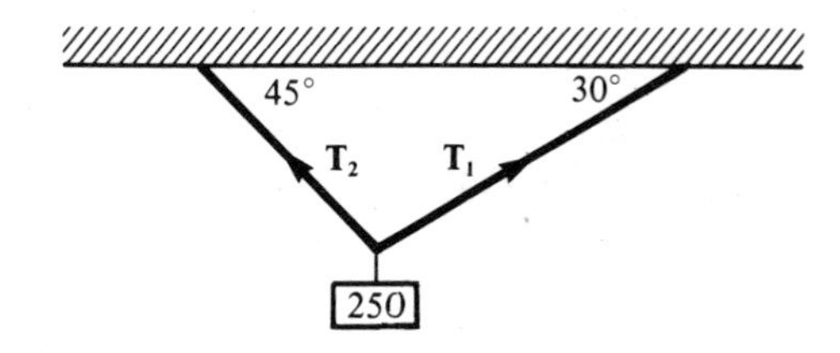

Solution Let $\mathbf{T}_1$ and $\mathbf{T}_2$ be the two tensions, as shown. Since the system is in equilibrium, the horizontal component of $\mathbf{T}_2$ must be the negative of the horizontal component of $\mathbf{T}_1$, and the sum of the vertical components must equal 250 newtons. Thus,

$$|\mathbf{T}_1| \cos 30° = |\mathbf{T}_2| \cos 45°$$

$$|\mathbf{T}_1| \sin 30° + |\mathbf{T}_2| \sin 45° = 250$$

or equivalently,

$$\frac{\sqrt{3}}{2} |\mathbf{T}_1| = \frac{\sqrt{2}}{2} |\mathbf{T}_2|$$

$$\frac{1}{2} |\mathbf{T}_1| + \frac{\sqrt{2}}{2} |\mathbf{T}_2| = 250$$

From the first equation we get

$$|\mathbf{T}_2| = \frac{\sqrt{3}}{\sqrt{2}} |\mathbf{T}_1|$$

which can be substituted into the second equation to give

$$\frac{1}{2} |\mathbf{T}_1| + \frac{\sqrt{3}}{2} |\mathbf{T}_1| = 250$$

$$|\mathbf{T}_1| (1 + \sqrt{3}) = 500$$

---

*A **newton** is the unit of force in the mks (meter-kilogram-second) system. It is the force necessary to produce an acceleration of 1 meter per second per second on a mass of 1 kilogram.

$$|\mathbf{T}_1| = \frac{500}{1 + \sqrt{3}} \approx 183 \text{ newtons}$$

$$|\mathbf{T}_2| = \frac{\sqrt{3}}{\sqrt{2}}(183.02) \approx 224 \text{ newtons}$$

Comment. We will give additional problems dealing with vectors in Chapter 6 when we are freed from the constraint of having to work only with right triangles.

EXERCISE SET 2.5

A In Problems 1–4 the horizontal and vertical components of a vector **V**, with initial point at the origin, are given. Find the magnitude of **V** and the angle α from the positive x axis to **V**.

1. $V_x = 7, V_y = 3$

2. $V_x = -5, V_y = 2$

3. $V_x = -2.10, V_y = -3.70$

4. $V_x = 30.4, V_y = -57.8$

In Problems 5–8 the magnitude of a vector **V** in standard position and its angle α from the positive x axis are given. Find V_x and V_y.

5. $|\mathbf{V}| = 25, \alpha = 32°$

6. $|\mathbf{V}| = 18.3, \alpha = 162.3°$

7. $|\mathbf{V}| = 2.03, \alpha = 342° \, 12'$

8. $|\mathbf{V}| = 102.6, \alpha = 212.53°$

In Problems 9–12 the coordinates of the initial point A and the terminal point B of a vector $\overrightarrow{AB}$ are given. Find the horizontal and vertical components of $\overrightarrow{AB}$.

9. $A = (3, 4), B = (6, 1)$

10. $A = (-2, 5), B = (0, 3)$

11. $A = (5, -2), B = (-7, -3)$

12. $A = (-4, -8), B = (-1, 3)$

In Problems 13–18 the terminal points of two vectors are given, and the initial point of each vector is the origin. Using graph paper draw the two vectors, and find the sum geometrically.

13. $(5, 1), (-2, 3)$

14. $(3, -2), (-1, -4)$

15. $(-8, -3), (-6, 4)$

16. $(6, -5), (-2, 4)$

17. $(3, 2), (6, 4)$

18. $(2, -1), (-4, 2)$

19. Let $\mathbf{U} = (2, -4)$ and $\mathbf{V} = (3, -1)$. Find:

a. $\mathbf{U} \cdot \mathbf{V}$. **b.** the angle between **U** and **V**.

20. Find, to the nearest minute, the angle between the vectors $\vec{R} = (3, 2)$ and $\vec{S} = (-4, 5)$.

21. Prove that **V** and **W** are perpendicular.

a. $\mathbf{V} = (-5, -3), \mathbf{W} = (-9, 15)$ **b.** $\mathbf{V} = (6, -10), \mathbf{W} = (25, 15)$

22. Find a so that $\mathbf{A} = (a, 3)$ and $\mathbf{B} = (4, 7)$ will be perpendicular.

23. Find b so that $\mathbf{U} = (3, -5)$ and $\mathbf{V} = (-2, b)$ will be perpendicular.

24. A force of 150 pounds is being exerted on a body acting horizontally to the right. A second force of 68 pounds exerted on the body is acting vertically upward. Find the magnitude and direction of the resultant force.

25. A pilot sets his heading at 120° and flies at an indicated airspeed of 195 miles per hour. There is a wind blowing from 210° at 62 miles per hour. What are the true course and speed of the airplane relative to the ground?

26. A pilot flew to a destination due south of her starting point. If the wind was from 300° at 80 miles per hour and her actual ground speed averaged 160 miles per hour, find her indicated heading and airspeed.

27. A boy can row 5 miles per hour in still water. He heads directly across a stream which has a current of 2 miles per hour. Find the resultant velocity of the boat, giving both magnitude and direction. If the river is 300 feet wide, how long will it take him to get across?

28. A 1,000 newton weight is resting on a plane making an angle of 28° with the horizontal. Find the components of the weight parallel to and perpendicular to the plane.

29. A hot-air balloon is rising vertically at the rate of 70 feet per second. If the wind is blowing horizontally at the rate of 30 miles per hour, find the actual path of the balloon and its speed along that path.

30. The pilot of a small plane plans a trip from town A to town B, 400 miles due south of A. He will cruise at an indicated airspeed of 180 miles per hour. The wind is from the west at 40 miles per hour. What should his heading be in order to compensate for the wind? How long will it take to make the trip?

B **31.** If $\mathbf{V}$ is a non-zero vector and c is a non-zero real number, then $c\mathbf{V}$ is a vector with magnitude $|c|\ |\mathbf{V}|$, directed in the same direction as $\mathbf{V}$ if $c > 0$ and in the opposite direction if $c < 0$. If $c = 0$, then $c\mathbf{V}$ is the zero vector. Let $\mathbf{V}$ and $\mathbf{W}$ be vectors in standard position, with terminal points $(3, -2)$ and $(2, 1)$, respectively. Find each of the following geometrically.

a. $2\mathbf{V}$ **b.** $-3\mathbf{W}$ **c.** $3\mathbf{V} + 2\mathbf{W}$ **d.** $2\mathbf{V} - 3\mathbf{W}$

32. Let $\mathbf{U}$, $\mathbf{V}$, and $\mathbf{W}$ be in standard position, with terminal points $(-1, 2)$, $(4, 5)$, and $(3, 1)$, respectively. By a geometric construction show that there are numbers a and b such that $\mathbf{U} = a\mathbf{V} + b\mathbf{W}$. (See Problem 31.)

33. Repeat Problem 32 if the terminal points of $\mathbf{U}$, $\mathbf{V}$, and $\mathbf{W}$ are $(-3, -2)$, $(7, 2)$, and $(-2, 1)$, respectively.

34. If $\mathbf{U}$ and $\mathbf{V}$ are vectors, $\mathbf{U} - \mathbf{V}$ is defined as $\mathbf{U} + (-\mathbf{V})$, where $-\mathbf{V} = (-1)\mathbf{V}$. (See Problem 31.) Show how $\mathbf{U} - \mathbf{V}$ can be described in terms of the parallelogram having $\mathbf{U}$ and $\mathbf{V}$ as adjacent sides.

35. For any vector $\mathbf{V} = (V_x, V_y)$ and any real number c, prove that $c\mathbf{V} = (cV_x, cV_y)$. (See Problem 31.)

36. Prove that if $\mathbf{V} = (V_x, V_y)$ and $\mathbf{W} = (W_x, W_y)$, then $\mathbf{V} + \mathbf{W} = (V_x + W_x, V_y + W_y)$.

37. Use Definition 2.1, together with the results of Problems 35 and 36 to prove the following properties of the dot product.

a. $\mathbf{V} \cdot \mathbf{W} = \mathbf{W} \cdot \mathbf{V}$ **b.** $\mathbf{U} \cdot (\mathbf{V} + \mathbf{W}) = \mathbf{U} \cdot \mathbf{V} + \mathbf{U} \cdot \mathbf{W}$

c. $c(\mathbf{U} \cdot \mathbf{V}) = (c\mathbf{U}) \cdot \mathbf{V} = \mathbf{U} \cdot (c\mathbf{V})$

38. Points A and B are on opposite sides of a lake, 1.5 kilometers apart, and the bearing of B from A is N 23° W. A swimmer wishes to swim from A to B, and she can swim at an average speed of 6 kilometers per hour. The predominant current is in the direction

S 67° W at 3 kilometers per hour. In what direction should she swim in order to actually go in a straight line from A to B? How long will it take her?

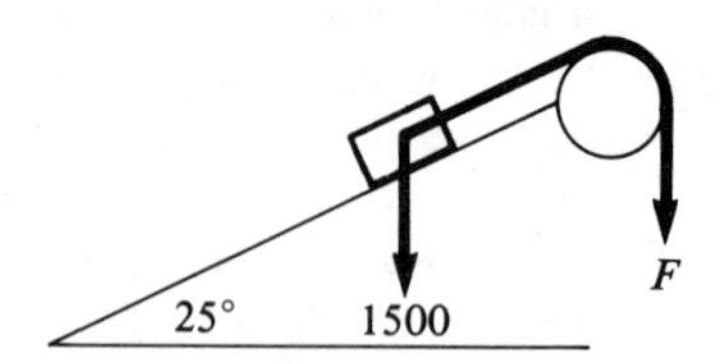

39. A weight of 1,500 pounds is being held in place on an inclined plane by a cable extending over a pulley as shown. What is the force **F** on the cable necessary to keep the weight from sliding? (Neglect friction.)

40. A weight of 3,000 newtons is being held in place by two cables as shown. Find the force in each cable.

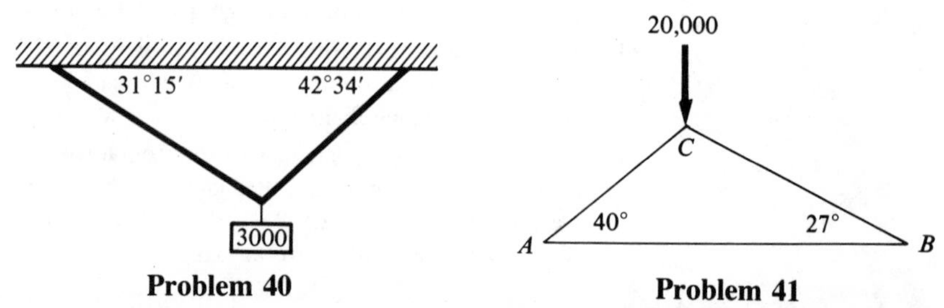

Problem 40 **Problem 41**

41. A section of a steel structure has the shape shown. The maximum weight to be supported at C is 20,000 pounds. Find the magnitude of the compressive force in each of the members AC and BC under the maximum load.

2.6 Review Exercise Set

A In Problems 1 and 2 use linear interpolation and the appropriate table to evaluate.

1. **a.** tan 36° 21′ **b.** sec 110° 35′ **c.** sin 76.34°
d. cos 215.82° **e.** cot(−132° 26′)

2. **a.** sin 1.243 **b.** cot 0.357 **c.** csc 3.586
d. cos(−2.149) **e.** tan 10.062

In Problems 3 and 4 find the acute angle A using tables and linear interpolation.

3. Find A to the nearest minute.
a. $\cos A = 0.3146$ **b.** $\tan A = 1.7850$ **c.** $\csc A = 2.3478$
d. $\sin A = 0.2496$ **e.** $\sec A = 3.5470$

4. Find A to the nearest thousandth of a radian.
a. $\sin A = 0.9684$ **b.** $\cot A = 0.3168$ **c.** $\cos A = 0.8865$
d. $\csc A = 1.2060$ **e.** $\tan A = 0.5496$

Use a hand calculator in Problems 5 and 6.

5. Find the specified function value correct to five significant figures.
   a. $\tan 205.3°$ b. $\sin 3.154$ c. $\cos 248° 37'$
   d. $\sec 2.756$ e. $\cot(-543.87°)$
6. Find the specified angle to the nearest hundredth of a degree.
   a. $\sin \theta = 0.89643$, θ acute
   b. $\cos \theta = -0.43215$, $90° \le \theta \le 180°$
   c. $\tan \theta = 3.7642$, $180° \le \theta \le 270°$
   d. $\csc \theta = -2.1543$, $-90° \le \theta \le 0°$
   e. $\sec \theta = 4.1587$, $270° \le \theta \le 360°$
7. Round off as specified.
   a. 2.0354998, nearest thousandth
   b. 0.0804501, three significant figures
   c. 19.535, four significant figures
   d. 1.8345, nearest thousandth
   e. 343.02, four significant figures

In Problems 8 and 9 perform the indicated operations and express answers to the appropriate accuracy. Use a calculator.

8. a. $\dfrac{(2.0348)(0.0473)}{0.159872}$ b. $\dfrac{12.53 \sin 32.4°}{25.648}$
9. a. $\sqrt{(25.2)^2 + (13.8)^2 - 2(25.2)(13.8) \cos 123° 32'}$
   Note. Treat 2 as exact.
   b. $\dfrac{254.5 \sin 34.81°}{\sin 56.23°}$

In Problems 10–23 solve the right triangles. The right angle is at C.

10. $A = 63.4°$, $b = 5.13$
11. $B = 32.7°$, $c = 19.1$
12. $B = 20° 13'$, $a = 4.05$
13. $A = 46° 11'$, $a = 12.4$
14. $B = 37° 35'$, $c = 1.374$
15. $A = 52° 12'$, $c = 52.30$
16. $a = 17.8$, $c = 24.3$
17. $a = 8$, $b = 15$
18. $A = 33.4°$, $c = 13.5$
19. $a = 40.31$, $b = 75.27$
20. $b = 102$, $c = 151$
21. $A = 67.5°$, $a = 12.6$
22. $B = 26.24°$, $c = 33.53$
23. $B = 72.3°$, $a = 2.53$
24. A 40 foot high building is at the edge of a river. Directly opposite on the other edge an observer finds that the angle of elevation of the top of the building is 33° 10′. How wide is the river at that point?
25. Two buildings are known to be 80 feet apart. From the roof of one of the buildings an observer finds that the angle of depression of the base of the other building is 67° 30′. How high is the building with the observer?
26. A ship leaves port and cruises for 3 hours on a course N 33.6° E at an average speed of 21.3 knots (nautical miles per hour). The course is then changed to S 56.4° E and the ship goes for 2 hours at an average speed of 24.7 knots. How far is the ship from the starting point at the end of this time? What is the bearing of the ship at the end of this time, as measured from the starting point?
27. A surveyor is at point A on the bank of a river, and he wishes to find the distance to point B directly across the river. With a transit he establishes a line of sight perpendic-

ular to line AB and measures 100 feet along this line to establish point C. From point C he measures the angle from CA to CB to be $41° 18'$. How far is it from A to B?

28. Find the angle between the vectors **A** and **B** if:
a. $\mathbf{A} = (3, 4)$, $\mathbf{B} = (-2, 1)$
b. $\mathbf{A} = (-6, 8)$, $\mathbf{B} = (-1, 3)$

29. Find t so that the vectors **U** and **V** will be perpendicular, where:
a. $\mathbf{U} = (-3, 2)$, $\mathbf{V} = (t, 4)$
b. $\mathbf{U} = (5, t)$, $\mathbf{V} = (3, -4)$

30. A girl and a boy in a canoe set out to row directly across a river 220 feet wide. They can row at an average speed of 18 miles per hour in still water. If the current is flowing at 6 miles per hour, how long will it take them to get across? How far downstream will they land from the point directly opposite their starting point?

31. The heading of an airplane is 234° and its indicated airspeed is 150 miles per hour. A wind is blowing from 324° at 65 miles per hour. What is the true course of the plane, and what is its true speed relative to the ground?

32. Forces of 32 pounds and 56 pounds are acting on an object as shown. Find the magnitude and direction of the resultant force.

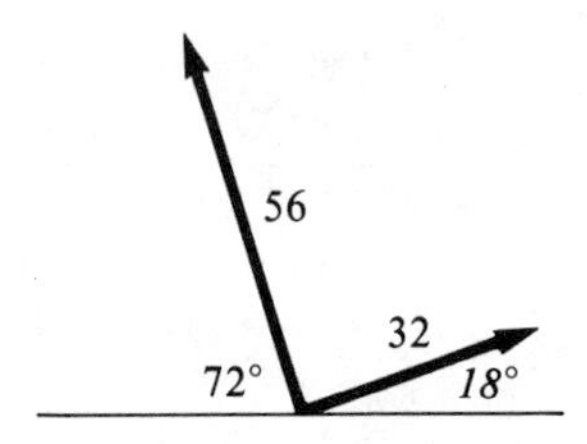

33. The magnitude of a force is 80 pounds, and its horizontal component is 35 pounds. Find its vertical component and the direction of the force.

34. A vector is in standard position and its angle from the positive x axis is 218.3°. If the magnitude of the vector is 16.32, find its horizontal and vertical components.

B **35.** Perform the following calculations, and express the answers to the appropriate degree of accuracy.

a. $\dfrac{(2.305 \times 10^{-4})\sqrt{198.2}}{4.96 \times 10^{-2}}$

b. $\dfrac{c \sin A}{\sqrt{b^2 + c^2 - 2bc \cos A}}$, where $b = 2.67$, $c = 3.02$, and $A = 68.37°$.

36. An observer is 100 feet from the base of a building. He finds the angles of elevation to the bottom and top of a flagpole on the roof of the building to be 59.3° and 62.4°, respectively. How high is the flagpole?

37. From the top of a tower the angles of depression of two points on the ground at the same level as the base of the tower and in line with it are $42° 38'$ and $61° 42'$. If the points are known to be 25 meters apart, find the height of the tower. How far is the closer point from the base of the tower?

38. Two guy wires on opposite sides of a telephone pole make angles of 57.3° and 46.8° with the horizontal. The distance between the points on the ground where the wires are attached is 80 feet. How high off the ground is the point on the pole to which the wires are attached?

39. Town B is 260 miles due east of town A, and town C is 175 miles due south of town B. The pilot of a small plane plans a trip from A to B, then to C, and back to A. The average airspeed on each leg of the trip is to be 125 miles per hour. A constant wind of 45 miles per hour is blowing from the north. Determine what the heading should be on each leg of the trip. What will be the estimated total flying time for the entire trip?

40. Let $\mathbf{V}_1$, $\mathbf{V}_2$ and $\mathbf{W}$ be vectors in standard position having terminal points $(-3, 1)$, $(4, 2)$, and $(8, -6)$, respectively. Find numbers a and b such that $\mathbf{W} = a\mathbf{V}_1 + b\mathbf{V}_2$.

41. Let $\mathbf{V}_1$, $\mathbf{V}_2$, and $\mathbf{V}_3$ be vectors in standard position having terminal points $(6, 0)$, $(0, 4)$, and $(-2, 3)$, respectively. Find the magnitude and direction of the vector $\mathbf{W} = \mathbf{V}_1 + \mathbf{V}_2 + \mathbf{V}_3$.

42. A weight of 500 newtons is supported by a brace as shown. Find the magnitude of the force each of the members AB and BC is subjected to. (Note that AB is under tension, and BC is under compression.)

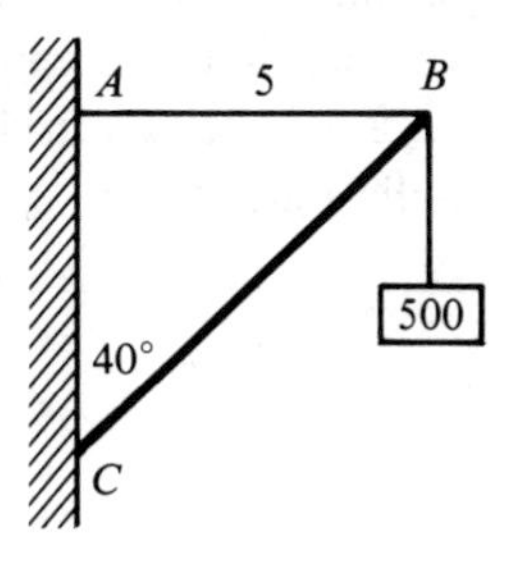

3 Analytical Trigonometry

Those of you who plan to take a calculus course will find that relationships between the trigonometric functions, known as **trigonometric identities**, make possible the solution of many otherwise intractable calculus problems. The use of such identities appears again and again in the part of calculus called *integration*. The following example illustrates this.*

Evaluate

$$\int \frac{dx}{(1 + x^2)^2}.$$

. . . We use the substitution $x = \tan \theta$, which gives $dx = \sec^2 \theta \, d\theta$. This leads to

$$\int \frac{dx}{(1 + x^2)^2} = \int \frac{\sec^2 \theta \, d\theta}{(1 + \tan^2 \theta)^2} = \int \frac{\sec^2 \theta \, d\theta}{\sec^4 \theta} = \int \frac{d\theta}{\sec^2 \theta}$$

$$= \int \cos^2 \theta \, d\theta = \frac{1}{2} \int (1 + \cos 2\theta) \, d\theta = \frac{1}{2} \theta + \frac{1}{4} \sin 2\theta + C$$

Now, from the original substitution, we have

$$\theta = \text{Arctan } x$$

and, from Figure 15.3 or trigonometric identities,

$$\sin \theta = \frac{x}{\sqrt{1 + x^2}} \quad \text{and} \quad \cos \theta = \frac{1}{\sqrt{1 + x^2}}$$

Figure 15.3

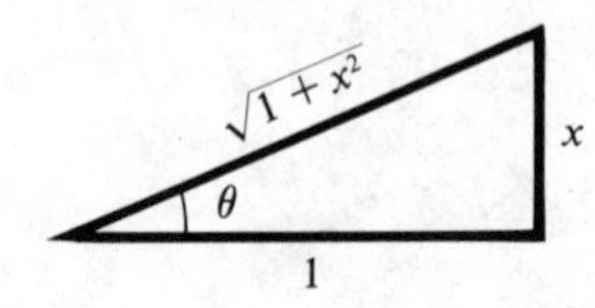

*From *Calculus and Analytic Geometry*, Third Edition, by Douglas F. Riddle. © 1979 by Wadsworth Publishing Company, Inc. Reprinted by permission of Wadsworth Publishing Company, Belmont, California 94002.

Thus

$$\sin 2\theta = 2 \sin \theta \cos \theta = \frac{2x}{1 + x^2}$$

This gives

$$\int \frac{dx}{(1 + x^2)^2} = \frac{1}{2}\theta + \frac{1}{4}\sin 2\theta + C$$
$$= \frac{1}{2}\text{Arctan } x + \frac{x}{2(1 + x^2)} + C$$

While this contains some mysterious symbols, it is loaded with notions that will be studied in this chapter. For example, the following relationships are employed that will be shown shortly:

$$1 + \tan^2 \theta = \sec^2 \theta \qquad \frac{1}{\sec \theta} = \cos \theta$$

$$\cos^2 \theta = \frac{1}{2}(1 + \cos 2\theta) \qquad \sin 2\theta = 2 \sin \theta \cos \theta$$

Also, the meaning of the notation "Arctan x" will be made clear in this chapter.

3.1 Trigonometric Functions of Real Numbers

So far in our definitions of trigonometric functions the domains have been sets of admissible angles. The values depend only on the angle and not whether it is measured in degrees or radians. The next stage of our development is an important one, although somewhat subtle. We are going to assign meanings to such expressions as $\sin x$, $\cos x$, and $\tan x$, where x stands for a *real number*.

Toward this end, let us recall that in the definition of the radian measure of an angle we used an arbitrary circle with center at the vertex of the angle. We find it useful in the present case to specify a circle of radius 1, with center at the origin, and we refer to this as the **unit circle** (Figure 3.1). If we are given an angle of θ radians, then the length of the subtended arc on a circle of radius r is $s = r\theta$, and so if $r = 1$, the arc length s equals θ. In other words, the linear measure (using the unit of measure for r) of arc length equals the angular measure of the central angle in radians. This relationship provides us with a way of defining the trigonometric functions of real numbers that is consistent with our previous definitions.

Given any real number θ (not at present to be thought of as the measure of an angle, but just a number), we mark off a distance of θ units on the unit circle, starting from $(1, 0)$ and moving counterclockwise when θ is positive and clockwise when θ is negative. Let $P(\theta)$ be the point arrived at in this manner, with coordinates (x, y) as shown in Figure 3.2. We now give the following definition.

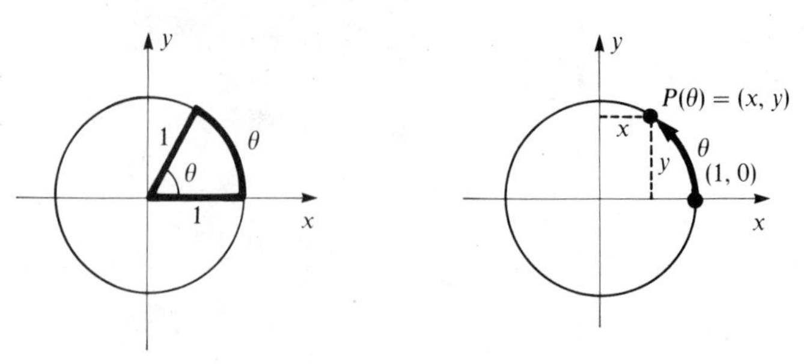

Figure 3.1 **Figure 3.2**

DEFINITION 3.1

$$\sin\theta = y \qquad \csc\theta = \frac{1}{y}$$

$$\cos\theta = x \qquad \sec\theta = \frac{1}{x}$$

$$\tan\theta = \frac{y}{x} \qquad \cot\theta = \frac{x}{y}$$

For the tangent and secant functions, θ cannot be any number for which $x = 0$; and for the cotangent and cosecant, θ cannot be any number for which $y = 0$.

The relationship between the functions defined in terms of numbers and those defined in terms of angles should now be clear. We simply consider the angle in standard position whose terminal side passes through $P(\theta)$, as shown in Figure 3.3.

Figure 3.3

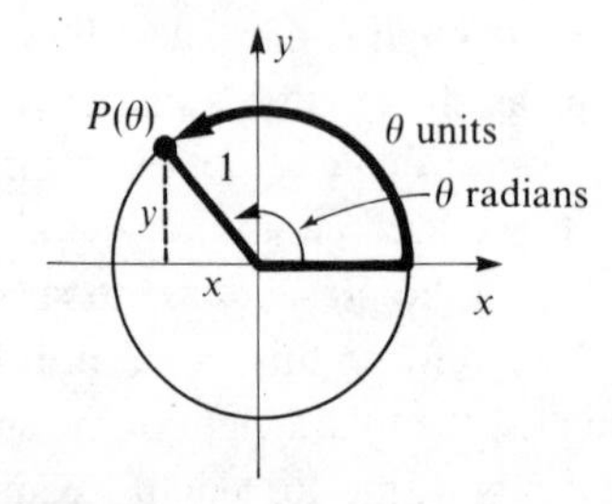

Then the radian measure of this angle is exactly θ, the same as the number of units of length of arc. Now by Definition 1.2 of the trigonmetric functions of the *angle* θ, we have

$$\sin\theta = \frac{y}{r} = \frac{y}{1} = y \qquad \csc\theta = \frac{r}{y} = \frac{1}{y}$$

$$\cos\theta = \frac{x}{r} = \frac{x}{1} = x \qquad \sec\theta = \frac{r}{x} = \frac{1}{x}$$

$$\tan\theta = \frac{y}{x} \qquad \cot\theta = \frac{x}{y}$$

Comparing these values with those given in Definition 3.1, we see that they are precisely the same. Suppose, then, that we are given the expression "sin 2." How shall we interpret it? Does it mean sine of the number 2 or sine of the angle whose radian measure is 2? The answer is that it does not matter, because the values are the same. Conceptually these are quite different, but sin 2, where 2 is a number, has the same value as sin 2, where 2 is the radian measure of an angle. More generally, *any trigonometric function of a real number x has the same value as the same trigonometric function of the angle of x radians*. Thus, insofar as values of the trigonometric functions are concerned, it makes no difference whether we think in terms of an angle measured in radians or in terms of the real number that equals the number of radians in the angle.

EXAMPLE 3.1 The x coordinate of the point $P(\theta)$ on the unit circle is $-\frac{4}{5}$, and $P(\theta)$ is in the second quadrant. Find the y coordinate of $P(\theta)$, and write all six trigonometric functions of θ.

Solution By the Pythagorean theorem, we have

$$\left(-\frac{4}{5}\right)^2 + y^2 = 1$$

or

$$y^2 = 1 - \frac{16}{25} = \frac{9}{25}$$

so that $y = \frac{3}{5}$. By Definition 3.1,

$$\sin\theta = y = \frac{3}{5} \qquad \csc\theta = \frac{1}{y} = \frac{5}{3}$$

$$\cos\theta = x = -\frac{4}{5} \qquad \sec\theta = \frac{1}{x} = -\frac{5}{4}$$

$$\tan\theta = \frac{y}{x} = -\frac{3}{4} \qquad \cot\theta = \frac{x}{y} = -\frac{4}{3}$$

EXAMPLE 3.2 Determine the value of θ if $0 \le \theta < 2\pi$ and

a. $P(\theta) = \left(-\frac{\sqrt{3}}{2}, -\frac{1}{2}\right)$ **b.** $P(\theta) = \left(-\frac{1}{\sqrt{2}}, \frac{1}{\sqrt{2}}\right)$

Solution

a. We observe that the reference angle of the angle in standard position with terminal side passing through $P(\theta)$ is 30° or $\pi/6$ radians (Figure 3.4). So the angle itself is $7\pi/6$ radians. Therefore $\theta = 7\pi/6$.

b. The reference angle in this case is 45°, or $\pi/4$ radians, and the angle is $3\pi/4$ radians (Figure 3.5). Therefore $\theta = 3\pi/4$.

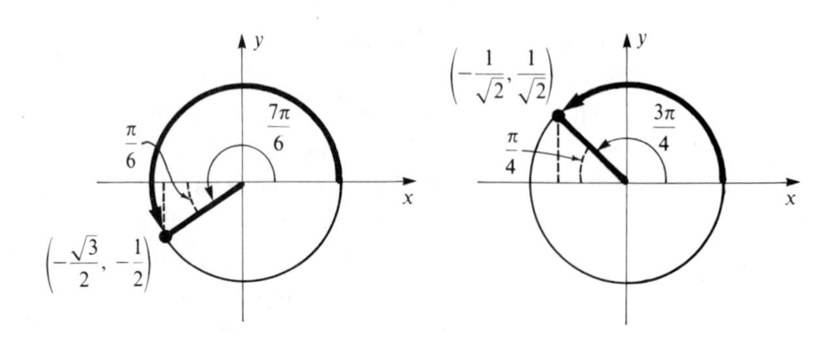

Figure 3.4

Figure 3.5

EXERCISE SET 3.1

A In Problems 1–10 find the values of all six trigonometric functions of θ.

1. $P(\theta) = \left(-\frac{1}{3}, \frac{-2\sqrt{2}}{3}\right)$

2. $P(\theta) = \left(-\frac{2}{\sqrt{5}}, \frac{1}{\sqrt{5}}\right)$

3. $P(\theta) = \left(\frac{5}{13}, -\frac{12}{13}\right)$

4. $P(\theta) = (-0.6, 0.8)$

5. The abscissa of $P(\theta)$ is $\frac{1}{4}$, and $P(\theta)$ is in the fourth quadrant.

6. The ordinate of $P(\theta) = -\frac{1}{2}$, and $\pi \le \theta \le 3\pi/2$. What is the value of θ?

7. The x coordinate of $P(\theta)$ is $-\frac{4}{5}$, and $\pi/2 \le \theta \le \pi$.

8. The y coordinate of $P(\theta)$ is $\sqrt{3}/2$, and its x coordinate is negative. What is the value of θ for which $0 \le \theta < 2\pi$?

9. The abscissa of $P(\theta)$ is $\frac{1}{3}$, and the ordinate is negative.

10. The ordinate of $P(\theta)$ is $-\frac{5}{13}$, and $3\pi \le \theta \le \frac{7\pi}{2}$.

In Problems 11 and 12 give the coordinates of $P(\theta)$ for the given value of θ.

11. **a.** $P\left(\frac{\pi}{2}\right)$ **b.** $P(\pi)$ **c.** $P\left(\frac{3\pi}{2}\right)$ **d.** $P\left(\frac{5\pi}{6}\right)$ **e.** $P\left(\frac{4\pi}{3}\right)$

12. **a.** $P\left(\frac{5\pi}{4}\right)$ **b.** $P\left(\frac{8\pi}{3}\right)$ **c.** $P\left(-\frac{7\pi}{4}\right)$ **d.** $P(5\pi)$ **e.** $P\left(\frac{31\pi}{6}\right)$

In Problems 13 and 14 indicate the approximate location of $P(\theta)$ on the unit circle for the given value of θ.

13. **a.** $\theta = 0.5$ **b.** $\theta = -2$ **c.** $\theta = 4$ **d.** $\theta = -15$

14. **a.** $\theta = 8.2$ **b.** $\theta = 12.75$ **c.** $\theta = -1.54$ **d.** $\theta = 4.72$

In Problems 15 and 16 give the value of θ if $0 \le \theta < 2\pi$.

15. **a.** $P(\theta) = \left(\frac{\sqrt{3}}{2}, \frac{1}{2}\right)$ **b.** $P(\theta) = \left(\frac{1}{\sqrt{2}}, \frac{1}{\sqrt{2}}\right)$ **c.** $P(\theta) = \left(\frac{1}{2}, \frac{\sqrt{3}}{2}\right)$

d. $P(\theta) = (0, 1)$ **e.** $P(\theta) = (1, 0)$

16. a. $P(\theta) = \left(-\frac{1}{2}, -\frac{\sqrt{3}}{2}\right)$ b. $P(\theta) = (0, -1)$ c. $P(\theta) = \left(-\frac{\sqrt{3}}{2}, \frac{1}{2}\right)$
    d. $P(\theta) = \left(\frac{\sqrt{2}}{2}, -\frac{\sqrt{2}}{2}\right)$ e. $P(\theta) = (-1, 0)$
17. Without using tables or a calculator, find all values of θ such that $0 \le \theta < 2\pi$ for which:
    a. $\sin \theta = 1$ b. $\cos \theta = 1$ c. $\tan \theta = 1$ d. $\sin \theta = 0$
    e. $\cos \theta = 0$ f. $\tan \theta = 0$ g. $\sin \theta = -1$ h. $\cos \theta = -1$
    i. $\tan \theta = -1$
18. Use the unit circle to show that:
    a. $\sin(-\theta) = -\sin \theta$ b. $\cos(-\theta) = \cos \theta$
19. Use the unit circle to show that:
    a. $\sin(\pi - \theta) = \sin \theta$ b. $\cos(\pi - \theta) = -\cos \theta$
    c. $\sin(\pi + \theta) = -\sin \theta$ d. $\cos(\pi + \theta) = -\cos \theta$
    e. $\tan(\pi + \theta) = \tan \theta$
20. Use a geometric argument to show that for $0 \le \theta \le \pi$, $\sin \theta \ge 0$. Give a corresponding inequality which is valid for $-\pi \le \theta \le 0$.

B 21. Show that for all integers n, each of the following is undefined for the stated values of θ.
    a. $\tan \theta$ for $\theta = \left(\frac{2n+1}{2}\right)\pi$ b. $\cot \theta$ for $\theta = n\pi$
    c. $\sec \theta$ for $\theta = \left(\frac{2n+1}{2}\right)\pi$ d. $\csc \theta$ for $\theta = n\pi$
22. Prove that the range of the secant and of the cosecant is the set $\{k \in \mathbb{R} : |k| \ge 1\}$.
23. Prove that the range of the tangent and of the cotangent is the set $\mathbb{R}$ of all real numbers.

3.2 Some Basic Trigonometric Identities

We have already noted the following reciprocal relationships among the trigonometric functions:

$$\sec \theta = \frac{1}{\cos \theta}, \qquad \csc \theta = \frac{1}{\sin \theta}, \qquad \cot \theta = \frac{1}{\tan \theta}$$

It is also a direct consequence of the definitions that

$$\tan \theta = \frac{\sin \theta}{\cos \theta} \quad \text{and} \quad \cot \theta = \frac{\cos \theta}{\sin \theta}$$

These are examples of what are called **trigonometric identities**, because they are true for all admissible values of θ. There are many other identities, and some are so basic that they should be committed to memory. These will be indicated.

The Pythagorean theorem is the basis for another important set of identities. Using the theorem, we have, for (x, y) on the unit circle (see Figure 3.6),

Figure 3.6

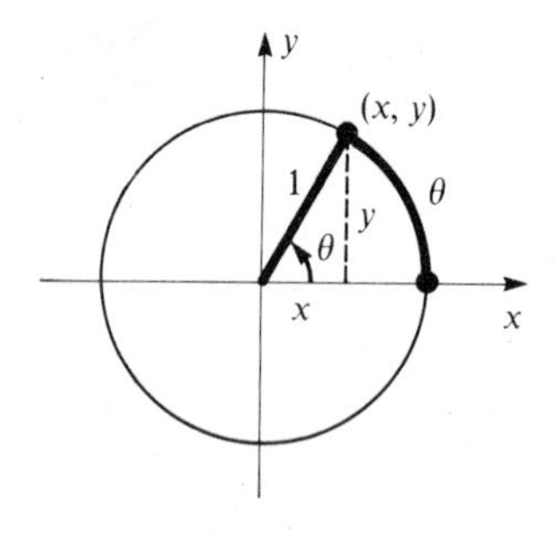

$$x^2 + y^2 = 1$$

But $x = \cos\theta$ and $y = \sin\theta$. So

$$\sin^2\theta + \cos^2\theta = 1$$

If we divide both sides of the first equation by x^2 (assuming $x \neq 0$), we obtain

$$1 + \left(\frac{y}{x}\right)^2 = \left(\frac{1}{x}\right)^2$$

which, according to our definitions, is

$$1 + \tan^2\theta = \sec^2\theta$$

If, however, we divide the first equation by y^2, we get

$$\left(\frac{x}{y}\right)^2 + 1 = \left(\frac{1}{y}\right)^2$$

or

$$\cot^2\theta + 1 = \csc^2\theta$$

We collect these results together:

$$\sin^2\theta + \cos^2\theta = 1$$
$$1 + \tan^2\theta = \sec^2\theta$$
$$1 + \cot^2\theta = \csc^2\theta$$

These are known as the **Pythagorean identities** and should be learned. Notice the similarity between the last two. In applying these, it may be necessary to write them in various equivalent forms, for example, $\cos^2\theta = 1 - \sin^2\theta$ or $\sec^2\theta - 1 = \tan^2\theta$, but it is probably best to concentrate on learning them in just one form, such as the one given. You can then mentally rewrite them in various ways.

Starting from any point on the unit circle, if we go 2π units along the circle either clockwise or counterclockwise, we wind up at the same point, since the circumference of the unit circle is 2π. Thus, for any θ, $P(\theta) = P(\theta + 2n\pi)$, where n is any integer, positive or negative. Since the values of the trigonometric functions depend only on the coordinates of $P(\theta)$, it follows that

$$\sin(\theta + 2n\pi) = \sin \theta$$
$$\cos(\theta + 2n\pi) = \cos \theta$$

and so on, for the remaining trigonometric functions. These can be expressed briefly by saying that each of the trigonometric functions is **periodic**, with **period 2π**.*

Consider next the relationships between the functions of θ and $-\theta$. By symmetry we see that if the coordinates of $P(\theta)$ are (x, y), then those of $P(-\theta)$ are $(x, -y)$ (see Figure 3.7). The sine of $-\theta$ is by definition the y coordinate of $P(-\theta)$, but this is the negative of the y coordinate of θ. Thus, $\sin(-\theta) = -\sin \theta$. Since the x coordinates of $P(\theta)$ and $P(-\theta)$ are the same, it follows that $\cos(-\theta) = \cos \theta$. Similar relationships for the other functions can now be obtained:

$\sin(-\theta) = -\sin \theta$	$\csc(-\theta) = -\csc \theta$
$\cos(-\theta) = \cos \theta$	$\sec(-\theta) = \sec \theta$
$\tan(-\theta) = -\tan \theta$	$\cot(-\theta) = -\cot \theta$

It is most important to learn just the first two of these since they occur more frequently than the others and since the others can be deduced easily from these two.

Figure 3.7

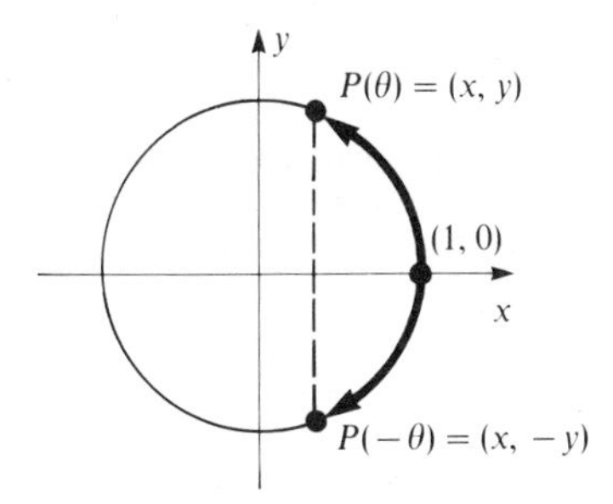

If f is any function (not necessarily a trigonometric function) for which $f(-x) = f(x)$ for all x in the domain of f, then f is said to be an **even** function. If $f(-x) = -f(x)$ for all x in the domain of f, then f is said to be an **odd** function. From what we have just seen, then, the cosine and secant are even functions, whereas the sine, tangent, cosecant, and cotangent are odd.

---

*In general, if $f(x + k) = f(x)$ for all x, then f is said to be *periodic with period k*. The smallest positive k for which this is true is called the **fundamental period**. For the sine, cosine, secant, and cosecant, the fundamental period is 2π, but for the tangent and cotangent it is π.

EXAMPLE 3.3 Make use of the basic identities of this section to find the values of the other five trigonometric functions of θ if $\tan \theta = -2$ and $\cos \theta < 0$.

Solution First we observe that $\cot \theta = 1/\tan \theta = -\frac{1}{2}$. Next, since $\sec^2 \theta = 1 + \tan^2 \theta = 1 + (-2)^2 = 5$, we have $\sec \theta = \pm\sqrt{5}$. But $\cos \theta$ is negative, so that $\sec \theta$, which is $1/\cos \theta$, is also negative. Thus, $\sec \theta = -\sqrt{5}$. Therefore, $\cos \theta = -1/\sqrt{5}$.

Now we use $\sin^2 \theta = 1 - \cos^2 \theta = 1 - \frac{1}{5} = \frac{4}{5}$. Since $\tan \theta$ and $\cos \theta$ are both negative, and $\tan \theta = \sin \theta/\cos \theta$, it follows that $\sin \theta$ is positive. Thus, $\sin \theta = 2/\sqrt{5}$. Finally, $\csc \theta = 1/\sin \theta = \sqrt{5}/2$.

Note. We could also have determined the signs of each of the functions by observing that for both $\tan \theta$, which is y/x, and $\cos \theta$, which is x, to be negative, we must have y positive, so that the point $P(\theta)$ is in the second quadrant or, equivalently, the angle of θ radians terminates in the second quadrant.

EXAMPLE 3.4 In each of the following use the basic identities to show that the first expression can be transformed into the second.

a. $\dfrac{\sin \theta}{\tan \theta}$, $\cos \theta$ **b.** $\dfrac{\cos \theta}{\tan \theta}$, $\csc \theta - \sin \theta$

Solution **a.** Since $\tan \theta = \sin \theta/\cos \theta$, we have

$$\frac{\sin \theta}{\tan \theta} = \frac{\sin \theta}{\dfrac{\sin \theta}{\cos \theta}} = \sin \theta \cdot \frac{\cos \theta}{\sin \theta} = \cos \theta$$

b. $$\frac{\cos \theta}{\tan \theta} = \cos \theta \cdot \frac{\cos \theta}{\sin \theta} = \frac{\cos^2 \theta}{\sin \theta}$$

Now we use the identities $\cos^2 \theta = 1 - \sin^2 \theta$ and $\csc \theta = 1/\sin \theta$ to get

$$\frac{\cos^2 \theta}{\sin \theta} = \frac{1 - \sin^2 \theta}{\sin \theta} = \frac{1}{\sin \theta} - \frac{\sin^2 \theta}{\sin \theta} = \csc \theta - \sin \theta$$

EXERCISE SET 3.2

A In Problems 1–8 find the values of the remaining five trigonometric functions of θ, using the basic identities of this section.

1. $\sin\theta = -\dfrac{3}{5}, \quad \tan\theta > 0$

2. $\sec\theta = \dfrac{13}{12}, \quad \csc\theta < 0$

3. $\tan\theta = -\dfrac{4}{3}, \quad \dfrac{\pi}{2} < \theta < \dfrac{3\pi}{2}$

4. $\cot\theta = \dfrac{5}{12}, \quad \pi < \theta < 2\pi$

5. $\cos\theta = \dfrac{2}{\sqrt{5}}, \quad \cot\theta < 0$

6. $\csc\theta = -\dfrac{17}{8}, \quad \tan\theta > 0$

7. $\sin\theta = \dfrac{1}{3}, \quad \dfrac{5\pi}{2} < \theta < \dfrac{7\pi}{2}$

8. $\sec\theta = 3, \quad 3\pi < \theta < 4\pi$

In Problems 9–18 show that the first expression can be transformed into the second, using basic identities.

9. $\dfrac{\tan\theta}{\sin\theta}, \quad \sec\theta$

10. $\dfrac{\cot\theta}{\csc\theta}, \quad \cos\theta$

11. $\dfrac{\sin\theta}{\cot\theta}, \quad \sec\theta - \cos\theta$

12. $\sec^2\theta\sin^2\theta, \quad \sec^2\theta - 1$

13. $\cot\theta\sec\theta, \quad \csc\theta$

14. $\tan\theta\csc\theta, \quad \sec\theta$

15. $\dfrac{\sec\theta}{\csc\theta}, \quad \tan\theta$

16. $1 - \dfrac{\sin\theta}{\csc\theta}, \quad \cos^2\theta$

17. $\dfrac{1}{\sec^2\theta} + \dfrac{1}{\csc^2\theta}, \quad 1$

18. $\sec^2\theta - \sin^2\theta\sec^2\theta, \quad 1$

B **19.** By making the substitution $t = \tan\theta$, where $-\pi/2 < \theta < \pi/2$, show that

$$\frac{\sqrt{1+t^2}}{t} = \csc\theta$$

20. By making the substitution $t = 2\sin\theta$, where $-\pi/2 < \theta < \pi/2$, show that

$$\frac{t}{\sqrt{4-t^2}} = \tan\theta$$

21. By making the substitution $t = \frac{3}{2}\sec\theta$, where $0 < \theta < \pi/2$ if $t > 0$ and $\pi < \theta < 3\pi/2$ if $t < 0$, show that

$$\frac{\sqrt{4t^2-9}}{t} = 2\sin\theta$$

22. If $\tan\theta = t$, prove that

$$\sin\theta\cos\theta = \frac{t}{1+t^2}$$

23. If $\cot\theta = t$, and $t \neq 0$, prove that

$$\sec\theta\csc\theta = \frac{1+t^2}{t}$$

3.3 Proving Identities

By using the basic identities of the last section, we can prove a multitude of others, but fortunately it is not particularly useful to try to memorize any of the results. By proving an identity, we mean to show that the given equation is true for all admissible values of the variable or variables involved. The general procedure is to work only on one side of the equation, and by use of the basic identities transform it so that in the final stage it is identical to the other side. Often the more complicated side is the better to work with, since it offers more obvious possibilities for alteration, but there are times in calculus when it is better to change a simpler expression into a more complicated one. We illustrate the procedure by means of several examples.

EXAMPLE 3.5 Prove the identity $\tan\theta + \cot\theta = \sec\theta\csc\theta$.

Solution Perhaps it would be better to word the instructions, "Prove that the following is an identity." For we cannot prove it is true by assuming it is true, and this is an important point of logic often missed. This precludes, for example, working on both sides of the equation, unless we verify that each step is reversible. A proper approach is to begin with the left-hand side and try to obtain the right-hand side:

$$\begin{aligned}\tan\theta + \cot\theta &= \frac{\sin\theta}{\cos\theta} + \frac{\cos\theta}{\sin\theta}\\ &= \frac{\sin^2\theta + \cos^2\theta}{\cos\theta\sin\theta}\\ &= \frac{1}{\cos\theta\sin\theta}\\ &= \frac{1}{\cos\theta}\cdot\frac{1}{\sin\theta}\\ &= \sec\theta\csc\theta\end{aligned}$$

and now the given equation is verified. It is true for all admissible values of θ, and in this case this means all values except 0, $\pi/2$, $3\pi/2$, and any angles coterminal with these since, at each of these values, two of the given functions are undefined. Usually we will understand that such exceptions are necessary without mentioning them explicitly.

EXAMPLE 3.6 Prove the identity

$$\frac{1}{\sec x + 1} = \cot x \csc x - \csc^2 x + 1$$

Solution It would be possible to transform the right-hand side into the left, and may even be easier, but we choose to work from the left-hand side because in later applications the right-hand side will be seen to be the more desirable final form. We begin with a commonly used trick, that of multiplying numerator and denominator by the same factor:

$$\frac{1}{\sec x + 1} = \frac{1}{\sec x + 1} \cdot \frac{\sec x - 1}{\sec x - 1} = \frac{\sec x - 1}{\sec^2 x - 1}$$

The object here is to bring $\sec^2 x - 1$ into the picture, because this is, by one of the Pythagorean identities, equal to $\tan^2 x$. So,

$$\begin{aligned}
\frac{1}{\sec x + 1} &= \frac{\sec x - 1}{\tan^2 x} = \frac{\sec x}{\tan^2 x} - \frac{1}{\tan^2 x} \\
&= \frac{1}{\cos x} \cdot \frac{\cos^2 x}{\sin^2 x} - \cot^2 x \\
&= \frac{\cos x}{\sin^2 x} - (\csc^2 x - 1) \\
&= \frac{\cos x}{\sin x} \cdot \frac{1}{\sin x} - \csc^2 x + 1 \\
&= \cot x \csc x - \csc^2 x + 1
\end{aligned}$$

EXAMPLE 3.7 Prove that

$$\frac{2 \sin^3 x}{1 - \cos x} = 2 \sin x(1 + \cos x)$$

Solution

$$\begin{aligned}
\frac{2 \sin^3 x}{1 - \cos x} &= \frac{2 \sin x(1 - \cos^2 x)}{1 - \cos x} \\
&= \frac{2 \sin x(1 + \cos x)(1 - \cos x)}{1 - \cos x} \\
&= 2 \sin x(1 + \cos x)
\end{aligned}$$

EXAMPLE 3.8 Prove the identity

$$\frac{\csc \theta - \cot \theta}{\sec \theta - 1} = \cot \theta$$

Solution

$$\frac{\csc \theta - \cot \theta}{\sec \theta - 1} = \frac{\dfrac{1}{\sin \theta} - \dfrac{\cos \theta}{\sin \theta}}{\dfrac{1}{\cos \theta} - 1}$$

$$= \frac{\dfrac{1-\cos\theta}{\sin\theta}}{\dfrac{1-\cos\theta}{\cos\theta}}$$

$$= \frac{1-\cos\theta}{\sin\theta} \cdot \frac{\cos\theta}{1-\cos\theta}$$

$$= \frac{\cos\theta}{\sin\theta}$$

$$= \cot\theta$$

A word is in order about the utility of proving identities. In calculus, as well as in more advanced courses, we are frequently confronted with an unwieldy expression involving trigonometric functions. Often, by use of the basic identities, these can be transformed into more manageable forms. So in practice we usually do not have a ready-made identity to verify, but rather an expression that is to be changed to some other, initially unknown, form. It would be more accurate to describe this procedure as deriving an identity. The value of proving identities already provided lies in the fact that it gives a goal to shoot for in transforming an expression. Without this and with little or no experience in such activity, a student may go in circles or change to a less desirable form.

EXERCISE SET 3.3

A In Problems 1–49 prove that the given equation is an identity by transforming the left-hand side into the right-hand side.

1. $\dfrac{1+\sin\theta}{\tan\theta} = \cos\theta + \cot\theta$

2. $\dfrac{\sec\theta - \cos\theta}{\tan\theta} = \sin\theta$

3. $\dfrac{\sin x \cot x}{\cos x \csc x} = \sin x$

4. $\dfrac{\sin^2 x - 1}{\cos^2 x - 1} = \csc^2 x - 1$

5. $(1 + \cot^2\theta)\tan^2\theta = \sec^2\theta$

6. $(\tan x + \cot x)^2 = \sec^2 x + \csc^2 x$

7. $\sec x - \sin x \tan x = \cos x$

8. $1 - \dfrac{\tan\theta\cos\theta}{\csc\theta} = \cos^2\theta$

9. $(\sin\phi - \cos\phi)^2 = 1 - 2\sin\phi\cos\phi$

10. $\dfrac{1-\sec^2\alpha}{\cos^2\alpha - 1} = \sec^2\alpha$

11. $\sin\theta\tan\theta = \sec\theta - \cos\theta$

12. $\sin\theta + \cos\theta\cot\theta = \csc\theta$

13. $(\tan^2\theta + 1)(\cos^2\theta - 1) = 1 - \sec^2\theta$

14. $\csc\theta - \cos\theta\cot\theta = \sin\theta$

15. $\sin^4 x - \cos^4 x = 2\sin^2 x - 1$

16. $\dfrac{(1-\tan\theta)^2}{\sec^2\theta} = 1 - 2\sin\theta\cos\theta$

17. $\dfrac{\tan\theta + 1}{\sec\theta + \csc\theta} = \sin\theta$

18. $\dfrac{\cos x \cot x}{\csc x} - 1 = -\sin^2 x$

19. $\dfrac{1}{\tan\theta + \cot\theta} = \sin\theta\cos\theta$

20. $\dfrac{\cos\theta\,(\tan\theta - \sec\theta)}{1 - \csc\theta} = \sin\theta$

21. $\dfrac{\sec^2 x}{1 + \cot^2 x} = \tan^2 x$

22. $\dfrac{\sec x + \tan x}{\sec^2 x} = \cos x(1 + \sin x)$

23. $(1 + \csc\theta)(\sec\theta - \tan\theta) = \cot\theta$

24. $\dfrac{1 - \cos x}{\sin x} + \dfrac{\sin x}{1 - \cos x} = 2\csc x$

25. $\tan^2 x - \sin^2 x = \tan^2 x \sin^2 x$

26. $\dfrac{1}{1 - \sin x} + \dfrac{1}{1 + \sin x} = 2\sec^2 x$

27. $\dfrac{\tan^2 x - 1}{\sin^2 x} = \sec^2 x - \csc^2 x$

28. $\cot^4\theta - 1 = \cot^2\theta\csc^2\theta - \csc^2\theta$

29. $\sec^4 x = \tan^2 x \sec^2 x + \sec^2 x$

30. $\dfrac{\tan^2\alpha - \sin^2\alpha}{\sec^2\alpha - 1} = \sin^2\alpha$

31. $\dfrac{1 + \tan x}{1 + \cot x} = \tan x$

32. $\dfrac{\sin x}{\cot x + \csc x} = 1 - \cos x$

33. $\dfrac{1}{\sec x - \tan x} = \sec x + \tan x$

34. $\tan^4 x = \sec^2 x \tan^2 x - \sec^2 x + 1$

35. $\dfrac{\sin x}{1 - \cos x} - \dfrac{1 - \cos x}{\sin x} = 2\cot x$

36. $\dfrac{(\tan^2\theta - 1)\cot\theta}{\sin\theta - \cos\theta} = \sec\theta + \csc\theta$

37. $\dfrac{1 - \sec^2 x}{1 - \csc^2 x} = \tan^2 x \sec^2 x - \sec^2 x + 1$

38. $(\cot\theta - \csc\theta)^2 = \dfrac{1 - \cos\theta}{1 + \cos\theta}$

B

39. $(1 - \cos^2\theta)(2 + \tan^2\theta) = \sec^2\theta - \cos^2\theta$

40. $\tan\theta\,(\sin\theta + \cos\theta)^2 + (1 - \sec^2\theta)\cot\theta = 2\sin^2\theta$

41. $2\cos x - \sec x\,(1 - 2\sin^2 x) = \sec x$

42. $\dfrac{\sin\theta}{1 - \cos\theta} = \csc\theta + \cot\theta$

43. $\dfrac{1}{1 - \sin\theta} = \sec^2\theta + \tan\theta\sec\theta$

44. $\dfrac{1 - \sin x}{1 + \sin x} = 2\sec^2 x - 2\sec x\tan x - 1$

45. $\dfrac{\sin\theta}{1 + \sin\theta} + \sec^2\theta = \tan\theta\sec\theta + 1$

46. $\dfrac{\sin^2 x \tan^2 x}{\tan x - \sin x} = \tan x + \sin x$

47. $\dfrac{\sec\theta - \cos\theta + \tan\theta}{\tan\theta + \sec\theta} = \sin\theta$

48. $\dfrac{\sin^3\theta - \sin^2\theta\cos\theta - \sin\theta\cos^2\theta + \cos^3\theta}{\sin\theta + \cos\theta} = (\sin\theta - \cos\theta)^2$

49. $\sqrt{\dfrac{1 - \cos\theta}{1 + \cos\theta}} = \dfrac{\sin\theta}{1 + \cos\theta} \quad (0 \le \theta < \pi)$

50. Make the substitution: $x = \sin\theta$ $(-\pi/2 < \theta < \pi/2)$ in the expression $x^2/\sqrt{1 - x^2}$ and show the result can be written in the form $\sec\theta - \cos\theta$.

51. Let $x = 2\tan\theta$ $(-\pi/2 < \theta < \pi/2)$ in $x/\sqrt{4 + x^2}$, and show the result is $\sin\theta$.

52. Substitute $x = 3 \sec \theta$, where θ lies either in the first or third quadrants, in $(x^2 - 9)^{3/2}/x$, and show the result can be written in the form $9(\sec \theta \tan \theta - \sin \theta)$.

53. Substitute $x = \frac{3}{2} \tan \theta \quad (-\pi/2 < \theta < \pi/2)$ in $x/\sqrt{4x^2 + 9}$ and show that the result is $\frac{1}{2} \sin \theta$.

3.4 The Addition Formulas for Sine and Cosine

In this section we will develop formulas for the sum and difference of two numbers (or of two angles). These are of fundamental importance in deriving other identities. We designate by α and β two numbers which we will initially restrict to be between 0 and 2π and for which $\alpha > \beta$. These restrictions will soon be removed.

In Figure 3.8 we have shown typical positions of $P(\alpha)$ and $P(\beta)$. Now the distance along the circle from (1, 0) to $P(\alpha)$ is α, and to $P(\beta)$ is β. So the arc length from $P(\beta)$ to $P(\alpha)$ is $\alpha - \beta$. If we think of sliding this latter arc along the circle

Figure 3.8

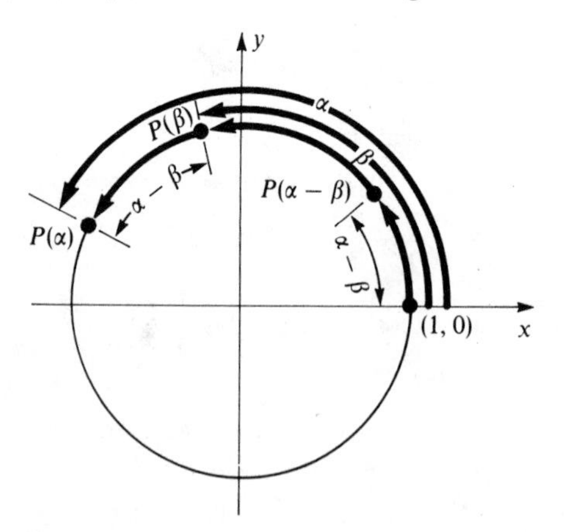

until the point originally at $P(\beta)$ coincides with (1, 0), then the distance $\alpha - \beta$ is in standard position and the endpoint of the arc is correctly labeled $P(\alpha - \beta)$. Since the x and y coordinates of a point $P(\theta)$ are $\cos \theta$ and $\sin \theta$, respectively, we can show the coordinates of the three points $P(\alpha)$, $P(\beta)$, and $P(\alpha - \beta)$ (Figure 3.9).

Figure 3.9

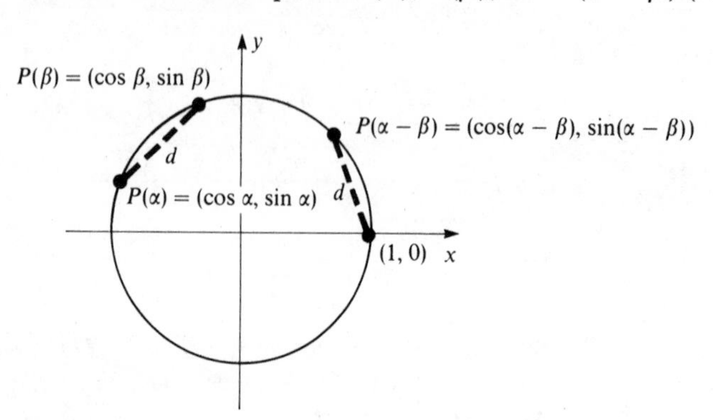

Now since the arc from (1, 0) to $P(\alpha - \beta)$ is equal in length to the arc from $P(\beta)$ to $P(\alpha)$, it follows that the chords connecting these respective pairs of points are also equal. We can get these lengths using the distance formula.

$$\text{Distance from } (1, 0) \text{ to } P(\alpha - \beta) = \sqrt{[\cos(\alpha - \beta) - 1]^2 + [\sin(\alpha - \beta) - 0]^2}$$

$$\text{Distance from } P(\beta) \text{ to } P(\alpha) = \sqrt{(\cos \alpha - \cos \beta)^2 + (\sin \alpha - \sin \beta)^2}$$

Since these distances are the same, their squares are the same; so we obtain the equation

$$[\cos(\alpha - \beta) - 1]^2 + [\sin(\alpha - \beta)]^2 = (\cos \alpha - \cos \beta)^2 + (\sin \alpha - \sin \beta)^2$$

On squaring and collecting terms, making use several times of the identity $\sin^2 \theta + \cos^2 \theta = 1$, this gives

$$\begin{aligned} \cos^2 (\alpha - \beta) - 2 \cos(\alpha - \beta) + 1 + \sin^2 (\alpha - \beta) &= \cos^2 \alpha - 2 \cos \alpha \cos \beta + \cos^2 \beta \\ &+ \sin^2 \alpha - 2 \sin \alpha \sin \beta + \sin^2 \beta \\ 2 - 2 \cos(\alpha - \beta) &= 2 - 2(\cos \alpha \cos \beta + \sin \alpha \sin \beta) \end{aligned}$$

or finally,

$$\boldsymbol{\cos(\alpha - \beta) = \cos \alpha \cos \beta + \sin \alpha \sin \beta} \qquad (3.1)$$

We next remove the restrictions imposed on the sizes of α and β. First, suppose $\beta > \alpha$ and both α and β are between 0 and 2π. Then, since $\cos \theta = \cos(-\theta)$,

$$\cos(\alpha - \beta) = \cos[-(\alpha - \beta)] = \cos(\beta - \alpha)$$

and now (3.1) applies, with the roles of α and β reversed. So

$$\begin{aligned} \cos(\alpha - \beta) &= \cos(\beta - \alpha) \\ &= \cos \beta \cos \alpha + \sin \beta \sin \alpha \\ &= \cos \alpha \cos \beta + \sin \alpha \sin \beta \end{aligned}$$

that is, (3.1) is true regardless of whether α or β is larger, as long as both are between 0 and 2π. The formula is also true if $\alpha = \beta$, for then we have

$$\begin{aligned} \cos(\alpha - \beta) &= \cos(\alpha - \alpha) \\ &= \cos \alpha \cos \alpha + \sin \alpha \sin \alpha \\ &= \cos^2 \alpha + \sin^2 \alpha \\ &= 1 \end{aligned}$$

which is true, since $\cos(\alpha - \alpha) = \cos 0 = 1$.

Now suppose α and β are not restricted to lie between 0 and 2π. Then we can always write

$$\alpha = \alpha_1 + 2n\pi \quad \text{and} \quad \beta = \beta_1 + 2m\pi$$

where α_1 and β_1 do lie between 0 and 2π and m and n are appropriately chosen integers. Because of periodicity we have that $\sin \alpha = \sin(\alpha_1 + 2n\pi) = \sin \alpha_1$, and similarly for β and β_1 with the cosine.

$$\begin{aligned}\cos(\alpha - \beta) &= \cos(\alpha_1 + 2n\pi - \beta_1 - 2m\pi)\\ &= \cos[(\alpha_1 - \beta_1) + (n - m)2\pi]\\ &= \cos(\alpha_1 - \beta_1)\\ &= \cos \alpha_1 \cos \beta_1 + \sin \alpha_1 \sin \beta_1\\ &= \cos \alpha \cos \beta + \sin \alpha \sin \beta\end{aligned}$$

The third step, namely, that

$$\cos[(\alpha_1 - \beta_1) + (n - m)2\pi] = \cos(\alpha_1 - \beta_1)$$

again follows from periodicity. This establishes (3.1) for *all* numbers α and β.

From (3.1) we are able to get a number of other results. Consider first $\cos(\alpha + \beta)$. We write $\alpha + \beta = \alpha - (-\beta)$ and apply (3.1), replacing β by $-\beta$:

$$\cos(\alpha + \beta) = \cos(\alpha - (-\beta)) = \cos \alpha \cos(-\beta) + \sin \alpha \sin(-\beta)$$

Since $\cos(-\beta) = \cos \beta$ and $\sin(-\beta) = -\sin \beta$,

$$\boldsymbol{\cos(\alpha + \beta) = \cos \alpha \cos \beta - \sin \alpha \sin \beta} \tag{3.2}$$

To get analogous formulas to (3.1) and (3.2) for the sine, we proceed as follows. In (3.1), replace α by $\pi/2$ and allow β to be arbitrary. Remember that $\sin(\pi/2) = 1$ and $\cos(\pi/2) = 0$. So,

$$\begin{aligned}\cos\left(\frac{\pi}{2} - \beta\right) &= \cos \frac{\pi}{2} \cos \beta + \sin \frac{\pi}{2} \sin \beta\\ &= 0 + \sin \beta\\ &= \sin \beta\end{aligned}$$

That is, for any number θ,

$$\cos\left(\frac{\pi}{2} - \theta\right) = \sin \theta \tag{3.3}$$

In particular, if we take $\theta = (\pi/2) - \alpha$, we obtain

$$\cos\left[\frac{\pi}{2} - \left(\frac{\pi}{2} - \alpha\right)\right] = \sin\left(\frac{\pi}{2} - \alpha\right)$$

or

$$\cos \alpha = \sin\left(\frac{\pi}{2} - \alpha\right) \tag{3.4}$$

for any number α. The letters θ and α used in (3.3) and (3.4) are independent of any relationships expressed between them in the derivation, that is, the formulas are self-contained. The letters are really just dummy variables, or place holders, and any other letters would express the same relationships. So we could write, for example,

$$\cos\left(\frac{\pi}{2} - x\right) = \sin x$$

and

$$\sin\left(\frac{\pi}{2} - x\right) = \cos x$$

If in the first of these we take x as $\alpha + \beta$, we have by (3.2)

$$\sin(\alpha + \beta) = \cos\left[\frac{\pi}{2} - (\alpha + \beta)\right] = \cos\left[\left(\frac{\pi}{2} - \alpha\right) - \beta\right]$$

$$= \cos\left(\frac{\pi}{2} - \alpha\right)\cos\beta + \sin\left(\frac{\pi}{2} - \alpha\right)\sin\beta$$

Now we use (3.3) and (3.4) again to get

$$\boldsymbol{\sin(\alpha + \beta) = \sin\alpha\cos\beta + \cos\alpha\sin\beta} \qquad (3.5)$$

This is valid for all numbers α and β. If we write $\alpha - \beta = \alpha + (-\beta)$, we can use (3.5) together with the facts that $\sin(-\theta) = -\sin\theta$ and $\cos(-\theta) = \cos\theta$ to get

$$\sin(\alpha - \beta) = \sin[\alpha + (-\beta)] = \sin\alpha\cos(-\beta) + \cos\alpha\sin(-\beta)$$

that is,

$$\boldsymbol{\sin(\alpha - \beta) = \sin\alpha\cos\beta - \cos\alpha\sin\beta} \qquad (3.6)$$

We summarize the formulas (3.1), (3.2), (3.5), and (3.6) as follows.

The Addition Formulas for Sine and Cosine

$$\sin(\alpha + \beta) = \sin\alpha\cos\beta + \cos\alpha\sin\beta$$
$$\sin(\alpha - \beta) = \sin\alpha\cos\beta - \cos\alpha\sin\beta$$
$$\cos(\alpha + \beta) = \cos\alpha\cos\beta - \sin\alpha\sin\beta$$
$$\cos(\alpha - \beta) = \cos\alpha\cos\beta + \sin\alpha\sin\beta$$

These are of fundamental importance and should be learned. Notice that for the sine, the sign between the terms on the right agrees with the sign between α and β on the left, whereas for the cosine these are reversed. So there really are just two basic patterns to learn.

EXAMPLE 3.9 Find the value of each of the following without the use of tables or a calculator.

a. $\sin\frac{5\pi}{12}$ **b.** $\sin\frac{\pi}{12}$ **c.** $\cos\frac{13\pi}{12}$ **d.** $\cos\left(-\frac{\pi}{12}\right)$

Solution **a.** We can write $5\pi/12$ as $2\pi/12 + 3\pi/12 = \pi/6 + \pi/4$. So

$$\sin\frac{5\pi}{12} = \sin\left(\frac{\pi}{6} + \frac{\pi}{4}\right)$$

$$= \sin\frac{\pi}{6}\cos\frac{\pi}{4} + \cos\frac{\pi}{6}\sin\frac{\pi}{4}$$

Now we know that we may treat $\pi/6$ and $\pi/4$ as if they are radian measures of angles. Thus,

$$\sin\frac{5\pi}{12} = \left(\frac{1}{2}\right)\left(\frac{\sqrt{2}}{2}\right) + \left(\frac{\sqrt{3}}{2}\right)\left(\frac{\sqrt{2}}{2}\right) = \frac{\sqrt{2}+\sqrt{6}}{4}$$

b. $$\sin\frac{\pi}{12} = \sin\left(\frac{3\pi}{12} - \frac{2\pi}{12}\right) = \sin\left(\frac{\pi}{4} - \frac{\pi}{6}\right)$$

$$= \sin\frac{\pi}{4}\cos\frac{\pi}{6} - \cos\frac{\pi}{4}\sin\frac{\pi}{6}$$

$$= \left(\frac{\sqrt{2}}{2}\right)\left(\frac{\sqrt{3}}{2}\right) - \left(\frac{\sqrt{2}}{2}\right)\left(\frac{1}{2}\right) = \frac{\sqrt{6}-\sqrt{2}}{4}$$

c. $$\cos\frac{13\pi}{12} = \cos\left(\frac{9\pi}{12} + \frac{4\pi}{12}\right) = \cos\left(\frac{3\pi}{4} + \frac{\pi}{3}\right)$$

$$= \cos\frac{3\pi}{4}\cos\frac{\pi}{3} - \sin\frac{3\pi}{4}\sin\frac{\pi}{3}$$

$$= \left(-\frac{\sqrt{2}}{2}\right)\left(\frac{1}{2}\right) - \left(\frac{\sqrt{2}}{2}\right)\left(\frac{\sqrt{3}}{2}\right) = -\frac{\sqrt{2}+\sqrt{6}}{4}$$

d. $\cos(-\pi/12) = \cos(\pi/12)$, since $\cos(-\theta) = \cos\theta$. So,

$$\cos\left(-\frac{\pi}{12}\right) = \cos\frac{\pi}{12} = \cos\left(\frac{3\pi}{12} - \frac{2\pi}{12}\right) = \cos\left(\frac{\pi}{4} - \frac{\pi}{6}\right)$$

$$= \cos\frac{\pi}{4}\cos\frac{\pi}{6} + \sin\frac{\pi}{4}\sin\frac{\pi}{6}$$

$$= \left(\frac{\sqrt{2}}{2}\right)\left(\frac{1}{2}\right) + \left(\frac{\sqrt{2}}{2}\right)\left(\frac{\sqrt{3}}{2}\right) = \frac{\sqrt{2}+\sqrt{6}}{4}$$

EXAMPLE 3.10 If $-\pi/2 \le x \le \pi/2$ and $0 \le y \le \pi$, $\sin x = \frac{3}{4}$, and $\cos y = -\frac{5}{6}$, find:

a. $\sin(x + y)$ **b.** $\cos(x - y)$

Solution We conclude that x is between 0 and $\pi/2$, since if it were between 0 and $-\pi/2$, $\sin x$ would be negative. Similarly, we must have y between $\pi/2$ and π in order for $\cos y$ to be negative. By the Pythagorean identity $\sin^2 x + \cos^2 x = 1$, we get

$$\cos^2 x = 1 - \sin^2 x$$

So

$$\cos x = \pm\sqrt{1 - \sin^2 x}$$

The positive sign must be chosen, since $0 \le x \le \pi/2$. So,

$$\cos x = \sqrt{1 - \left(\frac{3}{4}\right)^2} = \sqrt{1 - \frac{9}{16}} = \sqrt{\frac{7}{16}} = \frac{\sqrt{7}}{4}$$

Similarly,

$$\sin y = \sqrt{1 - \cos^2 y} = \sqrt{1 - \left(-\frac{5}{6}\right)^2} = \sqrt{1 - \frac{25}{36}} = \frac{\sqrt{11}}{6}$$

We now have all of the necessary ingredients for obtaining the solutions.

a. $\sin(x + y) = \sin x \cos y + \cos x \sin y = \frac{3}{4}\left(-\frac{5}{6}\right) + \left(\frac{\sqrt{7}}{4}\right)\left(\frac{\sqrt{11}}{6}\right)$

$$= \frac{-15 + \sqrt{77}}{24}$$

b. $\cos(x - y) = \cos x \cos y + \sin x \sin y = \frac{\sqrt{7}}{4}\left(-\frac{5}{6}\right) + \frac{3}{4}\left(\frac{\sqrt{11}}{6}\right)$

$$= \frac{-5\sqrt{7} + 3\sqrt{11}}{24}$$

EXERCISE SET 3.4

A In Problems 1–8 evaluate without using a calculator or tables.

1. $\sin(\alpha + \beta)$ and $\cos(\alpha + \beta)$, where $\alpha = \pi/4$ and $\beta = \pi/3$

2. $\sin(\alpha - \beta)$ and $\cos(\alpha - \beta)$, where $\alpha = \pi/4$ and $\beta = \pi/3$

3. **a.** $\sin\left(\frac{\pi}{3} - \frac{3\pi}{4}\right)$ **b.** $\cos\left(\frac{5\pi}{6} + \frac{\pi}{4}\right)$

4. **a.** $\sin\left(\frac{5\pi}{4} + \frac{11\pi}{6}\right)$ **b.** $\cos\left(\frac{4\pi}{3} - \frac{3\pi}{4}\right)$

5. **a.** $\cos\frac{7\pi}{12}$ **b.** $\sin\frac{17\pi}{12}$

6. **a.** $\sin\left(-\frac{\pi}{12}\right)$ **b.** $\cos\left(-\frac{5\pi}{12}\right)$

7. **a.** $\sin 75°$ **b.** $\cos 15°$

8. **a.** $\cos 255°$ **b.** $\sin 195°$

9. If $\sin \alpha = -\frac{3}{5}$, $\sin \beta = -\frac{5}{13}$, $P(\alpha)$ is in the third quadrant, and $P(\beta)$ is in the fourth quadrant, find the following.

a. $\sin(\alpha - \beta)$ **b.** $\cos(\alpha + \beta)$

10. If $\cos \alpha = \frac{12}{13}$, $\cos \beta = -\frac{4}{5}$, $P(\alpha)$ is the fourth quadrant, and $P(\beta)$ is in the second quadrant, find:

a. $\sin(\alpha + \beta)$ **b.** $\cos(\alpha - \beta)$

11. If $\sin\alpha < 0$, $\cos\alpha = \frac{1}{3}$, $\cos\beta < 0$, and $\sin\beta = -\frac{2}{3}$, find:

a. $\sin(\alpha + \beta)$ **b.** $\cos(\alpha - \beta)$

12. If $\tan x = -\frac{3}{4}$, $\csc x > 0$, $\sec y = \frac{13}{5}$, and $\cot y < 0$, find:

a. $\sin(x - y)$ **b.** $\cos(x + y)$

13. If $\sin\alpha = -2/\sqrt{5}$, $\tan\alpha > 0$, $\cos\beta = -\frac{8}{17}$, and $\csc\beta > 0$, find:

a. $\sin(\alpha - \beta)$ **b.** $\cos(\alpha + \beta)$

14. If $\cot A = -\frac{24}{7}$, $\sec A > 0$, $\tan B = \frac{4}{3}$, and $\sin B < 0$, find:

a. $\sin(A + B)$ **b.** $\cos(A - B)$

15. If $\sin x = -\frac{3}{5}$, $\tan x > 0$, $\sec y = -\frac{13}{5}$, and $\cot y < 0$, find:

a. $\csc(x + y)$ **b.** $\sec(x - y)$

Establish the formulas in Problems 16–20 by using addition formulas.

16. **a.** $\sin(\pi + \theta) = -\sin\theta$ **b.** $\cos(\pi + \theta) = -\cos\theta$

17. **a.** $\sin\left(\frac{\pi}{2} + \theta\right) = \cos\theta$ **b.** $\cos\left(\frac{\pi}{2} + \theta\right) = -\sin\theta$

18. **a.** $\sin\left(\frac{3\pi}{2} - \theta\right) = -\cos\theta$ **b.** $\cos\left(\frac{3\pi}{2} - \theta\right) = -\sin\theta$

19. **a.** $\sin(\pi - \theta) = \sin\theta$ **b.** $\cos(\pi - \theta) = -\cos\theta$

20. **a.** $\sin\left(\frac{\pi}{2} - \theta\right) = \cos\theta$ **b.** $\cos\left(\frac{\pi}{2} - \theta\right) = \sin\theta$

Prove the identities in Problems 21–29.

21. $\sin\alpha\cos\beta\,(\cot\alpha - \tan\beta) = \cos(\alpha + \beta)$

22. $\dfrac{\cos(\alpha + \beta)}{\cos(\alpha - \beta)} = \dfrac{1 - \tan\alpha\tan\beta}{1 + \tan\alpha\tan\beta}$

23. $\dfrac{\tan\alpha - \tan\beta}{\tan\alpha + \tan\beta} = \dfrac{\sin(\alpha - \beta)}{\sin(\alpha + \beta)}$

24. $\dfrac{\sin(\alpha + \beta) + \sin(\alpha - \beta)}{\cos(\alpha + \beta) + \cos(\alpha - \beta)} = \tan\alpha$

25. $\dfrac{\cos(\alpha - \beta) - \cos(\alpha + \beta)}{\sin(\alpha + \beta) - \sin(\alpha - \beta)} = \tan\alpha$

26. $\dfrac{\sin(\alpha + \beta)}{\sin\alpha\sin\beta} = \cot\alpha + \cot\beta$

27. $\dfrac{\cos(\alpha - \beta)}{\sin\alpha\sin\beta} = 1 + \cot\alpha\cot\beta$

28. $\sin(x + y)\cos y - \cos(x + y)\sin y = \sin x$

29. $\sin x\sin(x + y) + \cos x\cos(x + y) = \cos y$

B

30. Prove that, in general, $\sin(\alpha + \beta) \neq \sin\alpha + \sin\beta$.

31. Find all values of x for which $0 \le x < 2\pi$ and

$$\sin 5x\cos 4x = \cos 5x\sin 4x$$

32. Find all values of x for which $0 \le x < 2\pi$ and

$$2\cos 2x\cos x = 1 - 2\sin 2x\sin x$$

33. Derive a formula for

a. $\sin(\alpha + \beta + \gamma)$ **b.** $\cos(\alpha + \beta + \gamma)$

Hint. First use the associative property for addition.

Prove the identities in Problems 34–36.

34. $\sin(\alpha + \beta) \sin(\alpha - \beta) = \sin^2 \alpha - \sin^2 \beta$

35. $\cos(\alpha + \beta) \cos(\alpha - \beta) = \cos^2 \alpha - \sin^2 \beta$

36. $\sin(\alpha + \beta) \cos(\alpha - \beta) = \sin \alpha \cos \alpha + \sin \beta \cos \beta$

3.5 Double-Angle, Half-Angle, and Reduction Formulas

The importance of the addition formulas in Section 3.4 lies primarily in the fact that so many other identities can be derived from them. We will carry out the derivations for some of the most important of these.

The **double-angle formulas** are obtained from the addition formulas for sine and cosine of α and β by putting $\beta = \alpha$. If we denote the common value of α and β by θ, we obtain

$$\begin{aligned} \sin(\theta + \theta) &= \sin \theta \cos \theta + \cos \theta \sin \theta \\ &= 2 \sin \theta \cos \theta \end{aligned}$$

So,

$$\boldsymbol{\sin 2\theta = 2 \sin \theta \cos \theta}$$

Also,

$$\begin{aligned} \cos(\theta + \theta) &= \cos \theta \cos \theta - \sin \theta \sin \theta \\ &= \cos^2 \theta - \sin^2 \theta \end{aligned}$$

So,

$$\boldsymbol{\cos 2\theta = \cos^2 \theta - \sin^2 \theta} \tag{3.7}$$

Two other useful forms of $\cos 2\theta$ can be obtained by replacing, in turn, $\cos^2 \theta$ by $1 - \sin^2 \theta$ and $\sin^2 \theta$ by $1 - \cos^2 \theta$. This gives

$$\cos 2\theta = (1 - \sin^2 \theta) - \sin^2 \theta = 1 - 2 \sin^2 \theta$$

and

$$\cos 2\theta = \cos^2 \theta - (1 - \cos^2 \theta) = 2 \cos^2 \theta - 1$$

So we have

$$\boldsymbol{\cos 2\theta = 1 - 2 \sin^2 \theta} \tag{3.8}$$

and

$$\boldsymbol{\cos 2\theta = 2 \cos^2 \theta - 1} \tag{3.9}$$

Whether to use (3.7), (3.8), or (3.9) depends on the objective. We will shortly see some examples where a particular form is clearly preferable.

If we solve (3.8) for $\sin^2\theta$ and (3.9) for $\cos^2\theta$, we obtain

$$\sin^2\theta = \frac{1-\cos 2\theta}{2}$$

and

$$\cos^2\theta = \frac{1+\cos 2\theta}{2}$$

These forms are employed extensively in calculus.

By replacing θ by $\alpha/2$ in the last two equations and then taking square roots, we get the **half-angle formulas**:

$$\sin\frac{\alpha}{2} = \pm\sqrt{\frac{1-\cos\alpha}{2}}$$

$$\cos\frac{\alpha}{2} = \pm\sqrt{\frac{1+\cos\alpha}{2}}$$

The ambiguity of sign has to be resolved in each particular instance according to the quadrant in which $\alpha/2$ lies.

From the addition formulas we can obtain a class of identities sometimes called **reduction formulas**, of which (3.3) and (3.4) of Section 3.4 are special cases. Here are some others:

$$\begin{aligned}\sin(\pi+\theta) &= \sin\pi\cos\theta + \cos\pi\sin\theta\\ &= -\sin\theta\end{aligned}$$

and

$$\begin{aligned}\sin(\pi-\theta) &= \sin\pi\cos\theta - \cos\pi\sin\theta\\ &= \sin\theta\end{aligned}$$

Similarly,

$$\begin{aligned}\cos(\pi+\theta) &= \cos\pi\cos\theta - \sin\pi\sin\theta\\ &= -\cos\theta\end{aligned}$$

and

$$\begin{aligned}\cos(\pi-\theta) &= \cos\pi\cos\theta + \sin\pi\sin\theta\\ &= -\cos\theta\end{aligned}$$

In these, we have used the facts that $\sin\pi = 0$ and $\cos\pi = -1$. Since $\sin(3\pi/2) = -1$ and $\cos(3\pi/2) = 0$, we also have

$$\begin{aligned}\sin\left(\frac{3\pi}{2}+\theta\right) &= \sin\frac{3\pi}{2}\cos\theta + \cos\frac{3\pi}{2}\sin\theta\\ &= -\cos\theta\end{aligned}$$

and

$$\sin\left(\frac{3\pi}{2} - \theta\right) = \sin\frac{3\pi}{2}\cos\theta - \cos\frac{3\pi}{2}\sin\theta$$
$$= -\cos\theta$$

Likewise, there are similar formulas for $\cos[(3\pi/2) \pm \theta]$.

To generalize, it appears that we should consider two cases:

Case 1. Functions of $(n\pi \pm \theta)$

Case 2. Functions of $\left[\frac{(2n + 1)\pi}{2} \pm \theta\right]$

where n is an arbitrary integer. Note that in Case 2, an odd multiple of $\pi/2$ is involved since $(2n + 1)$ is always odd. The following identities can now be established.

Reduction Formulas for the Sine and Cosine

Case 1. $\sin(n\pi \pm \theta) = \pm\sin\theta$

$\cos(n\pi \pm \theta) = \pm\cos\theta$

Case 2. $\sin\left[\frac{(2n + 1)\pi}{2} \pm \theta\right] = \pm\cos\theta$

$\cos\left[\frac{(2n + 1)\pi}{2} \pm \theta\right] = \pm\sin\theta$

To determine the correct sign on the right, it suffices to determine what the sign is when θ is acute.

These are called reduction formulas since in each case the given expression is reduced to a function involving θ only. Note that in Case 1 the function on the right is the *same* as that on the left, whereas in Case 2 the function on the right is the *cofunction* of the one on the left.

We will prove the first of these formulas and leave the others as exercises. We note first that $\sin n\pi = 0$. Secondly, $\cos n\pi = (-1)^n$ since, when n is even, $n\pi$ is coterminal with 0 so that the cosine is $+1$ and, when n is odd, $n\pi$ is coterminal with π so that the cosine is -1. Using the addition formula for the sine, we have

$$\sin(n\pi + \theta) = \sin n\pi\cos\theta + \cos n\pi\sin\theta$$
$$= (0)\cos\theta + (-1)^n\sin\theta$$
$$= (-1)^n\sin\theta$$

This shows that $\sin(n\pi + \theta) = \pm\sin\theta$. Furthermore, the sign on the right is $(-1)^n$ which is independent of the size of θ. So if we determine the sign when θ is acute, it will be correct for all values of θ.

EXAMPLE 3.11 Reduce each of the following to a function of θ only.

a. $\sin(3\pi - \theta)$ **b.** $\cos\left(\frac{\pi}{2} + \theta\right)$ **c.** $\sin\left(-\frac{\pi}{2} + \theta\right)$

Solution **a.** $\sin(3\pi - \theta) = \pm\sin\theta$ (Case 1)

If θ is acute, $3\pi - \theta$ is in quadrant II, so $\sin(3\pi - \theta)$ is positive. Thus,

$$\sin(3\pi - \theta) = \sin\theta$$

b. $\cos\left(\frac{\pi}{2} + \theta\right) = \pm\sin\theta$ (Case 2)

For θ acute, $\pi/2 + \theta$ is in quadrant II, and the cosine is negative there. So

$$\cos\left(\frac{\pi}{2} + \theta\right) = -\sin\theta$$

c. $\sin\left(-\frac{\pi}{2} + \theta\right) = \pm\cos\theta$ (Case 2)

For θ acute, $-\pi/2 + \theta$ is in quadrant IV, where the sine is negative. So

$$\sin\left(-\frac{\pi}{2} + \theta\right) = -\cos\theta$$

EXERCISE SET 3.5

A In Problems 1–6 find $\sin 2\theta$ and $\cos 2\theta$.

1. $\sin\theta = \frac{4}{5}$ and θ terminates in quadrant II
2. $\cos\theta = -\frac{1}{3}$ and θ terminates in quadrant III
3. $\tan\theta = \frac{5}{12}$ and $\sin\theta < 0$
4. $\sec\theta = \sqrt{5}$ and $\csc\theta < 0$
5. $\cot\theta = -2$ and $0 \leq \theta \leq \pi$
6. $\csc\theta = \frac{17}{8}$ and $\frac{\pi}{2} \leq \theta \leq \frac{3\pi}{2}$
7. If $\sin\theta = x$ and $-\pi/2 \leq \theta \leq \pi/2$, find $\sin 2\theta$ and $\cos 2\theta$ in terms of x.
8. If $\cos\theta = x$ and $0 \leq \theta \leq \pi$, find $\sin 2\theta$ and $\cos 2\theta$ in terms of x.

In Problems 9–12 find $\sin\theta$ and $\cos\theta$.

9. $\cos 2\theta = \frac{1}{3}$ and $\pi \leq 2\theta \leq 2\pi$
10. $\sin 2\theta = -\frac{4}{5}$ and $\pi \leq 2\theta \leq \frac{3\pi}{2}$

11. $\sec 2\theta = \frac{25}{7}$ and $-\pi \leq 2\theta \leq 0$

12. $\tan 2\theta = -\frac{12}{5}$ and $3\pi \leq 2\theta \leq 4\pi$

13. Use the half-angle formulas to find

a. $\sin \frac{\pi}{8}$ **b.** $\cos \frac{\pi}{12}$ **c.** $\sin 75°$ **d.** $\cos 67.5°$

In Problems 14–17 find $\sin(\alpha/2)$ and $\cos(\alpha/2)$.

14. $\cos \alpha = \frac{7}{25}, \quad 0 \leq \alpha \leq \pi$

15. $\sin \alpha = -\frac{12}{13}, \quad \pi \leq \alpha \leq \frac{3\pi}{2}$

16. $\tan \alpha = -\frac{8}{15}, \quad 2\pi \leq \alpha \leq 3\pi$

17. $\sec \alpha = 3, \quad -\pi \leq \alpha \leq 0$

18. If $P(\theta) = \left(\frac{1}{\sqrt{5}}, -\frac{2}{\sqrt{5}}\right)$, find $P(2\theta)$.

19. If $0 \leq 2\theta \leq \pi$ and $P(2\theta) = \left(-\frac{7}{25}, \frac{24}{25}\right)$, find $P(\theta)$.

In Problems 20–22 reduce to a function involving θ only.

20. **a.** $\sin(2\pi - \theta)$ **b.** $\cos\left(\frac{3\pi}{2} - \theta\right)$ **c.** $\sin\left(\frac{5\pi}{2} + \theta\right)$

d. $\cos(-\pi + \theta)$ **e.** $\sin\left(-\frac{\pi}{2} - \theta\right)$

21. **a.** $\sin(\theta - \pi)$ **b.** $\cos\left(\frac{\pi}{2} + \theta\right)$ **c.** $\sin\left(-\frac{3\pi}{2} - \theta\right)$

d. $\cos(\theta - 3\pi)$ **e.** $\sin\left(\frac{\pi}{2} + \theta\right)$

22. **a.** $\cos\left(\frac{3\pi}{2} + \theta\right)$ **b.** $\sin(\pi + \theta)$ **c.** $\cos(\theta - 2\pi)$

d. $\sin\left(-\frac{\pi}{2} + \theta\right)$ **e.** $\cos\left(\theta - \frac{5\pi}{2}\right)$

In Problems 23–37 prove the identities.

23. $1 - \frac{\cos 2x - 1}{2\cos^2 x} = \sec^2 x$

24. $\sin^4 \theta \cot^2 \theta = \frac{1}{4}\sin^2 2\theta$

25. $\frac{\cos^2 \theta - \sin^2 \theta}{\sin \theta \cos \theta} = 2 \cot 2\theta$

26. $\frac{\sin^2 2x}{1 - \cos 2x} = 2\cos^2 x$

27. $(\sin x - \cos x)^2 = 1 - \sin 2x$

28. $\cos^4 x - \sin^4 x = \cos 2x$

29. $\frac{2}{1 + \cos 2x} = \sec^2 x$

30. $\frac{\sin 2\theta}{\sin^2 \theta - 1} + 3 \tan \theta = \tan \theta$

31. $\tan \alpha + \cot \alpha = 2 \csc 2\alpha$

32. $\frac{\cot \theta + \tan \theta}{\cot \theta - \tan \theta} = \sec 2\theta$

33. $\frac{\tan^2 \theta}{1 + \tan^2 \theta} = \frac{1 - \cos 2\theta}{2}$

34. $\sin^2 \frac{\theta}{2} \csc^2 \theta = \frac{1}{2(1 + \cos \theta)}$

35. $\dfrac{2\cos^2\dfrac{\theta}{2}}{\sin^2\theta} = \dfrac{1}{1-\cos\theta}$

36. $\dfrac{2\sin^2\dfrac{\theta}{2}+\cos\theta}{\sec\theta} = \cos\theta$

37. $\dfrac{1-\tan^2 x}{1+\tan^2 x} = \cos 2x$

B 38. Prove the reduction formula for each of the following:

a. $\sin(n\pi-\theta)$ **b.** $\cos(n\pi+\theta)$ **c.** $\cos(n\pi-\theta)$ **d.** $\sin\left[\dfrac{(2n+1)\pi}{2}+\theta\right]$

e. $\sin\left[\dfrac{(2n+1)\pi}{2}-\theta\right]$ **f.** $\cos\left[\dfrac{(2n+1)\pi}{2}+\theta\right]$ **g.** $\cos\left[\dfrac{(2n+1)\pi}{2}-\theta\right]$

39. If $\tan\theta = x$ and $-\pi/2 < \theta < \pi/2$, find $\sin 2\theta$ and $\cos 2\theta$ in terms of x.

40. If $\sec\theta = x$ and $0 \le \theta \le \pi$, $\theta \ne \pi/2$, find:

a. $\sin 2\theta$ **b.** $\cos 2\theta$ **c.** $\sin\dfrac{1}{2}\theta$ **d.** $\cos\dfrac{1}{2}\theta$

41. By writing $3\theta = 2\theta + \theta$, find formulas for $\sin 3\theta$ and $\cos 3\theta$ in terms of functions of θ.

42. Derive the formula $\sin 4\theta = 4\sin\theta\cos\theta - 8\sin^3\theta\cos\theta$.

43. Derive the formula $\cos 4\theta = 8\cos^4\theta - 8\cos^2\theta + 1$.

Prove the identities in Problems 44–50.

44. $\cos^4 x = \dfrac{3}{8} + \dfrac{\cos 2x}{2} + \dfrac{\cos 4x}{8}$

45. $\sin^6 x = \dfrac{5}{16} - \dfrac{\cos 2x}{2} + \dfrac{3\cos 4x}{16} + \dfrac{\sin^2 2x\cos 2x}{8}$

46. $\dfrac{1-\cos 2nx}{\sin 2nx} = \tan nx$

47. $\dfrac{\sin 3\theta}{\sin\theta} - \dfrac{\cos 3\theta}{\cos\theta} = 2$

48. $\sin 3x + \sin x = 4\sin x\cos^2 x$

49. $\dfrac{\sin^3 x+\cos^3 x}{\sin x+\cos x} = 1 - \dfrac{1}{2}\sin 2x$

50. By calculating $\sin(\pi/12)$ in two ways, using the half-angle formulas and using the addition formulas, prove that

$$\sqrt{2-\sqrt{3}} = \frac{1}{2}(\sqrt{6}-\sqrt{2})$$

3.6 Further Identities

To obtain addition formulas for the tangent, it is necessary only to use the fact that $\tan\theta = \sin\theta/\cos\theta$. Thus,

$$\tan(\alpha+\beta) = \frac{\sin(\alpha+\beta)}{\cos(\alpha+\beta)} = \frac{\sin\alpha\cos\beta+\cos\alpha\sin\beta}{\cos\alpha\cos\beta-\sin\alpha\sin\beta}$$

This can be improved by dividing numerator and denominator by $\cos\alpha\cos\beta$:

$$\tan(\alpha + \beta) = \frac{\dfrac{\sin \alpha \cos \beta}{\cos \alpha \cos \beta} + \dfrac{\cos \alpha \sin \beta}{\cos \alpha \cos \beta}}{\dfrac{\cos \alpha \cos \beta}{\cos \alpha \cos \beta} - \dfrac{\sin \alpha \sin \beta}{\cos \alpha \cos \beta}}$$

or

$$\boldsymbol{\tan(\alpha + \beta) = \frac{\tan \alpha + \tan \beta}{1 - \tan \alpha \tan \beta}} \tag{3.10}$$

And in a similar manner,

$$\boldsymbol{\tan(\alpha - \beta) = \frac{\tan \alpha - \tan \beta}{1 + \tan \alpha \tan \beta}} \tag{3.11}$$

The double-angle formula for the tangent is obtained from (3.10) by taking $\alpha = \beta$. If we call this common value θ, we have

$$\tan(\theta + \theta) = \frac{\tan \theta + \tan \theta}{1 - \tan \theta \tan \theta}$$

So,

$$\boldsymbol{\tan 2\theta = \frac{2 \tan \theta}{1 - \tan^2 \theta}}$$

We can derive half-angle formulas for the tangent as follows:

$$\tan \tfrac{1}{2}\alpha = \frac{\sin \frac{1}{2}\alpha}{\cos \frac{1}{2}\alpha} = \frac{\sin \frac{1}{2}\alpha}{\cos \frac{1}{2}\alpha} \cdot \frac{2 \cos \frac{1}{2}\alpha}{2 \cos \frac{1}{2}\alpha}$$

$$= \frac{2 \sin \frac{1}{2}\alpha \cos \frac{1}{2}\alpha}{2 \cos^2 \frac{1}{2}\alpha}$$

Since $2 \sin(\alpha/2) \cos(\alpha/2) = \sin 2(\alpha/2) = \sin \alpha$ and

$$2 \cos^2 \frac{1}{2}\alpha = 2\left(\sqrt{\frac{1 + \cos \alpha}{2}}\right)^2 = 1 + \cos \alpha$$

we obtain

$$\boldsymbol{\tan \frac{1}{2}\alpha = \frac{\sin \alpha}{1 + \cos \alpha}}$$

If in this derivation we had multiplied numerator and denominator by $2 \sin(\alpha/2)$ instead of $2 \sin(\alpha/2)$, we would have obtained the equivalent formula

$$\boldsymbol{\tan \frac{1}{2}\alpha = \frac{1 - \cos \alpha}{\sin \alpha}}$$

(See Problem 35, Exercise Set 3.6.)

We could obtain analogous formulas for the cotangent, secant, and cosecant in a similar way, but these are so seldom used tht we will not clutter up our already formidable list with them.

We do choose to list one more group of identities known as the **sum and product formulas** for the sine and cosine. These can be obtained from the addition formulas (see Problems 32 and 33 in Exercise Set 3.6). Although these are used less frequently than the other identities we have considered, they are indispensable at times.

Sum Formulas

$$\sin A + \sin B = 2 \sin \frac{A + B}{2} \cos \frac{A - B}{2}$$

$$\sin A - \sin B = 2 \cos \frac{A + B}{2} \sin \frac{A - B}{2}$$

$$\cos A + \cos B = 2 \cos \frac{A + B}{2} \cos \frac{A - B}{2}$$

$$\cos A - \cos B = -2 \sin \frac{A + B}{2} \sin \frac{A - B}{2}$$

Product Formulas

$$\sin \alpha \cos \beta = \frac{1}{2} [\sin(\alpha + \beta) + \sin(\alpha - \beta)]$$

$$\cos \alpha \sin \beta = \frac{1}{2} [\sin(\alpha + \beta) - \sin(\alpha - \beta)]$$

$$\sin \alpha \sin \beta = \frac{1}{2} [\cos(\alpha - \beta) - \cos(\alpha + \beta)]$$

$$\cos \alpha \cos \beta = \frac{1}{2} [\cos(\alpha + \beta) + \cos(\alpha - \beta)]$$

EXERCISE SET 3.6

A

1. Use (3.10) and (3.11) to find:

   a. $\tan \frac{5\pi}{12}$ **b.** $\tan \frac{\pi}{12}$

2. Find $\tan 2\theta$ if $P(\theta)$ is the point $(-\frac{1}{3}, 2\sqrt{2}/3)$.
3. Find $\tan 2\theta$ if $\sin \theta = -\frac{3}{5}$ and $\cos \theta = -\frac{4}{5}$.
4. Find $\tan(\alpha + \beta)$ if $\sin \alpha = \frac{4}{5}$, α is in the second quadrant, $\cos \beta = -\frac{5}{13}$, and β is in the third quadrant.
5. Find $\tan(\alpha - \beta)$ if $\sec \alpha = 3$, $\csc \alpha < 0$, $\csc \beta = -\sqrt{5}$, and $\cos \beta < 0$.
6. If $\sec \alpha = \frac{13}{5}$, $\sin \alpha < 0$, $\csc \beta = \frac{17}{8}$, and $\cos \beta < 0$, find $\tan(\alpha + \beta)$.
7. For α and β as in Problem 6, find $\tan 2\alpha$, $\tan 2\beta$, $\tan \frac{1}{2}\alpha$, and $\tan \frac{1}{2}\beta$.
8. Find the value of each of the following without using tables or a calculator.

a. $\tan 105°$ **b.** $\tan \dfrac{\pi}{12}$ **c.** $\tan \dfrac{11\pi}{12}$ **d.** $\tan 195°$

e. $\tan \dfrac{\pi}{8}$

9. Find $\tan 2\theta$ and $\tan \frac{1}{2}\theta$ if $\csc \theta = \frac{5}{3}$ and $\sec \theta < 0$.

10. For the angles shown in the figures, find:

a. $\tan(\alpha + \beta)$ **b.** $\tan 2\alpha$ **c.** $\tan(\alpha - \beta)$ **d.** $\tan 2\beta$

e. $\tan \dfrac{1}{2}\alpha$ **f.** $\tan \dfrac{1}{2}\beta$

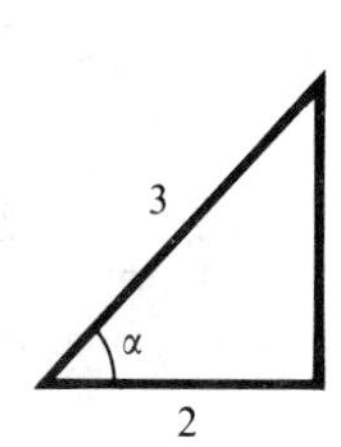

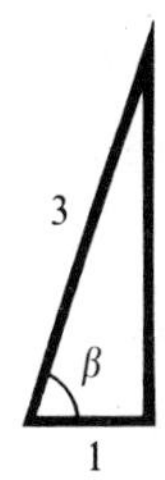

In Problems 11–14 evaluate by use of the sum and product formulas.

11. **a.** $\sin \dfrac{5\pi}{12} + \sin \dfrac{\pi}{12}$ **b.** $\cos \dfrac{7\pi}{12} - \cos \dfrac{\pi}{12}$

12. **a.** $\sin \dfrac{5\pi}{12} \cos \dfrac{7\pi}{12}$ **b.** $\cos \dfrac{3\pi}{8} \cos \dfrac{\pi}{8}$

13. **a.** $\sin 105° - \sin 15°$ **b.** $\cos 165° + \cos 75°$

14. **a.** $\sin 105° \sin 15°$ **b.** $\cos 165° \sin 75°$

15. **a.** Write as a product: $\sin 5x + \sin 3x$
b. Write as a sum: $\sin 5x \cos 3x$

16. **a.** Write as a product: $\cos 7x - \cos 5x$
b. Write as a sum: $\sin 7x \sin 5x$

17. Derive the formula

$$\cot(\alpha + \beta) = \frac{\cot \alpha \cot \beta - 1}{\cot \alpha + \cot \beta}$$

18. If $\cos \theta = x$ and $\sin \theta > 0$, find $\tan 2\theta$ and $\tan \frac{1}{2}\theta$ in terms of x.

In Problems 19–29 prove the identities.

19. $\dfrac{2}{\tan 2x} = \cot x - \tan x$

20. $\csc \theta - \tan \dfrac{\theta}{2} = \cot \theta$

21. $\dfrac{1}{1 - \tan \theta} - \tan 2\theta = \dfrac{1}{1 + \tan \theta}$

22. $1 + \tan \alpha \tan \dfrac{\alpha}{2} = \sec \alpha$

23. $\cot 2\theta = \dfrac{\cot^2 \theta - 1}{2 \cot \theta}$

24. $\dfrac{2 \sin \theta - \sin 2\theta}{2 \sin \theta + \sin 2\theta} = \tan^2 \dfrac{\theta}{2}$

25. $\dfrac{\sin A - \sin B}{\cos A + \cos B} = \tan \dfrac{1}{2}(A - B)$

26. $\dfrac{\sin 7\theta - \sin 5\theta}{\cos 7\theta + \cos 5\theta} = \tan \theta$

27. $\dfrac{\cos 3\theta + \cos \theta}{\sin 3\theta + \sin \theta} = \cot 2\theta$

28. $\sin\left(\dfrac{3\pi}{4} + \theta\right) \sin\left(\dfrac{\pi}{4} - \theta\right) = \dfrac{1 - \sin 2\theta}{2}$

29. $2 \cos \dfrac{3\theta}{2} \sin \dfrac{\theta}{2} = \sin 2\theta - \sin \theta$

B 30. Prove that

a. $\tan(n\pi \pm \theta) = \pm\tan \theta$ b. $\tan\left[\dfrac{(2n + 1)\pi}{2} \pm \theta\right] = \pm\cot \theta$

where the sign on the right-hand side can be determined by considering θ to be acute.

31. Prove the identity

$$\frac{\sin \theta + \sin 2\theta + \sin 3\theta}{\cos \theta + \cos 2\theta + \cos 3\theta} = \tan 2\theta$$

32. Use the addition formulas for the sine and cosine to derive the product formulas as follows:

a. Add the formulas for $\sin(\alpha + \beta)$ and $\sin(\alpha - \beta)$ to get the formula for $\sin \alpha \cos \beta$.
b. Subtract the formula for $\sin(\alpha - \beta)$ from that for $\sin(\alpha + \beta)$ to get the formula for $\cos \alpha \sin \beta$.
c. Add the formulas for $\cos(\alpha + \beta)$ and $\cos(\alpha - \beta)$ to get the formula for $\cos \alpha \cos \beta$.
d. Subtract the formula for $\cos(\alpha + \beta)$ from that for $\cos(\alpha - \beta)$ to get the formula for $\sin \alpha \sin \beta$.

33. In the addition formulas for the sine and cosine make the following substitutions: $\alpha + \beta = A$ and $\alpha - \beta = B$. Solve these two equations simultaneously for α and β in terms of A and B, and obtain the sum formulas.

34. Prove the identity

$$\frac{\sin(x + h) - \sin x}{h} = \frac{\sin \frac{h}{2}}{\frac{h}{2}} \cos(x + \frac{h}{2})$$

35. Derive the formula

$$\tan \frac{\alpha}{2} = \frac{1 - \cos \alpha}{\sin \alpha}$$

3.7 Summary of Identities

We summarize below the identities given in Sections 3.2–3.6.

PYTHAGOREAN IDENTITIES

$\sin^2 \theta + \cos^2 \theta = 1$ $1 + \tan^2 \theta = \sec^2 \theta$ $1 + \cot^2 \theta = \csc^2 \theta$

RECIPROCAL RELATIONS

$$\csc \theta = \frac{1}{\sin \theta} \qquad \sec \theta = \frac{1}{\cos \theta} \qquad \cot \theta = \frac{1}{\tan \theta}$$

TANGENT AND COTANGENT IN TERMS OF SINE AND COSINE

$$\tan \theta = \frac{\sin \theta}{\cos \theta} \qquad \cot \theta = \frac{\cos \theta}{\sin \theta}$$

ADDITION FORMULAS

$$\sin(\alpha \pm \beta) = \sin \alpha \cos \beta \pm \cos \alpha \sin \beta$$

$$\cos(\alpha \pm \beta) = \cos \alpha \cos \beta \mp \sin \alpha \sin \beta$$

$$\tan(\alpha \pm \beta) = \frac{\tan \alpha \pm \tan \beta}{1 \mp \tan \alpha \tan \beta}$$

DOUBLE-ANGLE FORMULAS

$$\sin 2\theta = 2 \sin \theta \cos \theta$$

$$\cos 2\theta = \cos^2 \theta - \sin^2 \theta = 2 \cos^2 \theta - 1 = 1 - 2 \sin^2 \theta$$

$$\tan 2\theta = \frac{2 \tan \theta}{1 - \tan^2 \theta}$$

SIN² θ AND COS² θ IN TERMS OF COS 2θ

$$\sin^2 \theta = \frac{1 - \cos 2\theta}{2} \qquad \cos^2 \theta = \frac{1 + \cos 2\theta}{2}$$

HALF-ANGLE FORMULAS

$$\sin \frac{\alpha}{2} = \pm \sqrt{\frac{1 - \cos \alpha}{2}} \qquad \cos \frac{\alpha}{2} = \pm \sqrt{\frac{1 + \cos \alpha}{2}}$$

$$\tan \frac{\alpha}{2} = \frac{\sin \alpha}{1 + \cos \alpha} = \frac{1 - \cos \alpha}{\sin \alpha}$$

SUM FORMULAS

$$\sin A + \sin B = 2 \sin \frac{A + B}{2} \cos \frac{A - B}{2}$$

$$\sin A - \sin B = 2 \cos \frac{A + B}{2} \sin \frac{A - B}{2}$$

$$\cos A + \cos B = 2 \cos \frac{A + B}{2} \cos \frac{A - B}{2}$$

$$\cos A - \cos B = -2 \sin \frac{A + B}{2} \sin \frac{A - B}{2}$$

PRODUCT FORMULAS

$$\sin \alpha \cos \beta = \frac{1}{2}[\sin(\alpha + \beta) + \sin(\alpha - \beta)]$$

$$\cos \alpha \sin \beta = \frac{1}{2}[\sin(\alpha + \beta) - \sin(\alpha - \beta)]$$

$$\sin \alpha \sin \beta = \frac{1}{2}[\cos(\alpha - \beta) - \cos(\alpha + \beta)]$$

$$\cos \alpha \cos \beta = \frac{1}{2}[\cos(\alpha + \beta) + \cos(\alpha - \beta)]$$

REDUCTION FORMULAS

Let f designate any trigonometric function and cf its cofunction. Then for any integer n,

$$f(n\pi \pm \theta) = \pm f(\theta)$$

$$f\left[\frac{(2n + 1)\pi}{2} \pm \theta\right] = \pm cf(\theta)$$

The sign on the right can be determined by considering θ to be acute.

Remark. The particular letters used in these formulas are unimportant. The occurrence of α and β at some times, θ sometimes, and A and B at other times is only because of customary usage.

3.8 Trigonometric Equations

The solutions of most trigonmetric equations cannot be obtained exactly, and we have to settle for approximate solutions obtained through some numerical procedure, usually with the aid of a calculator. There are, however, enough such equations for which elementary techniques will yield exact answers that some time devoted to these techniques is justified.

In order to make clear the sorts of solutions we are looking for, let us consider the very simple equation

$$2 \sin \theta - 1 = 0$$

The problem is to discover all real numbers θ for which this equation is true. We write the equation in the equivalent form

$$\sin\,\theta = \frac{1}{2}$$

and then rely on our knowledge of special angles to conclude that this is satisfied if, and only if,

$$\theta = \frac{\pi}{6}, \frac{5\pi}{6}$$

or any other number obtained from these by adding integral multiples of 2π to each. To see this, remember that the sine of a number θ is the y coordinate of the point on the unit circle that is θ units along the arc from $(1, 0)$. Thus, we are seeking those numbers θ for which the y coordinate of $P(\theta) = \frac{1}{2}$. There are only two points on the unit circle with y coordinate $\frac{1}{2}$, and our knowledge of the 30°–60° right triangle tells us that these points are $P(\pi/6)$ and $P(5\pi/6)$; see Figure 3.10. Since $P(\theta) = P(\theta + 2n\pi)$, we conclude that all solutions of the equation are given by $(\pi/6) + 2n\pi$ or $(5\pi/6) + 2n\pi$.

Figure 3.10

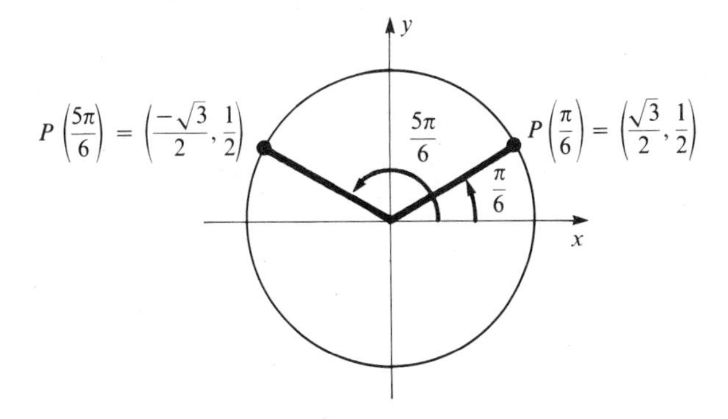

Usually, it is sufficient to give only the **primary** solutions, that is, those lying between 0 and 2π. We would know then that all other solutions are obtainable from these by adding multiples of 2π.

We can usually condense the above reasoning as follows:

1. Determine the appropriate reference angle.
2. By the sign of the function, locate all primary angles having this reference angle.
3. Write the answer as the radian measure of these angles.

In our example, since we know that $\sin 30° = \frac{1}{2}$, the reference angle is 30°. The sine is positive in quadrants I and II, so the angles are 30° and 150°. In radians these are $\pi/6$ and $5\pi/6$.

It might be useful to review at this time the values of the functions of 30°, 45°, and 60°.

$$\sin 30° = \sin\frac{\pi}{6} = \frac{1}{2} \qquad \sin 45° = \sin\frac{\pi}{4} = \frac{1}{\sqrt{2}} \qquad \sin 60° = \sin\frac{\pi}{3} = \frac{\sqrt{3}}{2}$$

$$\cos 30° = \cos\frac{\pi}{6} = \frac{\sqrt{3}}{2} \qquad \cos 45° = \cos\frac{\pi}{4} = \frac{1}{\sqrt{2}} \qquad \cos 60° = \cos\frac{\pi}{3} = \frac{1}{2}$$

$$\tan 30° = \tan\frac{\pi}{6} = \frac{1}{\sqrt{3}} \qquad \tan 45° = \tan\frac{\pi}{4} = 1 \qquad \tan 60° = \tan\frac{\pi}{3} = \sqrt{3}$$

Also,

$$\sin 0° = \sin 0 = 0 \qquad \sin 90° = \sin\frac{\pi}{2} = 1$$

$$\cos 0° = \cos 0 = 1 \qquad \cos 90° = \cos\frac{\pi}{2} = 0$$

$$\tan 0° = \tan 0 = 0 \qquad \tan 90° = \tan\frac{\pi}{2} \text{ is not defined}$$

(The cotangent, secant, and cosecant are not listed because they occur less frequently and can always be obtained by the reciprocal relations.) If you feel comfortable with radian measure by now, you can omit the degree measure entirely.

If an equation can be worked around to a form such as $\sin\theta = \pm a$, $\cos\theta = \pm b$, or $\tan\theta = \pm c$, for instance, where a and b are any of the numbers 0, 1, $\frac{1}{2}$, $1/\sqrt{2}$, or $\sqrt{3}/2$, and c is any of the numbers 0, 1, $1/\sqrt{3}$, or $\sqrt{3}$, then we can obtain all solutions in the manner outlined above. For other values of a, b, and c, a calculator or tables can be used.

EXAMPLE 3.12 Find all values of θ, for which $0 \le \theta < 2\pi$, satisfying $2\cos^2\theta - \cos\theta - 1 = 0$.

Solution We treat this first as a quadratic equation in $\cos\theta$ and factor:

$$(2\cos\theta + 1)(\cos\theta - 1) = 0$$

This is true if and only if $2\cos\theta + 1 = 0$ or $\cos\theta = 1$, that is, if and only if

$$\cos\theta = -\frac{1}{2} \quad \text{or} \quad \cos\theta = 1$$

Since $\cos 60° = \frac{1}{2}$, the reference angle in the first of these is 60°. Since the cosine is negative in quadrants II and III, it follows that we are seeking the angles 120° and 240°. In radians these are $2\pi/3$ and $4\pi/3$. The only primary angle for which $\cos\theta = 1$ is $\theta = 0°$, or 0 radians. Thus, the primary solution set for the equation is $\{0, 2\pi/3, 4\pi/3\}$.

Remark. In many equations involving trigonometric functions the variable must be treated as a real number, which is the reason for writing the above answers as the radian measures of the angles. Recall that the values of $\sin\theta$, $\cos\theta$, and the other trigonometric functions are unchanged whether we consider θ a real number or the radian measure of an angle. An example of a mixed algebraic and trigonometric equation will perhaps help to make this point clearer. The equation

$$2 \sin x - x = 0$$

is clearly satisfied when $x = 0$, but by trial and error, using a calculator (or by some more sophisticated numerical procedure), we can find that another solution is

$$x \approx 1.8955$$

If we think of this x as the radian measure of an angle, then the degree measure of the angle is approximately 108.6°. Now it would *not* be true that $x = 108.6°$; in fact, it would not even make sense, for we would have

$$2 \sin 108.6° - 108.6° = 0$$

which is a totally meaningless statement.

EXAMPLE 3.13 Find all primary solutions of $\sin 2x + \cos x = 0$.

Solution This time we will omit most discussion and go only through the steps that would be expected in doing the problem.

$$\sin 2x + \cos x = 0$$

$$2 \sin x \cos x + \cos x = 0$$

$$\cos x(2 \sin x + 1) = 0$$

$\cos x = 0$	$2 \sin x = -1$
$x = \frac{\pi}{2}, \frac{3\pi}{2}$	$\sin x = -\frac{1}{2}$
	$x = \frac{7\pi}{6}, \frac{11\pi}{6}$

The solution set is $\{\pi/2, 7\pi/6, 3\pi/2, 11\pi/6\}$.

EXAMPLE 3.14 Find all primary solutions of $\tan \alpha - 3 \cot \alpha = 0$.

Solution

$$\tan \alpha - 3 \cot \alpha = 0$$

$$\tan \alpha - \frac{3}{\tan \alpha} = 0$$

$$\tan^2 \alpha - 3 = 0$$

$$\tan \alpha = \pm\sqrt{3}$$

The reference angle is 60°, or $\pi/3$ radians. Therefore,

$$\alpha = \frac{\pi}{3}, \frac{2\pi}{3}, \frac{4\pi}{3}, \frac{5\pi}{3}$$

Since we multiplied by an unknown expression, namely, $\tan \alpha$, it is essential that we check to see that this was not zero for the values of α found, which is clearly the case here, since $\tan \alpha = \pm\sqrt{3} \neq 0$. So the solution set is $\{\pi/3, 2\pi/3, 4\pi/3, 5\pi/3\}$.

EXAMPLE 3.15 Find all primary solutions of $\sqrt{3} \cos x = 2 + \sin x$.

Solution The fact that $\cos^2 x = 1 - \sin^2 x$ suggests that if we square both sides of the given equation, it will be easier to work with.

$$3 \cos^2 x = 4 + 4 \sin x + \sin^2 x$$

$$3(1 - \sin^2 x) = 4 + 4 \sin x + \sin^2 x$$

$$4 \sin^2 x + 4 \sin x + 1 = 0$$

$$(2 \sin x + 1)^2 = 0$$

$$\sin x = -\frac{1}{2}$$

$$x = \frac{7\pi}{6}, \frac{11\pi}{6}$$

We again must check our answers, because squaring does not necessarily lead to an equivalent equation; it may introduce extraneous roots. So we check in the original equation.

When $x = 7\pi/6$, we get

$$\sqrt{3}\left(-\frac{\sqrt{3}}{2}\right) \stackrel{?}{=} 2 + \left(-\frac{1}{2}\right)$$

$$-\frac{3}{2} \neq \frac{3}{2}$$

So $7\pi/6$ is not a solution.

When $x = 11\pi/6$,

$$\sqrt{3}\left(\frac{\sqrt{3}}{2}\right) \stackrel{?}{=} 2 + \left(-\frac{1}{2}\right)$$

$$\frac{3}{2} = \frac{3}{2}$$

So $x = 11\pi/6$ is the only primary solution.

EXAMPLE 3.16 Find all primary solutions of $2 \sin 3\theta - \sqrt{3} \tan 3\theta = 0$.

Solution

$$2 \sin 3\theta - \frac{\sqrt{3} \sin 3\theta}{\cos 3\theta} = 0$$

$$\sin 3\theta(2 \cos 3\theta - \sqrt{3}) = 0 \text{ (We multiplied by } \cos 3\theta.)$$

$$\sin 3\theta = 0 \quad \Big| \quad \cos 3\theta = \frac{\sqrt{3}}{2} \text{ (So we did not multiply by zero.)}$$

Now we want all values of θ lying between 0 and 2π. We must therefore find all values of 3θ lying between 0 and 6π. In general, if $n\theta$ is involved, in order to find all values of θ between 0 and 2π, we find all values of $n\theta$ between 0 and $2n\pi$, and then divide by n. We have from $\sin 3\theta = 0$ that

$$3\theta = 0, \quad \pi, \quad 2\pi, \quad 3\pi, \quad 4\pi, \quad 5\pi$$

and

$$\theta = 0, \quad \frac{\pi}{3}, \quad \frac{2\pi}{3}, \quad \pi, \quad \frac{4\pi}{3}, \quad \frac{5\pi}{3}$$

From $\cos 3\theta = \sqrt{3}/2$, we have

$$3\theta = \frac{\pi}{6}, \quad \frac{11\pi}{6}, \quad \frac{13\pi}{6}, \quad \frac{23\pi}{6}, \quad \frac{25\pi}{6}, \quad \frac{35\pi}{6}$$

$$\theta = \frac{\pi}{18}, \quad \frac{11\pi}{18}, \quad \frac{13\pi}{18}, \quad \frac{23\pi}{18}, \quad \frac{25\pi}{18}, \quad \frac{35\pi}{18}$$

In both cases we found the first two values of 3θ, then added 2π once, and then 2π again. The complete solution set is

$$\left\{0, \quad \frac{\pi}{18}, \quad \frac{\pi}{3}, \quad \frac{11\pi}{18}, \quad \frac{2\pi}{3}, \quad \frac{13\pi}{18}, \quad \pi, \quad \frac{23\pi}{18}, \quad \frac{4\pi}{3}, \quad \frac{25\pi}{18}, \quad \frac{5\pi}{3}, \quad \frac{35\pi}{18}\right\}$$

EXAMPLE 3.17 Find all real solutions to $\tan x - \cot x = 2$.

Solution

$$\tan x - \cot x = 2$$

$$\frac{\sin x}{\cos x} - \frac{\cos x}{\sin x} = 2$$

$$\sin^2 x - \cos^2 x = 2 \sin x \cos x$$

$$-\cos 2x = \sin 2x$$

$$\tan 2x = -1$$

$$2x = \frac{3\pi}{4}, \quad \frac{7\pi}{4}, \quad \frac{11\pi}{4}, \quad \frac{15\pi}{4} \quad \text{(We "went around" twice.)}$$

$$x = \frac{3\pi}{8}, \frac{7\pi}{8}, \frac{11\pi}{8}, \frac{15\pi}{8}$$

As none of these values of x makes $\sin x$, $\cos x$, or $\cos 2x$ equal to zero, we did not multiply by zero. We must also check to see that we did not lose any roots when dividing by $\cos 2x$; that is, we must check to see if any of the roots of $\cos 2x = 0$ are roots of the original equation. Since $\cos 2x = 0$ when $2x = \pi/2, 3\pi/2, 5\pi/2, 7\pi/2$, and so when $x = \pi/4, 3\pi/4, 5\pi/4, 7\pi/4$, we check and find that none of these is a root of the original equation. So we have found all primary solutions. To get *all* solutions, we add arbitrary multiples of 2π to each of these. So the complete solution set is

$$\left\{\frac{3\pi}{8}, \frac{7\pi}{8}, \frac{11\pi}{8}, \frac{15\pi}{8}, \frac{19\pi}{8}, \frac{23\pi}{8}, \frac{27\pi}{8}, \frac{31\pi}{8}, \frac{35\pi}{8}, \frac{39\pi}{8}, \frac{43\pi}{8}, \frac{47\pi}{8}, \ldots\right\}$$

or we could write

$$\{3\pi/8 + 2n\pi, \quad 7\pi/8 + 2n\pi, \quad 11\pi/8 + 2n\pi, \quad 15\pi/8 + 2n\pi: \\ n = 0, \pm 1, \pm 2, \ldots\}$$

We could continue with examples, each possessing its own special features, but these illustrate the main techniques. It might be useful to summarize some of the points that the examples were meant to bring out:

1. Use algebraic techniques such as factoring, multiplying by the LCD, and squaring in conjunction with basic trigonometric identities to simplify so as to obtain one or more elementary equations of the form $\sin x = \pm a$, $\cos x = \pm b$, and so on.
2. Find the radian measure of all primary angles (that is, $0 \leq x < 2\pi$) satisfying the elementary equation in 1. If *all* solutions are desired, add arbitrary multiples of 2π to each of these to obtain the complete solution set.
3. If in arriving at the elementary equations both sides of an equation were squared, or if both sides were multiplied by an expression containing a variable, it is necessary to check the answers. (In the case of multiplying by an unknown, it is sufficient to check that the multiplier was not zero.) If both sides of an equation are divided by an expression containing a variable, then the values of the variable for which this expression equals zero mut be checked in the original equation to see if they are solutions. (In general, such division can be avoided by factoring.)
4. If the final elementary equation(s) is of the form $\sin nx = \pm a$, $\cos nx = \pm b$, and so on, then to get all primary values of x, find all values of nx between 0 and $2n\pi$, and divide by n. This amounts to finding the angles in the first revolution and then adding 2π a total of $n - 1$ separate times and finally dividing everything by n.

EXERCISE SET 3.8

A Find all primary solutions unless otherwise specified.

1. $2\cos x = \sqrt{3}$

2. $\cot x + 1 = 0$

3. $\sec x = 2$

4. $\sin^2 x = 1$ (Give all solutions.)

5. $\cos x - 2\cos^2 x = 0$

6. $\sin 2\theta - \cos\theta = 0$

7. $\cos 2x + \cos x = 0$

8. $2\tan\psi - \tan\psi\sec\psi = 0$

9. $\sin^2 t - \cos^2 t = 1$

10. $\sec^2\theta = 2\tan\theta$

11. $2\cos x - \cot x = 0$

12. $2\cos^2 x + \sin x - 1 = 0$

13. $4\tan^2\alpha = 3\sec^2\alpha$

14. $\sin^2 x = 2(\cos x - 1)$

15. $\cos^2 x - \sin x + 5 = 0$

16. $\tan 2x - \cos 2x + \sec 2x = 0$

17. $\dfrac{1-\cos x}{\sin x} = \sin x$

18. $2\cos^2 2\theta - 3\cos 2\theta + 1 = 0$

19. $\sin 6\theta + \sin 3\theta = 0$

20. $\sin^2 4\theta - \sin 4\theta - 2 = 0$

21. $\tan\dfrac{x}{2} + \cos x = 1$

22. $\sin^2\dfrac{x}{2} - \sin^2 x = 0$

23. $2(\sin^2 2\theta - \cos^2 2\theta) + \sqrt{3} = 0$

24. $2\cos x - \cot x\csc x = 0$

25. $2\tan^2\theta + \sec\theta + 2 = 0$

B **26.** $\sin\theta + \cos\theta = 1$ (Give all solutions.)

27. $\cos 2x\tan 3x + \sqrt{3}\cos 2x = 0$

28. $\dfrac{\sin x}{1-\cos x} = \dfrac{\cos x}{1+\sin x}$

29. $\sin 2\theta - 4\sin\theta = 3(2-\cos\theta)$

30. $\sqrt{3}\tan\theta = 2\sec\theta - 1$

In Problems 31–36 use a calculator or tables to find all primary solutions correct to three significant figures.

31. $12\sin^2 x - \sin x - 6 = 0$

32. $\cos 2x + 2\cos x = 0$

33. $2\sec^2 x = 5\tan x$

34. $3\sin x - \tan\dfrac{1}{2}x = 0$

35. $2\sin x - \csc x + 3 = 0$

36. $\sec x - 2 = 10\cos x$

In Problems 37–39 find all primary solutions without using a calculator or tables.

37. $\tan^3 x + \tan^2 x - 3\tan x - 3 = 0$

38. $8\sin^4\theta - 2\sin^2\theta - 3 = 0$

39. $2\sin^3 x + 3\sin^2 x - 1 = 0$

3.9 Review Exercise Set

A **1.** If $P(\theta) = (\frac{2}{3}, -\sqrt{5}/3)$ find:

a. $\sec\theta$ **b.** $\cot\theta$ **c.** $\cos 2\theta$ **d.** $\tan 2\theta$

2. Evaluate without using tables or a calculator.

a. $\tan \frac{7\pi}{12}$ **b.** $\sin \frac{23\pi}{12}$ **c.** $\cos \frac{\pi}{8}$ **d.** $\cos \frac{13\pi}{12}$

3. Reduce each of the following to a function involving θ only.

a. $\sin\left(\frac{3\pi}{2} - \theta\right)$ **b.** $\tan(\theta + \pi)$ **c.** $\sec(4\pi - \theta)$ **d.** $\cos\left(\theta - \frac{\pi}{2}\right)$

e. $\csc\left(\frac{9\pi}{2} + \theta\right)$

4. If $\sec \theta = \frac{5}{4}$ and $-\pi < \theta < 0$, find:

a. $\cos 2\theta$ **b.** $\tan 2\theta$ **c.** $\cos \frac{1}{2}\theta$ **d.** $\sin \frac{1}{2}\theta$

5. If $\sin \theta = \frac{4}{5}$, $\tan \theta < 0$ and $0 \le \theta \le 2\pi$, find:

a. $\sin 2\theta$ **b.** $\cos 2\theta$ **c.** $\sin \frac{1}{2}\theta$ **d.** $\tan \frac{1}{2}\theta$

6. If $\sin \alpha = -\frac{3}{5}$, $\cos \alpha > 0$, and $\cos \beta = -\frac{5}{13}$, $\sin \beta > 0$, find:

a. $\sin(\alpha + \beta)$ **b.** $\cos(\alpha - \beta)$ **c.** $\tan(\alpha + \beta)$

7. If $\tan \alpha = -\frac{4}{3}$, $\sec \alpha > 0$, and $\cos \beta = -\frac{8}{17}$, $\csc \beta > 0$, find:

a. $\sin(\alpha - \beta)$ **b.** $\cos(\alpha + \beta)$ **c.** $\tan(\alpha - \beta)$

8. If $\tan 2\theta = -\frac{24}{7}$, and $\pi/2 < 2\theta < \pi$, find $\sin \theta$ and $\cos \theta$.

9. If $\tan \theta = \frac{15}{8}$ and $\pi < \theta < 3\pi/2$, find:

a. $\sin \frac{\theta}{2}$ **b.** $\cos \frac{\theta}{2}$ **c.** $\tan \frac{\theta}{2}$

10. If $\sec \theta = \frac{5}{3}$ and $3\pi/2 < \theta < 2\pi$, find:

a. $\sin 2\theta$ **b.** $\cos \frac{1}{2}\theta$ **c.** $\tan 2\theta$ **d.** $\tan \frac{1}{2}\theta$

11. Use the sum and product formulas to evaluate the following:

a. $\cos \frac{5\pi}{12} + \cos \frac{\pi}{12}$ **b.** $\sin \frac{5\pi}{8} \sin \frac{\pi}{8}$ **c.** $\sin \frac{7\pi}{12} \cos \frac{5\pi}{12}$

In Problems 12–25 prove that the given equations are identities.

12. $\csc \theta - \cos \theta \cot \theta = \sin \theta$

13. $\dfrac{2}{\tan \theta + \cot \theta} = \sin 2\theta$

14. $\dfrac{\cos \theta + \cot \theta}{1 + \csc \theta} = \cos \theta$

15. $\dfrac{\cos \theta}{\sec \theta - \tan \theta} = 1 + \sin \theta$

16. $\dfrac{\cot \theta - \tan \theta}{\cot \theta + \tan \theta} = \cos 2\theta$

17. $\sin \theta \tan \theta + \cos \theta = \sec \theta$

18. $\dfrac{\tan \theta}{\sec \theta + 1} + \dfrac{\sec \theta - 1}{\tan \theta} = \dfrac{2 \tan \theta}{\sec \theta + 1}$

19. $\tan \theta (\cos 2\theta + 1) = \sin 2\theta$

20. $\sin \theta + \cos \theta \cot \theta = \csc \theta$

21. $\sin 2\theta + (\cos \theta - \sin \theta)^2 = 1$

22. $1 + \dfrac{\tan^2 \theta}{\sec \theta + 1} = \sec \theta$

23. $\cot \theta - \tan \theta = 2 \cot 2\theta$

24. $\dfrac{1 + \tan^2 \theta}{1 - \tan^2 \theta} = \sec 2\theta$

25. $\dfrac{\sin 2\theta}{1 - \cos 2\theta} = \cot \theta$

In Problems 26–31 find all primary solutions.

26. $\sin \theta - 2 \sin^2 \theta = 0$

27. $3 \tan^3 \theta - \tan \theta = 0$

28. $\sin 2\theta + \cos \theta = 0$

29. $\cos 2\theta - \sin \theta = 0$

30. $2 \cos^2 3\theta - 3 \cos 3\theta - 2 = 0$

31. $2 \cos^2 \theta = 1 - \sin \theta$

B **32.** Derive the following formula for $\tan 3\theta$:

$$\tan 3\theta = \frac{3 \tan \theta - \tan^3 \theta}{1 - 3 \tan^2 \theta}$$

33. If $\sec \theta = x$ and $\pi/2 < \theta < \pi$, find the following in terms of x.

a. $\cos 2\theta$ **b.** $\sin 2\theta$ **c.** $\cos \frac{1}{2} \theta$ **d.** $\tan \frac{1}{2} \theta$

34. If $\sin \alpha = 1/\sqrt{5}$, $\cos \alpha < 0$, and $\cot \beta = \frac{4}{3}$, $\pi \le \beta \le 2\pi$, find:

a. $\sin(2\alpha - \beta)$ **b.** $\cos\left(\alpha + \frac{1}{2}\beta\right)$ **c.** $\tan\left(2\alpha - \frac{1}{2}\beta\right)$

In Problems 35–40 prove the identities.

35. $\dfrac{\sin \theta + \cos 2\theta - 1}{\cos \theta - \sin 2\theta} = \tan \theta$

36. $\sin^2 \theta + \dfrac{2 - \tan^2 \theta}{\sec^2 \theta} - 1 = \cos 2\theta$

37. $\dfrac{\sec \theta - 1}{\sec \theta + 1} = 2 \csc^2 \theta - 2 \cot \theta \csc \theta - 1$

38. $\sec \theta \csc \theta - 2 \cos \theta \csc \theta = \tan \theta - \cot \theta$

39. $\dfrac{\sin 8\theta - \sin 6\theta}{\cos 8\theta + \cos 6\theta} = \tan \theta$

40. $\dfrac{\cos \theta}{1 - \sin \theta} = \tan \theta + \sec \theta$

In Problems 41–45 find all primary solutions.

41. $\tan^2 2\theta + 3 \sec 2\theta + 3 = 0$

42. $\cos 2\theta \sec \theta + \sec \theta = 1$

43. $\tan 2\theta = 2 \cos \theta$

44. $1 + \sin \theta = \cos \theta$

45. $\sqrt{2} \sec \theta + \tan \theta = 1$

Use a calculator or tables in Problems 46–49 to find the primary solutions correct to three significant figures.

46. $12 \sin^2 \theta + 5 \sin \theta - 3 = 0$

47. $2 \sec^2 x + 3 \tan x = 4$

48. $2 \cos^2 \frac{1}{2} x - 3 \cos^2 x = 0$

49. $3 \tan^2 x = 2 \sec x + 5$

4 Graphs and Inverses of the Trigonometric Functions

Graphs of the trigonometric functions, especially of the sine and cosine, are an integral part of the study of wave phenomena. The following excerpt from a chemistry textbook illustrates how various sine curves arise in the study of crystal structure.* As usual, you are not expected to understand the meaning of all of the terms, but you can still observe the use of trigonometric graphs in this context.

> The net scattering from an hkl plane requires the summing up, allowing for the phase differences, of the scattering amplitudes from all atoms in the unit cell. The summation of waves, all with the same frequency but with different amplitudes and phases, can be accomplished by representing the amplitude and phase of each wave by the diagrams. . . .

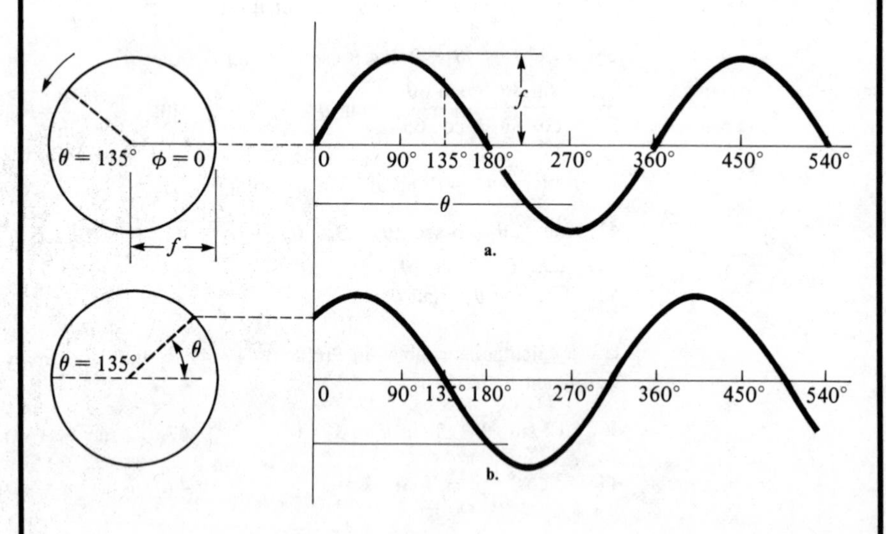

> If two scattered waves are considered, . . . the square of the amplitude is obtained as

*Gordon M. Barrow, *Physical Chemistry*, 4th ed. (New York: McGraw-Hill Book Company, 1979), pp. 518–519. Reprinted by permission.

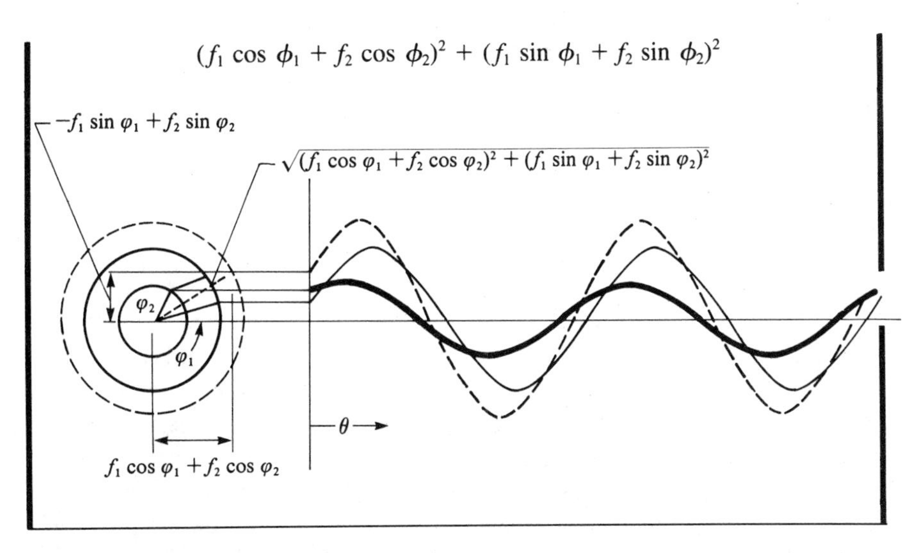

We will learn in this chapter how to graph each of the trigonometric functions, and we will learn the meaning of the terms "amplitude," "frequency," and "phase angle." We will also see how to graphically add sine waves that are out of phase, as in the last figure of the example.

4.1 Graphs of the Sine and Cosine

To obtain the graphs of $y = \sin x$ and $y = \cos x$, we make use of Definition 3.1 in which the sine is the ordinate and the cosine the abscissa of a point on the unit circle, as shown in Figure 4.1. Because we want to use the letter x as the independent variable and y as the dependent variable for the graphs of $y = \sin x$ and $y = \cos x$, we have departed from the usual notation in Figure 4.1 and designated the horizontal and vertical axes by capital letters. We have used x to designate the arc length on the unit circle from the point $(1, 0)$ to $P(x)$, rather than the customary θ.

Figure 4.1

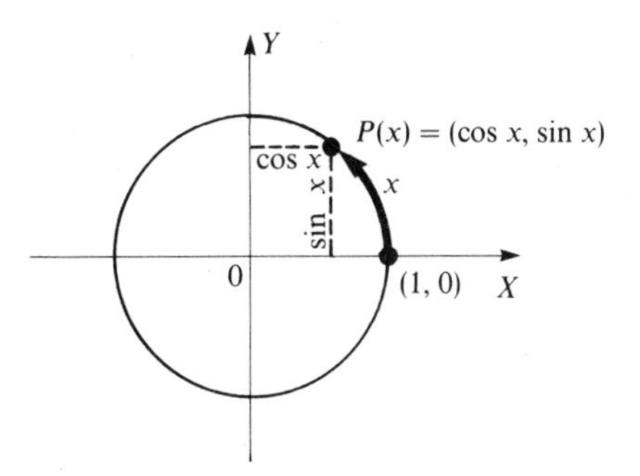

The following table indicates how $\sin x$ and $\cos x$ vary as x varies from 0 to 2π.

As x goes from	$\sin x$ goes from	$\cos x$ goes from
0 to $\pi/2$	0 to 1	1 to 0
$\pi/2$ to π	1 to 0	0 to -1
π to $3\pi/2$	0 to -1	-1 to 0
$3\pi/2$ to 2π	-1 to 0	0 to 1

These facts can be verified by visualizing the point $P(x)$ moving around the circle in a counterclockwise direction, starting at (1, 0). Furthermore, it should be clear from considering the figure that the ordinate and abscissa of $P(x)$ (that is, $\sin x$ and $\cos x$) vary in a uniform way, with no breaks or sudden changes, as $P(x)$ moves around the circle.

This information, together with our knowledge of $\sin x$ and $\cos x$ for $x = \pi/6$, $\pi/4$, $\pi/3$, and related values in the other quadrants, enable us to draw the graphs with reasonable accuracy. Since $\sin x$ and $\cos x$ each has period 2π, the graphs repeat every 2π units. So once we know the graphs in the interval from 0 to 2π, we can extend them indefinitely in either direction. The graphs are shown in Figures 4.2 and 4.3.

Figure 4.2

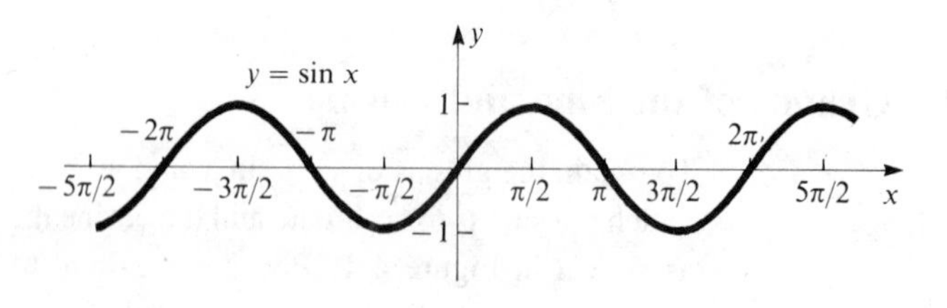

Figure 4.3

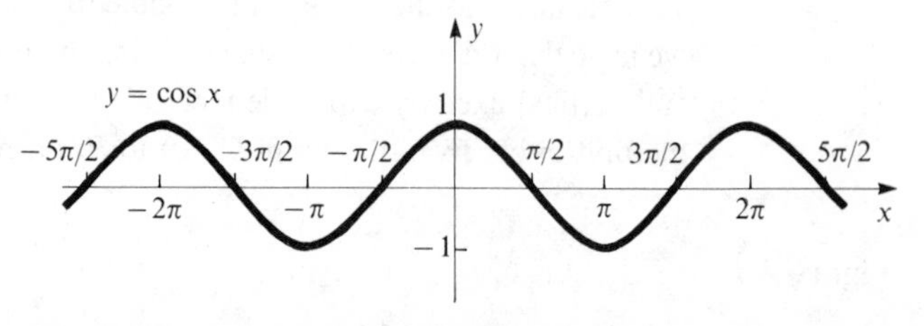

The maximum height attained by each of these curves is called the **amplitude**. So for both $y = \sin x$ and $y = \cos x$ the amplitude is 1. When x varies from 0 to 2π we say that the graphs of $\sin x$ and $\cos x$ complete one **cycle.** We will refer to the graphs shown in Figures 4.2 and 4.3 as the **basic** sine and cosine curves. In Figure 4.4 we repeat one cycle of each of these basic curves, showing the most important features.

We will examine below the effects of introducing various constants into the equations. We will see that the basic shapes remain the same, but they may be stretched, compressed, or shifted.

Figure 4.4

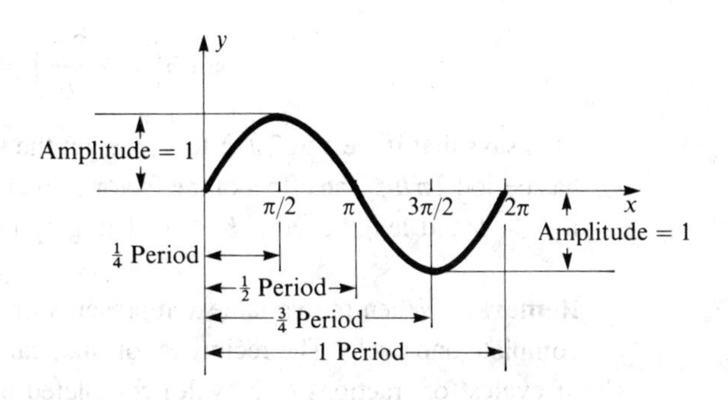

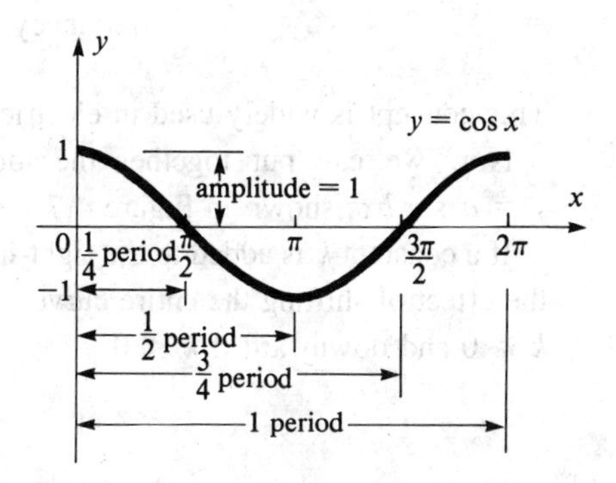

Consider first the graph of $y = a \sin x$, where $a > 0$. Introducing the factor a has the effect of multiplying every y value of the basic sine curve by a. So when the basic curve is at its maximum height 1, the new curve will be a units high. Thus, the amplitude becomes a. The period is unchanged. Figure 4.5 shows the graph.

Figure 4.5

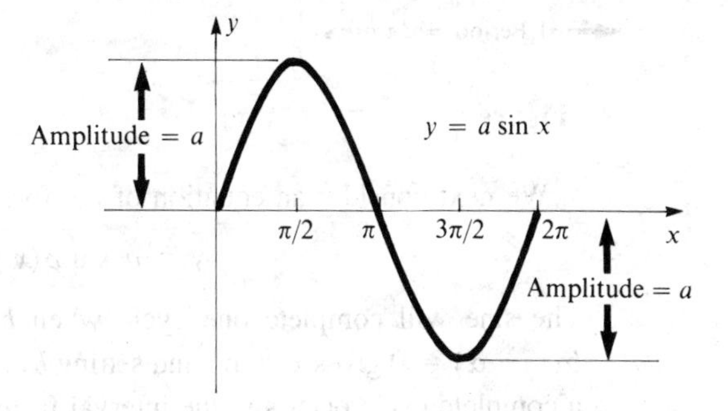

Next, let us consider $y = \sin bx$, where $b > 0$. We know that the sine curve completes one cycle when bx goes from 0 to 2π. But then x goes from 0 to $2\pi/b$. So we conclude that the period of $y = \sin bx$ is $2\pi/b$. This can also be seen in another way. Since the sine curve repeats itself when the angle is increased by 2π, we have $\sin(bx + 2\pi) = \sin bx$. But $\sin(bx + 2\pi) = \sin b[x + (2\pi/b)]$ so that

$$\sin b\left(x + \frac{2\pi}{b}\right) = \sin bx$$

This says that if we add $2\pi/b$ to x, we get the same value as $\sin bx$. That is, $\sin bx$ has period $2\pi/b$. The effect of the factor b, then, is to alter the period. It is shortened if $b > 1$ and lengthened if $b < 1$. The graph is shown in Figure 4.6.

Remark. When the variable x represents time, the period is the time required to complete one cycle. The reciprocal of this, called the **frequency,** gives the number of cycles (or fractions of a cycle) completed per unit of time. Thus,

$$\text{frequency} = \frac{1}{\text{period}} = \frac{b}{2\pi}$$

This concept is widely used in electrical engineering and electronics.

Now we can put together the ideas discussed above to get the graph of $y = a \sin bx$, shown in Figure 4.7.

If a constant k is added to the right-hand side, giving $y = a \sin bx + k$, this has the effect of shifting the entire curve of Figure 4.7 vertically by k units, upward if $k > 0$ and downward if $k < 0$.

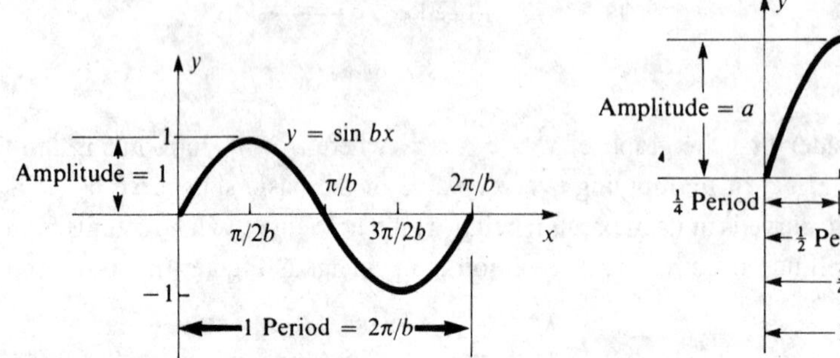

Figure 4.6

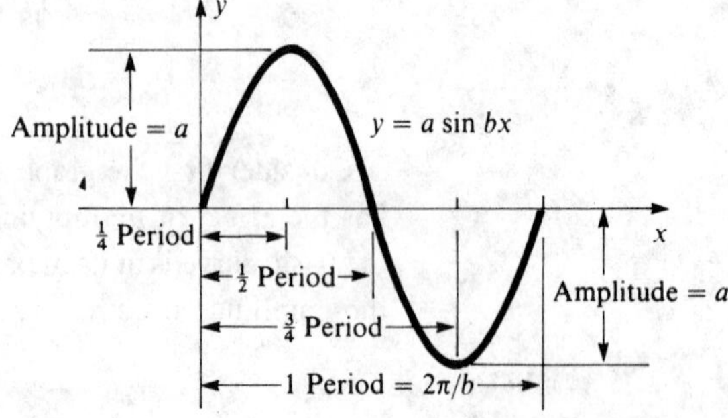

Figure 4.7

We next consider an equation of the form

$$y = a \sin b(x - \alpha) \tag{4.1}$$

The sine will complete one cycle when $b(x - \alpha)$ goes from 0 to 2π. Setting $b(x - \alpha) = 0$ gives $x = \alpha$, and setting $b(x - \alpha) = 2\pi$ gives $x = 2\pi/b + \alpha$. So a complete cycle occurs in the interval from α to $2\pi/b + \alpha$. This is a distance of $2\pi/b$, and so the period of $2\pi/b$ remains unchanged, but the curve is shifted α units horizontally. If $\alpha > 0$, the shift is to the right, and if $\alpha < 0$, it is to the left. We call α the **phase shift** and say the curve whose equation is given by (4.1) is $|\alpha|$ units out of phase with the curve $y = a \sin bx$. This is illustrated in Figure 4.8.

Finally, we note the effect on the graph of $y = a \sin bx$ if either a or b is negative. If $a < 0$, every y value is the negative of what it would have been if a

Figure 4.8

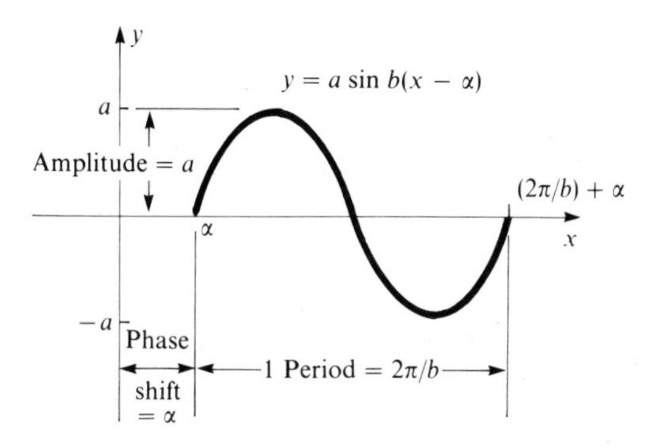

were positive. For example, in $y = -2 \sin x$, every y value is the negative of the corresponding y value in $y = 2 \sin x$. So the effect is to flip the graph of $y = 2 \sin x$ about the x axis. If $b < 0$, we make use of $\sin(-\theta) = -\sin \theta$. For example, we would write $y = \sin(-2x)$ as $y = -\sin 2x$, and proceed as above.

The graph of $y = a \cos b(x - \alpha) + k$ can be analyzed in a similar manner to that of the sine. Note, however, that since $\cos(-\theta) = \cos \theta$, the graph of $y = \cos(-bx)$ is the same as that of $y = \cos bx$. Figure 4.9 shows the graph of one cycle of $y = a \cos b(x - \alpha)$ for a, b, and α positive.

Figure 4.9

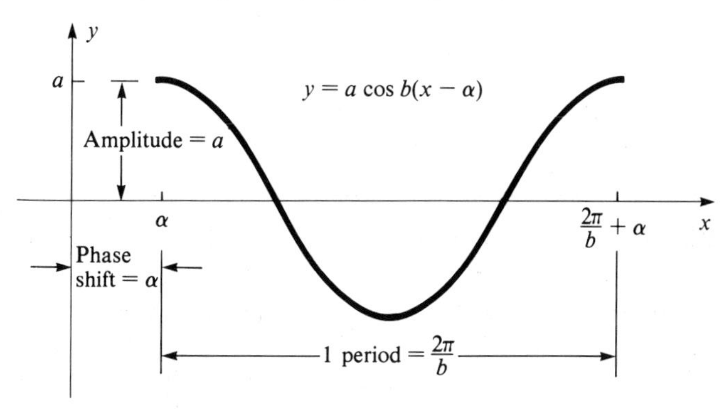

EXAMPLE 4.1 Sketch the graph of $y = 3 \sin \frac{1}{2}x$.

Solution The amplitude is 3, and the period is $2\pi \div \frac{1}{2} = 4\pi$. The graph is shown in Figure 4.10.

Figure 4.10

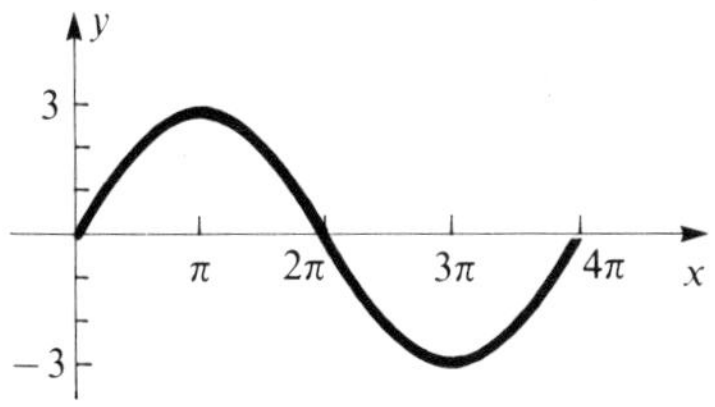

EXAMPLE 4.2 Sketch the graph of $y = 2\sin[3x - (\pi/4)]$.

Solution We first factor out the 3 to put the equation in the form (4.1).

$$y = 2\sin 3\left(x - \frac{\pi}{12}\right)$$

This is therefore a sine curve with amplitude 2, period $2\pi/3$, and phase shift $\pi/12$ units to the right. Figure 4.11 shows the graph.

Figure 4.11

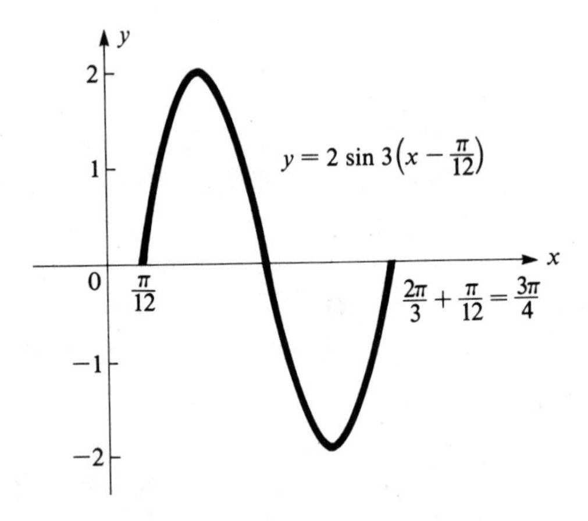

EXAMPLE 4.3 Sketch the graph of $y = \cos[2x + (4\pi/3)] - 1$.

Solution We first factor out the coefficient of x.

$$y = \cos 2\left(x + \frac{2\pi}{3}\right) - 1 = \cos 2\left[x - \left(-\frac{2\pi}{3}\right)\right] - 1$$

The amplitude is 1, the period π, and the phase shift $2\pi/3$ units to the left. The effect of the -1 is to lower the entire curve by 1 unit. Figure 4.12 shows the graph.

Figure 4.12

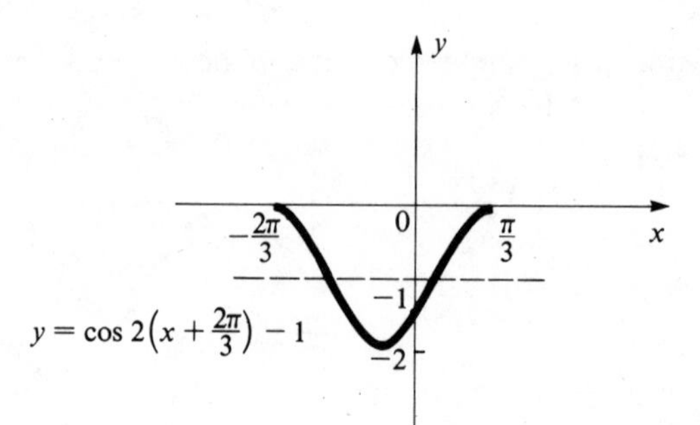

EXERCISE SET 4.1

A Sketch one cycle of each of the following curves, giving amplitude, period, and phase shift when appropriate.

1. $y = \sin 2x$
2. $y = \cos 3x$
3. $y = 2 \sin x$
4. $y = 2 \sin 3x$
5. $y = 3 \cos \dfrac{\pi x}{2}$
6. $y = 4 \sin \dfrac{\pi x}{3}$
7. $y = \dfrac{1}{2} \cos \pi x$
8. $y = \dfrac{3}{2} \sin \dfrac{x}{2}$
9. $y = 2 \sin(-3x)$
10. $y = 3 \cos(-\pi x)$
11. $y = -\sin \dfrac{x}{2}$
12. $y = -2 \cos x$
13. $y = \sin 2x + 1$
14. $y = 2 \cos x - 3$
15. $y = 2 \sin 3x - 1$
16. $y = \cos \pi x + 2$
17. $y = \sin\left(x - \dfrac{\pi}{3}\right)$
18. $y = \cos\left(x - \dfrac{\pi}{4}\right)$
19. $y = \sin 2\left(x + \dfrac{\pi}{4}\right)$
20. $y = \cos 3\left(x + \dfrac{\pi}{3}\right)$
21. $y = 2 \sin\left(3x + \dfrac{\pi}{2}\right)$
22. $y = 2 \cos\left(2x - \dfrac{\pi}{3}\right)$
23. $y = \sin(\pi x - 3\pi)$
24. $y = 3 \cos(2\pi x + \pi)$

B

25. $y = 2 \sin(3x + 2)$
26. $y = 3 \cos(3 - 2x)$
27. $y = 1 + 2 \sin\left(\pi x - \dfrac{\pi}{3}\right)$
28. $y = 2 + 2 \cos\left(3x - \dfrac{\pi}{2}\right)$
29. $y = -\dfrac{3}{2} \sin\left(2x - \dfrac{4}{3}\right) + 1$
30. $2(y + 1) = 3 \cos(4 - 2x)$
31. $y - 1 = \dfrac{3}{4} \cos\left(\pi x + \dfrac{\pi}{3}\right)$
32. $3(y + 2) = -4 \sin(5 - 3x)$

4.2 Graphs of the Other Trigonometric Functions

To obtain the graph of $y = \tan x$ we make use of the identity $\tan x = \sin x/\cos x$. Referring again to the point $P(x)$ on the unit circle (Figure 4.13), we consider the ratio of its ordinate to its abscissa as the point traces out the circle, starting from (1,0).

When $x = 0$, $\tan x = 0/1 = 0$, and as x increases toward $\pi/2$, the ratio of the ordinate of $P(x)$ to its abscissa steadily increases, taking on the value 1 at $x = \pi/4$. For x near $\pi/2$ the ratio becomes very large, and gets larger without limit as x approaches $\pi/2$. At $x = \pi/2$ the abscissa is 0, so the ratio is undefined. For x slightly larger than $\pi/2$, the ratio is negative, since the ordinate is positive and the

Figure 4.13

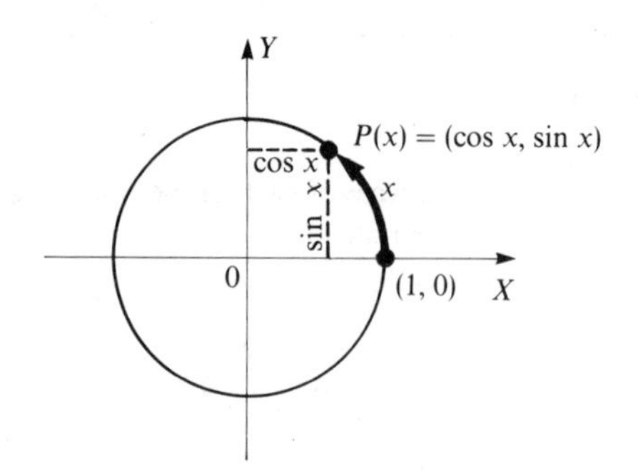

abscissa negative, but its absolute value is large. At $x = 3\pi/4$ the ratio is -1, and at π it is back to 0. Since

$$\tan(x + \pi) = \frac{\tan x + \tan \pi}{1 - \tan x \tan \pi} = \frac{\tan x + 0}{1 - (\tan x) \cdot 0} = \tan x,$$

it follows that $\tan x$ has period π. So we can draw the graph between 0 and π and duplicate this in each succeeding interval of length π. Similarly, we can extend it to the left. With this information and our knowledge of $\tan x$ for $x = \pi/6$, $\pi/3$ and related angles in the other quadrants, we can draw the graph, as shown in Figure 4.14. The vertical lines at odd multiples of $\pi/2$ are called **asymptotes.**

Figure 4.14

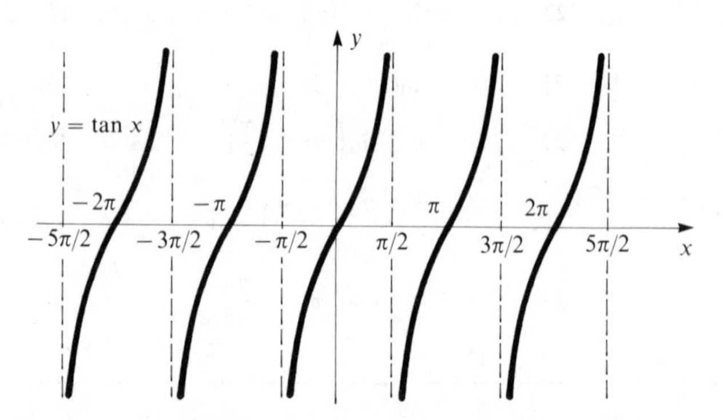

While no amplitude is defined for the tangent curve, it is still true that the graph of $y = a \tan x$ is obtained from the basic tangent curve by multiplying each ordinate by a. This effect can be seen particularly well when $x = \pm\, \pi/4$. Since the period of $y = \tan x$ is π, it follows that for a positive constant b, the period of $y = \tan bx$ is π/b. If $b < 0$, we use the fact that $\tan bx = -\tan(-bx)$.

As with the sine and cosine, the effect of adding a constant k, so that

$$y = a \tan bx + k$$

is to shift the entire curve k units in a vertical direction, upward if $k > 0$ and downward if $k < 0$. Also, the graph of

$$y = a \tan k(x - \alpha)$$

represents a phase shift of α units, to the right if α is positive and to the left if α is negative.

A similar analysis holds for the cotangent curve. We omit the details but show the basic graph in Figure 4.15. The asymptotes are at multiples of π.

Figure 4.15

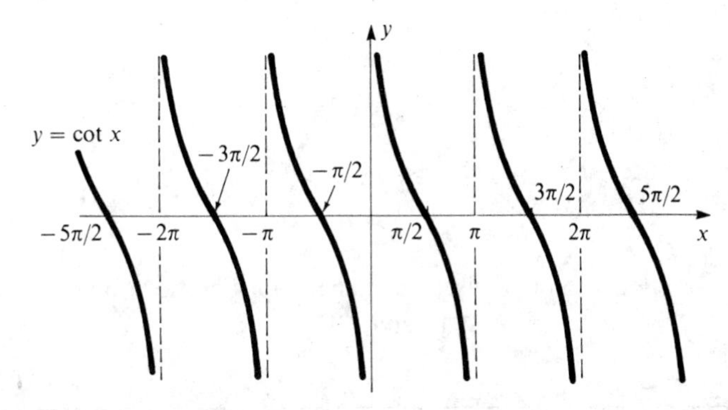

Since $\sec x = 1/\cos x$ and $\csc x = 1/\sin x$, we can obtain the graphs of the secant and cosecant from those of the cosine and sine, respectively, by taking the reciprocal of the y value at each point. This can be done except when it involves division by 0. We know, for example, that $\cos(\pi/2) = 0$. Hence $\sec(\pi/2)$ is undefined. When x is smaller than $\pi/2$ but close to it, $\cos x$ is positive and close to 0. Therefore, $\sec x$ is positive and large. The closer x gets to $\pi/2$, the larger $\sec x$ becomes. When x is slightly larger than $\pi/2$, the cosine is negative and small in absolute value. Therefore the secant is negative and large in absolute value. The vertical line $x = \pi/2$ is an asymptote for the secant curve. Similarly, asymptotes occur at $x = 3\pi/2, 5\pi/2, -\pi/2$, and at every odd multiple of $\pi/2$.

The graph of $y = \csc x$ is obtained similarly. Since $\sin n\pi = 0$ for all integral values of n, it follows as above that the vertical lines $x = n\pi$ are asymptotes for the cosecant curve. Both the basic secant and cosecant functions have period 2π. No amplitude is defined for them. Notice, however, that since both the sine and cosine always lie between $+1$ and -1, the secant and cosecant are always greater than or equal to $+1$ or less than or equal to -1. The effects of multipliers, additive constants, and phase shifts are similar to those for the sine and cosine and so will

Figure 4.16

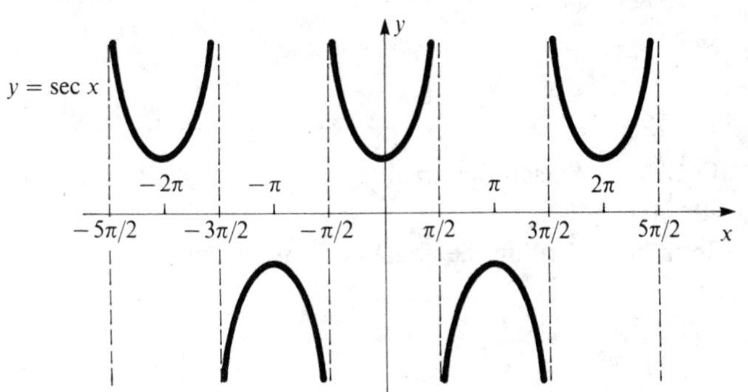

Figure 4.17

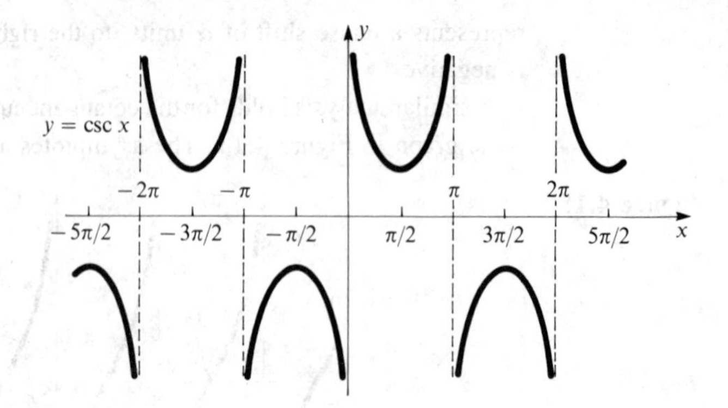

not be discussed. The graphs of the basic secant and cosecant curves are shown in Figures 4.16 and 4.17.

EXAMPLE 4.4 Sketch the graph of $y = 2\tan(x/2)$.

Solution The period is $\pi \div \frac{1}{2} = 2\pi$. Asymptotes occur at odd multiples of π, where $\tan(x/2)$ is undefined. At $x = \pi/2$, $\tan(x/2) = 1$, so $y = 2$. Using the basic curve as a guide, the graph can now be obtained, as shown in Figure 4.18.

Figure 4.18

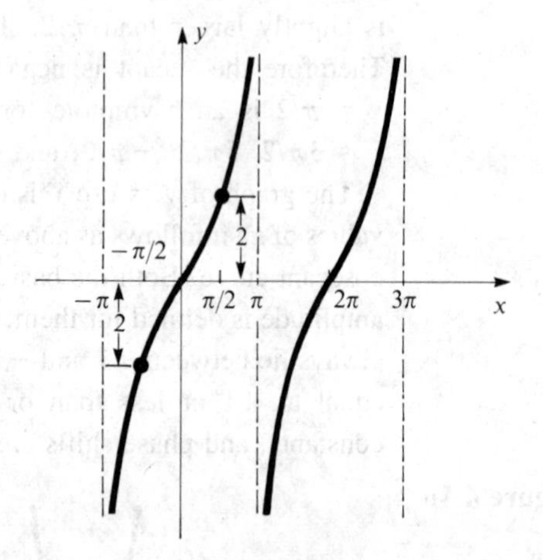

EXAMPLE 4.5 Sketch the graph of $y = 1 + \frac{1}{2}\sec[2x - (\pi/3)]$.

Solution Writing the equation in the form

$$y = \frac{1}{2}\sec 2\left(x - \frac{\pi}{6}\right) + 1$$

we see that the period is π and the phase shift is $\pi/6$ units to the right. Also, the curve is shifted one unit upward from the basic position. The minimum points on the upper branch and maximum points on the lower branch are $\frac{1}{2}$ unit from the shifted axis. The graph is shown in Figure 4.19.

Figure 4.19

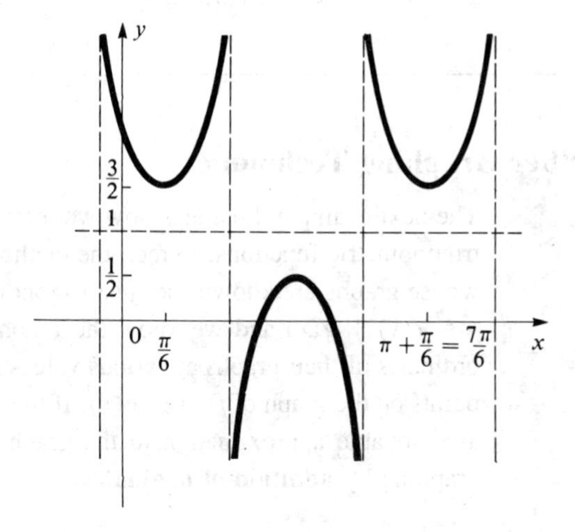

EXERCISE SET 4.2

A Sketch the graphs of each of the following.

1. $y = 2 \tan x$

2. $y = \cot 2x$

3. $y = 3 \tan \dfrac{x}{2}$

4. $y = 2 \cot 3x$

5. $y = \dfrac{1}{2} \sec x$

6. $y = \csc 2x$

7. $y = \dfrac{3}{2} \csc \dfrac{x}{2}$

8. $y = 2 \sec \dfrac{2x}{3}$

9. $y = \cot(-2x)$

10. $y = \tan\left(\dfrac{-\pi x}{2}\right)$

11. $y = \tan\left(x + \dfrac{\pi}{4}\right)$

12. $y = \cot\left(2x - \dfrac{\pi}{3}\right)$

13. $y = \csc\left(\dfrac{x}{2} - \dfrac{\pi}{8}\right)$

14. $y = \dfrac{1}{2} \sec 2\left(x + \dfrac{\pi}{3}\right)$

15. $y = \dfrac{1}{2} \tan\left(\dfrac{x}{2} - \dfrac{\pi}{4}\right)$

16. $y = 2 \cot \pi\left(x - \dfrac{1}{3}\right)$

17. $y = -2 \cot\left(\pi x + \dfrac{\pi}{4}\right)$

18. $y = -\dfrac{1}{2} \sec\left(2x - \dfrac{\pi}{2}\right)$

19. $y = 1 + \tan \dfrac{x}{2}$

20. $y = -\csc(-2x) - 2$

B **21.** $y = 2 - 3 \tan(2x + 3)$

22. $y = \dfrac{1}{3} \csc\left(\dfrac{1}{2} - \dfrac{x}{3}\right)$

23. $y = 1 - 2\sec(3x - 1)$

24. $2(y + 3) = \cot(2 - \pi x)$

25. $3(y - 2) = 2\tan(3 - 2x)$

26. $y = 2 - \frac{1}{2}\csc(3x + 4)$

4.3 Further Graphing Techniques

The next example illustrates how we may approximate the graph of the sum of two trigonometric functions. In fact, the method works for the sum of any two functions whose graphs are known, not just trigonometric functions. The basic idea is that if $y = f(x) + g(x)$ and we know the graphs of f and g, then we can measure the ordinates of their graphs at various values of x and add them together to get several points on the graph of $f(x) + g(x)$. If this is done for a sufficient number of points, a reasonable approximation to the graph can be obtained. This method is called graphing by **addition of ordinates.**

Remark. The method works for subtraction as well, since we may write

$$f(x) - g(x) = f(x) + [-g(x)]$$

and so can add the ordinates of $f(x)$ and $-g(x)$.

EXAMPLE 4.6 Draw the graph of $y = \sin 2x + \cos x$ by the method of addition of ordinates.

Solution We first draw the graphs of $\sin 2x$ and $\cos x$ individually, as shown by the dashed lines in Figure 4.20. Then, as described above, we select various x values and measure the ordinate to each curve, adding these together to find a point on the desired curve. The addition must be done carefully, taking into account whether the two ordinates are like in sign, or unlike in sign.

Figure 4.20

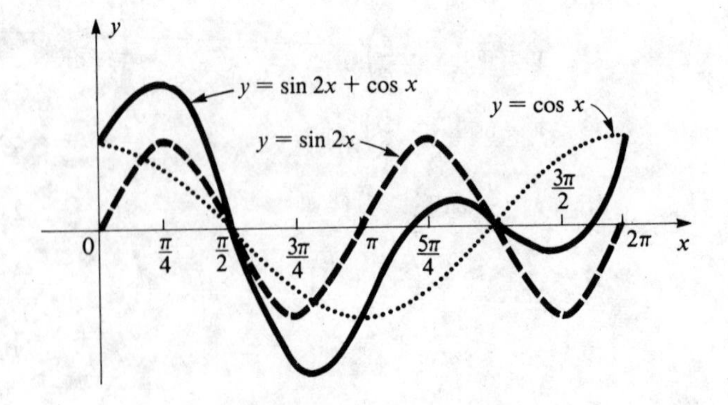

One special case of the sum of two trigonometric functions deserves to be singled out, namely the sum (or difference) of a sine and cosine function having the same period. Thus, we want to consider a function of the form

$$y = a \sin kx + b \cos kx \tag{4.2}$$

As an alternative to adding ordinates to obtain the graph, we proceed as follows. Multiply and divide the right-hand side of (4.2) by $\sqrt{a^2 + b^2}$, getting

$$y = \sqrt{a^2 + b^2}\left(\frac{a}{\sqrt{a^2 + b^2}} \sin kx + \frac{b}{\sqrt{a^2 + b^2}} \cos kx\right) \tag{4.3}$$

Let ϕ be the angle in standard position whose terminal side passes through the point (a, b), as in Figure 4.21. Then we have

$$\frac{a}{\sqrt{a^2 + b^2}} = \cos \phi$$

$$\frac{b}{\sqrt{a^2 + b^2}} = \sin \phi$$

Figure 4.21

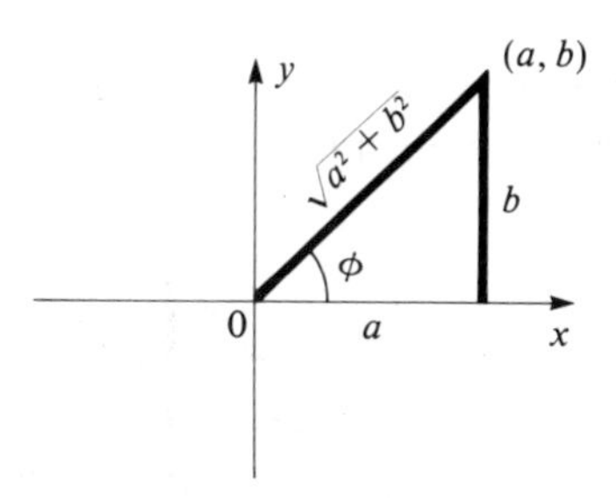

Equation (4.3) can therefore be put in the following form.

$$\begin{aligned} y &= \sqrt{a^2 + b^2}\,(\cos \phi \sin kx + \sin \phi \cos kx) \\ &= \sqrt{a^2 + b^2}\,\sin(kx + \phi) \\ &= \sqrt{a^2 + b^2}\,\sin k\left(x + \frac{\phi}{k}\right) \end{aligned}$$

We recognize this as a sine curve with amplitude $\sqrt{a^2 + b^2}$, period $2\pi/k$, and phase shift $\alpha = -\phi/k$.

Remark. This technique will work regardless of the signs of the coefficients a and b. It is probably best, however, to consider a and b positive and to take care of the signs separately. In this way ϕ will always be an acute angle. For example, we can show in a way similar to the above that

$$a \sin kx - b \cos kx = \sqrt{a^2 + b^2} \sin k\left(x - \frac{\phi}{k}\right)$$

$$-a \sin kx + b \cos kx = -\sqrt{a^2 + b^2} \sin k\left(x - \frac{\phi}{k}\right)$$

and

$$-a \sin kx - b \cos kx = -\sqrt{a^2 + b^2} \sin k\left(x + \frac{\phi}{k}\right)$$

EXAMPLE 4.7 Draw the graph of $y = \sin x + \cos x$ by the method described above.

Solution Multiplying and dividing by $\sqrt{a^2 + b^2} = \sqrt{1^2 + 1^2} = \sqrt{2}$ gives

$$y = \sqrt{2}\left(\frac{1}{\sqrt{2}} \sin x + \frac{1}{\sqrt{2}} \cos x\right)$$

We recognize in this case that the angle $\phi = \pi/4$ (Figure 4.22). So we have

$$y = \sqrt{2}\left(\sin x \cos \frac{\pi}{4} + \cos x \sin \frac{\pi}{4}\right)$$

$$= \sqrt{2} \sin\left(x + \frac{\pi}{4}\right)$$

The amplitude is $\sqrt{2}$, the period is 2π, and the phase shift is $-\pi/4$. The graph is shown in Figure 4.23.

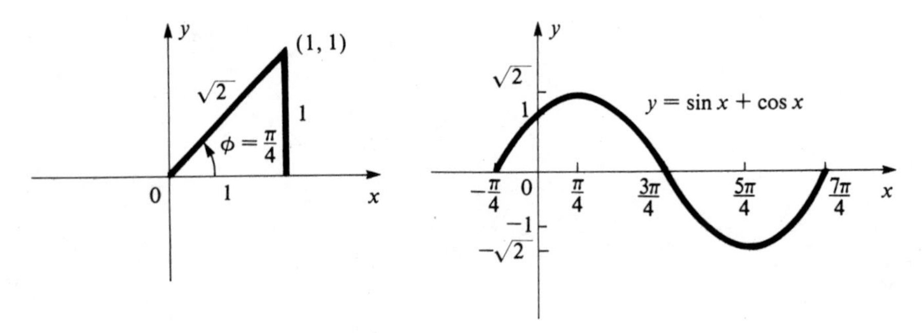

Figure 4.22

Figure 4.23

EXAMPLE 4.8 Write $y = 2 \sin 2x - 3 \cos 2x$ as a sine function and draw its graph.

Solution Following the procedure described above, we get

$$y = \sqrt{13}\left(\frac{2}{\sqrt{13}} \sin 2x - \frac{3}{\sqrt{13}} \cos 2x\right)$$

$$= \sqrt{13}\,(\sin 2x \cos \phi - \cos 2x \sin \phi)$$

$$= \sqrt{13} \sin(2x - \phi)$$

$$= \sqrt{13} \sin 2\left(x - \frac{\phi}{2}\right)$$

where ϕ is defined by Figure 4.24. From a calculator or tables we find ϕ in radians, as an angle with tangent equal to 1.5. We find that ϕ is approximately 0.98. So the phase shift of $\phi/2$ is approximately 0.49 units to the right. The period is $2\pi/2 = \pi$, and the amplitude is $\sqrt{13}$. The graph is shown in Figure 4.25.

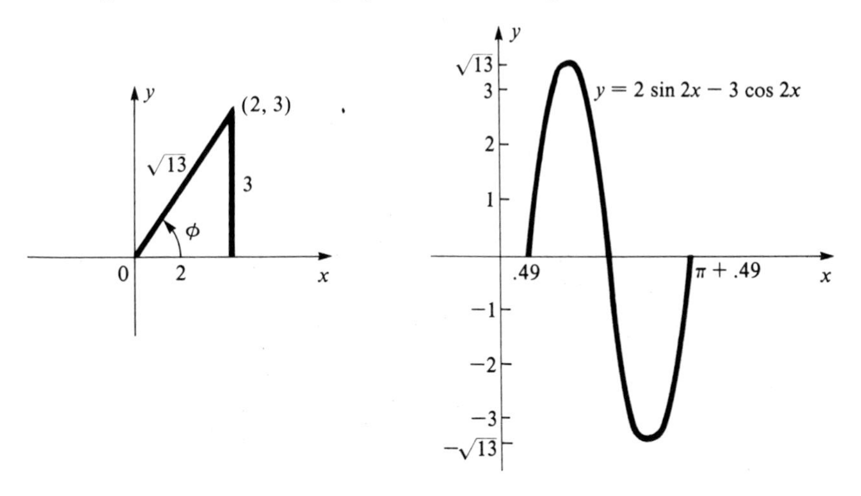

Figure 4.24 **Figure 4.25**

When $y = f(x) \cdot g(x)$ and the graphs of f and g are known, then it is possible to use **multiplication of ordinates** to obtain points on the graph of the product function. In general, this is more difficult than addition of ordinates, since addition can be done graphically with the aid of a ruler or a pair of dividers, but multiplication cannot easily be done graphically. We will consider two special cases, however, in which the technique is effective and can be done with relative ease. The two types of products we have in mind are of the form $f(x) \sin kx$ and $f(x) \cos kx$, where $f(x) > 0$ for all x.

Since both $\sin kx$ and $\cos kx$ oscillate between $+1$ and -1, it follows that each of the curves $y = f(x) \sin kx$ and $y = f(x) \cos kx$ oscillates between $f(x)$ and $-f(x)$. Furthermore, in the case of $y = f(x) \sin kx$ the product curve touches $f(x)$ when $\sin kx = 1$, touches $-f(x)$ when $\sin kx = -1$, and is 0 when $\sin kx = 0$. Similar remarks apply to $y = f(x) \cos kx$. This suggests drawing the graphs of both $f(x)$ and $-f(x)$ as upper and lower boundaries to the product curve. Plotting the high and low points, together with the points where the curve crosses the x axis, is usually sufficient for a sketch. We illustrate this in the next example.

EXAMPLE 4.9 Use multiplication of ordinates to sketch the graph of $y = x \sin x$ for $x \geq 0$.

Solution We draw the graphs of $y = x$, $y = -x$, and $y = \sin x$. The product curve will touch $y = x$ when $\sin x = 1$, will touch $y = -x$ when $\sin x = -1$, and will be zero when $\sin x = 0$. These points enable us to sketch the curve with reasonable accuracy, as shown in Figure 4.26.

Figure 4.26

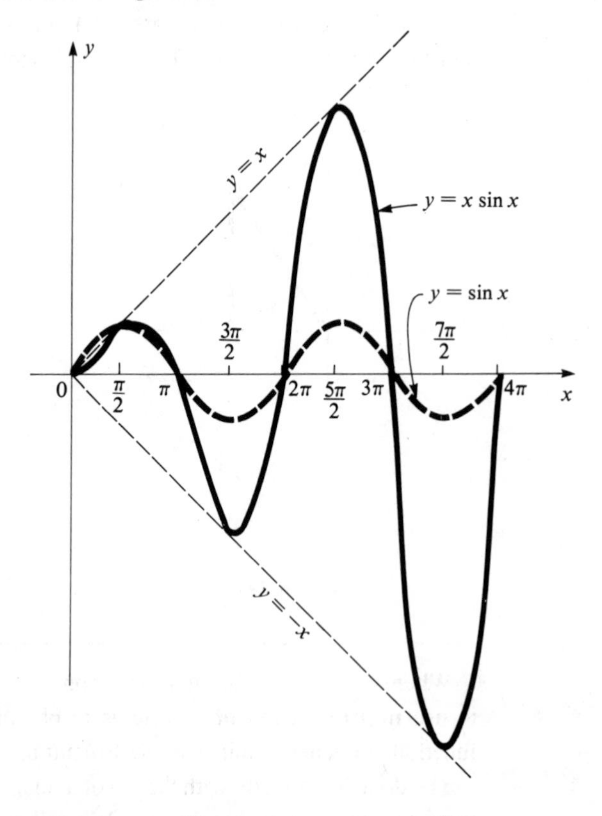

EXERCISE SET 4.3

A In Problems 1–10 use addition of ordinates to sketch the graph.

1. $y = 2 \sin x + \cos x$

2. $y = \sin x - \cos x$

3. $y = \sin x + \cos 2x$

4. $y = \sin 2x + \cos 3x$

5. $y = \cos \frac{x}{2} - \sin x$

6. $y = \frac{1}{2} \sin x + \frac{3}{2} \cos 2x$

7. $y = 2 \sin x + \sin 2x$

8. $y = \cos x + \cos \frac{1}{2} x$

9. $y = x + \sin x$

10. $y = \frac{x}{2} - \cos x$

In problems 11–18 write the given function as a sine function, give the amplitude, period, phase shift, and draw the graph.

11. $y = \sin x - \cos x$

12. $y = \sqrt{3} \sin x + \cos x$

12. $y = \sin 2x + \sqrt{3} \cos 2x$

14. $y = \sin 3x + \cos 3x$

15. $y = 3 \sin x - 4 \cos x$

16. $y = 5 \sin x + 12 \cos x$

17. $y = \cos x - 2 \sin x$

18. $y = -\sin 2x - \cos 2x$

In Problems 19–22 sketch the graph using multiplication of ordinates.

19. $y = \frac{x}{2} \cos x, \quad x \geq 0$

20. $y = \frac{x^2}{4} \sin x$

21. $y = x \sin 2x, \quad x \geq 0$

22. $y = \frac{1}{x} \cos x, \quad x > 0$

B Use addition of ordinates to sketch the graphs in Problems 23–25.

23. $y = \sin x + \csc x$

24. $y = \cos x + \tan x$

25. $y = \sec x + \csc x$

Use multiplication of ordinates to sketch the graphs in Problems 26 and 27.

26. $y = 2^{-x} \cos x$

27. $y = x \sin \frac{1}{x}, \quad x \neq 0$

28. Show that $a \sin kx + b \cos kx$ can be written in the form

$$\sqrt{a^2 + b^2} \cos(kx - \theta)$$

29. Prove that $y = a \sin kx + b \cos lx$ is a periodic function of period $2\pi/m$, where m is the greatest common divisor of k and l.

4.4 Inverse Functions

Let f be a function, and set $y = f(x)$. The definition of a function requires that each value of x in the domain of the function correspond to exactly one value of y in the range. Graphically, this amounts to the fact that each vertical line intersects the curve in at most one point (when x is in the domain, a vertical line through x will intersect the graph in exactly one point, and when x is not in the domain, a vertical line through x will not intersect the curve). It is entirely possible, however, for two or more different x values to correspond to the same y value. For example, if $y = x^2$, then $y = 4$ when $x = 2$ and when $x = -2$. Also, if $y = \sin x$, then there are infinitely many values of x corresponding to each y value between $+1$ and -1. Graphically, there are horizontal lines in these examples that intersect the curves in more than one point.

A special class of functions consists of those for which each y value in the range corresponds to one and only one x value in the domain. These functions are called one-to-one functions, written 1-1. If f is 1-1, then each vertical line and also each horizontal line intersects the graph of $y = f(x)$ in at most one point. An important property of 1-1 functions is that each such function has an *inverse*. We explain below what this means.

Roughly speaking, the inverse of a function has just the opposite effect as the function itself. If applying a function f to a value x in its doman yields the value y, then applying the inverse function to y yields the value x that we started with. When f has an inverse, we designate it by f^{-1}. The precise definition is as follows.

DEFINITION 4.1 Let f be a function with domain D and range R. Then f^{-1} is the **inverse** of f provided

(i) domain of $f^{-1} = R$, range of $f^{-1} = D$

(ii) $f^{-1}(f(x)) = x$ for each x in D

(iii) $f(f^{-1}(y)) = y$ for each y in R

It is easy to see that if f^{-1} is the inverse of f, then f is the inverse of f^{-1} (i.e. $(f^{-1})^{-1} = f$). So we say f and f^{-1} are inverses of each other. If we put $y = f(x)$ then by (ii) in the definition, $x = f^{-1}(y)$. This suggests that to find the inverse of a function f we should solve the equation $y = f(x)$ for x in terms of y. For example, suppose

$$f(x) = \frac{2x - 3}{5}$$

We set $y = f(x)$ and then solve for x.

$$y = \frac{2x - 3}{5}$$

$$5y = 2x - 3$$

$$x = \frac{5y + 3}{2}$$

So

$$f^{-1}(y) = \frac{5y + 3}{2}$$

Since it is customary to use the letter x for an arbitrary element of the domain (the **independent variable**) and y for the corresponding element of the range (the **dependent variable**), we usually reverse the roles of x and y after we have solved for x. So in the above example we can say

$$f^{-1}(x) = \frac{5x + 3}{2}$$

Unfortunately, the procedure shown in this example cannot always be applied, since we cannot always solve the equation for x. Even so, it is still possible to obtain essential information about f^{-1}, including its graph, when we know the graph of f.

In summary, we can say that a function f has an inverse, f^{-1}, if and only if f is 1-1 on its domain. This amounts to the requirement that each vertical line and each horizontal line intersect the graph of $y = f(x)$ in at most one point. If f has an inverse and it is possible to solve $y = f(x)$ for x in terms of y, then $y = f^{-1}(x)$ is found by reversing the roles of x and y in this solution.

If a point (a, b) lies on the graph of $y = f(x)$, then the point (b, a) lies on the graph of $y = f^{-1}(x)$, from which it follows that the two graphs are symmetrically placed with respect to the 45° line $y = x$. This is illustrated in Figure 4.27.

Some functions which are not 1-1 can be restricted to a subset of their domains on which they are 1-1, so that inverses exist on the restricted domains. For example, consider $f(x) = x^2$. The graph of $y = x^2$ is a parabola, as shown in Figure 4.28, and this is clearly not a 1-1 function (Why?) However, if we restrict the domain to be the set of non-negative real numbers, then f is 1-1 and has an inverse. For x in this restricted domain, we can solve $y = x^2$ for the unique solution $x = \sqrt{y}$. Interchanging x and y, we see that $f^{-1}(x) = \sqrt{x}$.

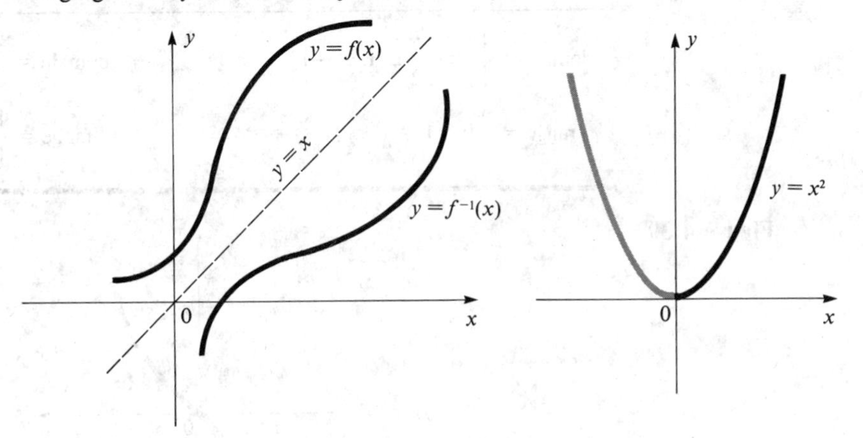

Figure 4.27 **Figure 4.28**

The situation with the trigonometric functions is similar to this last example, in that their domains must be restricted before an inverse can exist. Consider first $y = \sin x$ (Figure 4.29.). We make the restriction $-\pi/2 \leq x \leq \pi/2$, and the por-

Figure 4.29

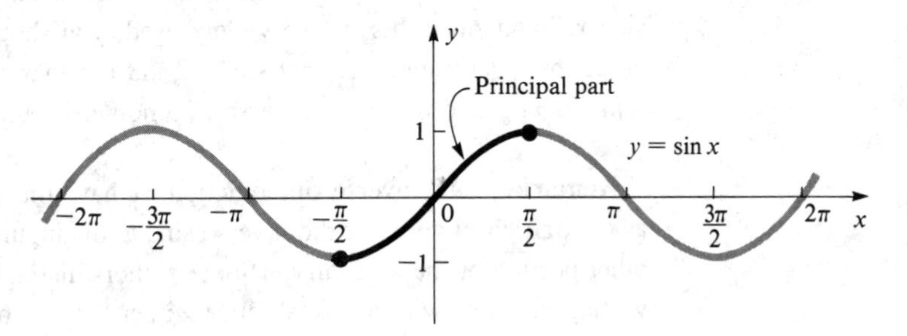

tion of the curve corresponding to this restricted domain (shown as a heavy line) is called the **principal part** of the sine curve. On this portion of the curve a horizontal line crosses the curve at most one time, so that an inverse exists. To find the inverse we would like to solve the equation $y = \sin x$ for x in terms of y and then reverse the roles of x and y. Unfortunately, there is no way we can do this. Yet we know an inverse exists. What we do, then, is to invent a symbol, taking our cue from the notation f^{-1} for the inverse of a general function f. We "solve" $y = \sin x$ for x by saying $x = \sin^{-1} y$ and read this "x is the inverse sine of y." Interchanging x and y, we have $y = \sin^{-1} x$. (*Caution*: $\sin^{-1} x$ does not mean $1/\sin x$.) The graph of this function is obtained by reflecting the principal part of the sine curve in the line $y = x$, resulting in Figure 4.30. The relationship between the domains and ranges of $\sin x$ and $\sin^{-1} x$ are as follows:

Principal part of $y = \sin x$	$y = \sin^{-1} x$
domain $= \left\{x: -\frac{\pi}{2} \le x \le \frac{\pi}{2}\right\}$	domain $= \{x: -1 \le x \le 1\}$
range $= \{y: -1 \le y \le 1\}$	range $= \left\{y: -\frac{\pi}{2} \le y \le \frac{\pi}{2}\right\}$

Figure 4.30

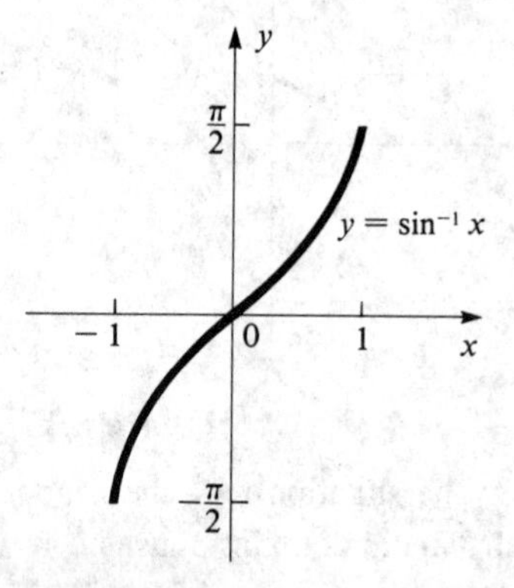

There is another notation for the inverse sine function, namely **arcsin x**, read "arc sine x," and since this is also widely used, you should become familiar with both notations. Just remember that $\sin^{-1} x$ and $\arcsin x$ mean the same thing. Similar remarks apply to the other inverse trigonometric functions discussed below.

Remark. The inverse sine function we have illustrated is sometimes referred to as the **principal value** of the inverse sine to distinguish it from the inverse of some other portion of the sine curve. Some authors distinguish this particular inverse by writing its name with a capital letter, either $\text{Sin}^{-1} x$ or $\text{Arcsin}\, x$. Since we will not consider the inverse of any portion of the sine curve other than what we have called the principal part, this distinction is unnecessary. In other words, the principal value of the inverse sine (as well as of the other trigonometric functions) is all we will consider.

In seeking the value of $\sin^{-1} x$ for a particular x we ask the question, "What is the number between $-\pi/2$ and $\pi/2$ whose sine is x?" Equivalently, we can ask, "What is the radian measure of the angle between $-\pi/2$ and $\pi/2$ whose sine is x?" For example, if we want to find $\sin^{-1} \frac{1}{2}$, we ask what angle in the interval from $-\pi/2$ to $\pi/2$ has a sine of $\frac{1}{2}$. We know the answer is $\pi/6$. So $\sin^{-1} \frac{1}{2} = \pi/6$. Similarly, $\sin^{-1}(-\frac{1}{2}) = -\pi/6$. Notice that the answer here is *not* $11\pi/6$, even though $-\pi/6$ and $11\pi/6$ are coterminal angles, because $11\pi/6$ does not lie in the interval from $-\pi/2$ to $\pi/2$.

EXAMPLE 4.10 Evaluate each of the following.

a. $\sin^{-1}\left(-\frac{\sqrt{3}}{2}\right)$

b. $\arcsin 0$

c. $\sin^{-1}(-1)$

d. $\cos\left[\sin^{-1}\left(-\frac{3}{5}\right)\right]$

e. $\sin\left[2 \sin^{-1} \frac{2}{3}\right]$

Solution

a. $\sin^{-1}\left(-\frac{\sqrt{3}}{2}\right) = -\frac{\pi}{3}$

b. $\arcsin 0 = 0$

c. $\sin^{-1}(-1) = -\frac{\pi}{2}$

d. Let $\theta = \sin^{-1}(-\frac{3}{5})$. Then as an angle, θ is negative and in the fourth quadrant, as shown in Figure 4.31. So

$$\cos\left[\sin^{-1}\left(-\frac{3}{5}\right)\right] = \cos \theta = \frac{4}{5}$$

e. Again let $\theta = \sin^{-1} \frac{2}{3}$. Then θ is as shown in Figure 4.32.

$$\sin\left(2 \sin^{-1} \frac{2}{3}\right) = \sin 2\theta = 2 \sin \theta \cos \theta = 2\left(\frac{2}{3}\right)\left(\frac{\sqrt{5}}{3}\right) = \frac{4\sqrt{5}}{9}$$

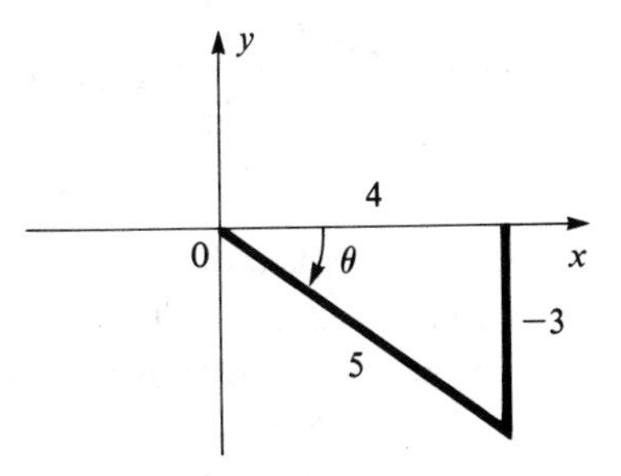

Figure 4.31

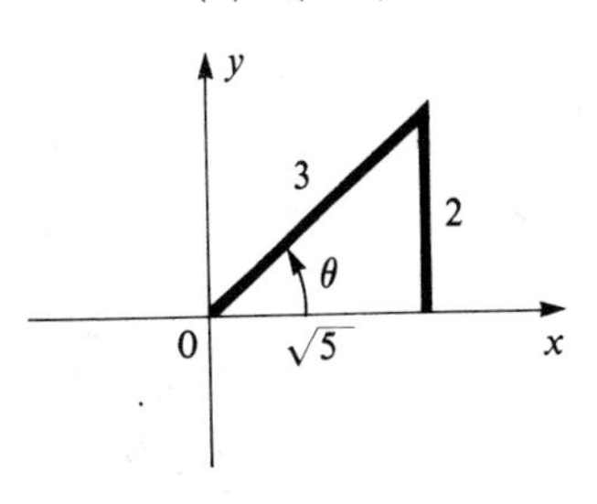

Figure 4.32

We next consider the inverse of the tangent function, since it has much in common with the inverse sine. We call that part of the tangent curve lying in the interval $-\pi/2 < x < \pi/2$ the principal part, or **principal branch,** of the tangent curve. It can be seen from Figure 4.33 that this portion of the curve is 1-1 and hence has an inverse, designated by $\tan^{-1} x$ (or arctan x). The graph of $y = \tan^{-1} x$ is shown in Figure 4.34. Its domain is the set of all real numbers, and its range is the set of y values such that $-\pi/2 < y < \pi/2$. Just as with $\sin^{-1} x$, when $x > 0$, $\tan^{-1} x$ is positive and lies between 0 and $\pi/2$. And when $x < 0$, $\tan^{-1} x$ is negative and lies between $-\pi/2$ and 0. For example,

$$\tan^{-1}(-1) = -\frac{\pi}{4} \quad \text{and} \quad \tan^{-1}\sqrt{3} = \frac{\pi}{3}$$

Figure 4.33

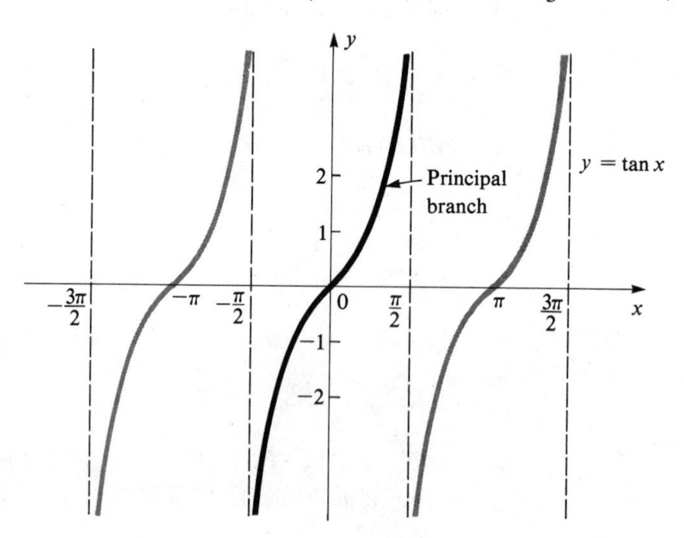

Figure 4.34

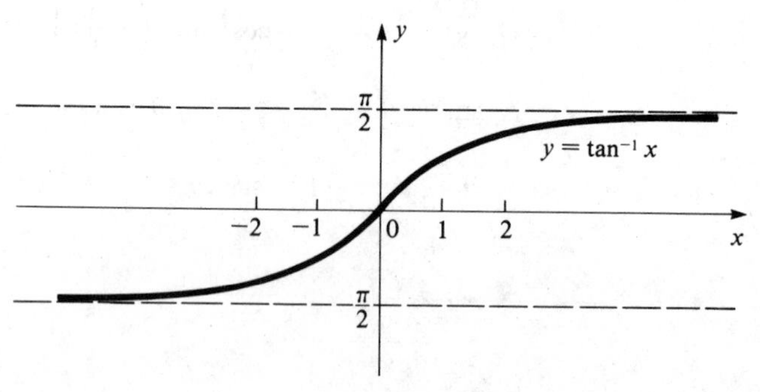

The cosine function must be handled differently since, as Figure 4.35 shows, it is not 1-1 on the interval from $-\pi/2$ to $\pi/2$. However, it is 1-1 on the interval from 0 to π, and this is the portion we designate as the principal part. We invert this to get $y = \cos^{-1} x$, whose graph is shown in Figure 4.36. When $x > 0$, $\cos^{-1} x$ is between 0 and $\pi/2$, but when $x < 0$, $\cos^{-1} x$ is between $\pi/2$ and π. For example, $\cos^{-1}(-\frac{1}{2}) = 2\pi/3$ and $\cos^{-1}(-1) = \pi$.

Figure 4.35

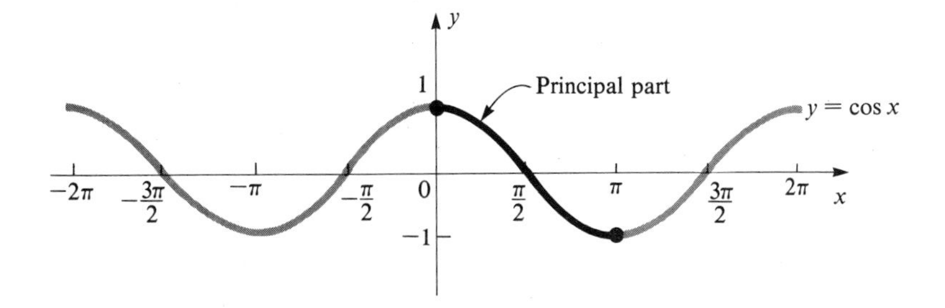

As with the cosine, the principal branch of the cotangent curve is the part lying in the interval $0 < x < \pi$. On interchanging the roles of x and y, we get the graph of $y = \cot^{-1} x$ shown in Figure 4.37.

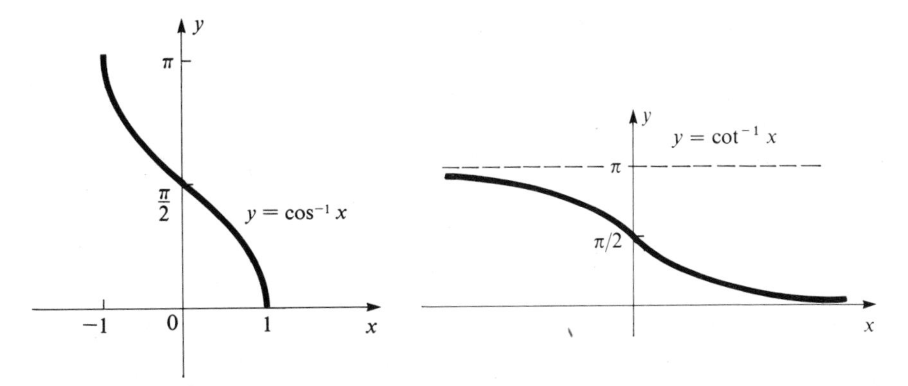

Figure 4.36 **Figure 4.37**

The secant and cosecant present certain problems, and there is no general agreement on which parts of the curves to invert, that is, which portions to define as the principal branches. Fortunately, this lack of agreement is not serious. The inverse cosecant is seldom used and, since it can always be circumvented, we will omit it entirely. The inverse secant is sufficiently useful to warrant consideration. For purposes of later use in calculus, there is some justification for the choice of principal parts made below. Let us consider first the graph of $y = \sec x$ (Figure 4.38). No

Figure 4.38

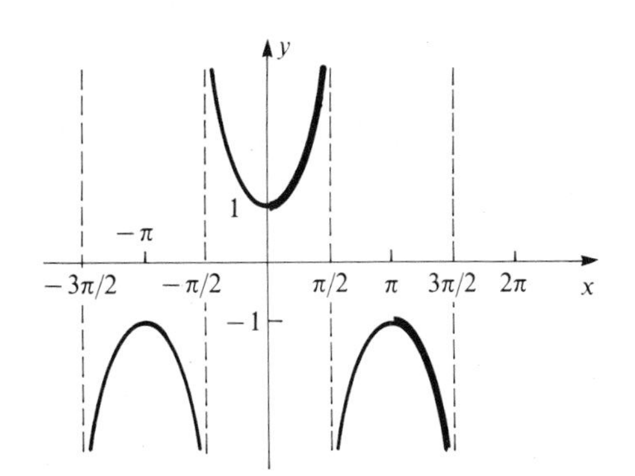

single branch of the graph is suitable for defining the inverse function. While the reason is not apparent, we select the portion shown in heavy lines in Figure 4.38 to define the inverse function. Thus, when $y \geq 1$, we take $0 \leq x < \pi/2$, and when $y \leq -1$, we take $\pi \leq x < 3\pi/2$. When we interchange the roles of x and y, we obtain the graph shown in Figure 4.39.

Figure 4.39

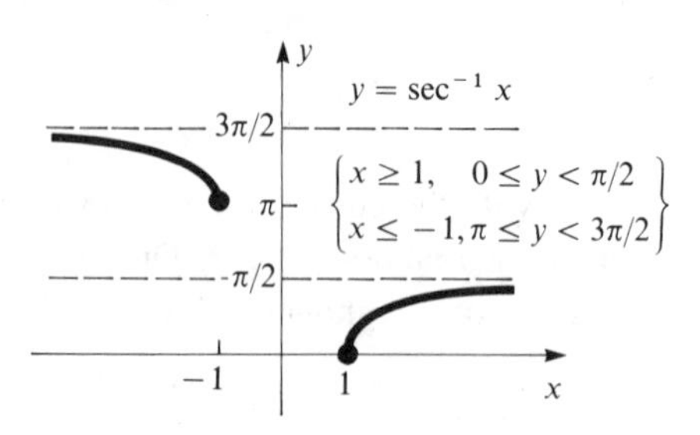

EXERCISE SET 4.4

A In Problems 1–36 give the value.

1. **a.** $\sin^{-1}\left(-\frac{\sqrt{3}}{2}\right)$ **b.** $\arccos\left(-\frac{\sqrt{3}}{2}\right)$
2. **a.** $\tan^{-1}(-1)$ **b.** $\text{arcsec}(-2)$
3. **a.** $\text{arccot}(-1)$ **b.** $\sin^{-1} 1$
4. **a.** $\arccos 1$ **b.** $\tan^{-1} 1$
5. **a.** $\text{arcsec } 1$ **b.** $\sin^{-1}\left(-\frac{1}{2}\right)$
6. **a.** $\arctan 0$ **b.** $\cos^{-1}(-1)$
7. **a.** $\arccos 0$ **b.** $\sec^{-1}\left(-\frac{2}{\sqrt{3}}\right)$
8. **a.** $\cos\alpha$ if $\alpha = \sin^{-1}\left(-\frac{2}{3}\right)$ **b.** $\tan\theta$ if $\theta = \arccos\left(-\frac{1}{4}\right)$
9. **a.** $\sin x$ if $x = \sec^{-1}(-3)$ **b.** $\sec\theta$ if $\theta = \tan^{-1}\frac{3}{4}$
10. **a.** $\sin\left[\cos^{-1}\left(-\frac{1}{3}\right)\right]$ **b.** $\sec\left[\sin^{-1}\left(-\frac{1}{4}\right)\right]$
11. **a.** $\tan\left(\arccos\frac{3}{5}\right)$ **b.** $\cos\left[\tan^{-1}\left(-\frac{4}{3}\right)\right]$
12. **a.** $\sin\left(\arctan\frac{5}{12}\right)$ **b.** $\csc\left[\arccos\left(-\frac{2}{3}\right)\right]$
13. **a.** $\cos\left[\sin^{-1}\left(-\frac{1}{3}\right)\right]$ **b.** $\cot\left[\cos^{-1}\left(-\frac{7}{25}\right)\right]$
14. **a.** $\sin^{-1}\left(\sin\frac{\pi}{8}\right)$ **b.** $\cos^{-1}\left(\cos\frac{\pi}{12}\right)$

15. **a.** $\tan^{-1}\left(\tan \frac{\pi}{5}\right)$ **b.** $\sin^{-1}\left(\cos \frac{\pi}{3}\right)$

16. **a.** $\cos^{-1}\left(\sin \frac{\pi}{6}\right)$ **b.** $\tan^{-1}\left(\cot \frac{\pi}{8}\right)$

17. **a.** $\cos^{-1}\left[\cos\left(-\frac{\pi}{4}\right)\right]$ **b.** $\sin^{-1}\left(\sin \frac{2\pi}{3}\right)$

18. **a.** $\arctan\left(\tan \frac{3\pi}{4}\right)$ **b.** $\arcsin\left(\sin \frac{5\pi}{3}\right)$

19. $\sin(\alpha - \beta)$ if $\alpha = \sin^{-1} \frac{3}{5}$ and $\beta = \cos^{-1}(-\frac{5}{13})$

20. $\cos(\alpha + \beta)$ if $\alpha = \arctan(-\frac{1}{2})$ and $\beta = \sec^{-1} \frac{5}{3}$

21. $\tan(\alpha + \beta)$ if $\alpha = \sin^{-1}(-\frac{4}{5})$ and $\beta = \cos^{-1}(-\frac{12}{13})$

22. $\sin 2\theta$ if $\theta = \cos^{-1}(-\frac{3}{5})$

23. $\cos 2\theta$ if $\theta = \sin^{-1}(-\frac{1}{3})$

24. $\tan 2\theta$ if $\theta = \sin^{-1}(-\frac{5}{13})$

25. $\cos\left[2 \cos^{-1}\left(-\frac{2}{3}\right)\right]$

26. $\tan\left(2 \tan^{-1} \frac{1}{2}\right)$

27. $\sin\left[2 \sin^{-1}\left(-\frac{3}{7}\right)\right]$

28. $\cos\left[2 \sin^{-1}\left(-\frac{2}{3}\right)\right]$

29. $\sin\left[\frac{1}{2} \cos^{-1} \frac{7}{25}\right]$

30. $\cos\left[\frac{1}{2} \arcsin\left(-\frac{3}{5}\right)\right]$

31. $\tan \frac{1}{2}\left[\tan^{-1}\left(-\frac{12}{5}\right)\right]$

32. $\sin \theta$ and $\cos \theta$ if $2\theta = \tan^{-1}\left(-\frac{24}{7}\right)$

33. $\sin \frac{\alpha}{2}$, $\cos \frac{\alpha}{2}$, and $\tan \frac{\alpha}{2}$ if $\alpha = \cos^{-1} \frac{3}{5}$

34. $\sin\left(\sin^{-1} \frac{12}{13} + \cos^{-1} \frac{4}{5}\right)$

35. $\cos\left[\cos^{-1} \frac{8}{17} - \sin^{-1}\left(-\frac{7}{25}\right)\right]$

36. $\tan\left(\tan^{-1} \frac{2}{3} + \tan^{-1} \frac{1}{4}\right)$

In Problems 37–40 use a calculator or tables to find the value correct to four significant figures.

37. **a.** $\sin^{-1} 0.6650$ **b.** $\cos^{-1} 0.4081$

38. **a.** $\tan^{-1} 1.473$ **b.** $\sec^{-1} 2.135$

39. **a.** $\sin^{-1}(-0.9943)$ **b.** $\tan^{-1}(-0.5472)$

40. **a.** $\cos^{-1}(-0.5702)$ **b.** $\sec^{-1}(-1.125)$

B In Problems 41 and 42 find the value without using tables or a calculator.

41. $\tan^{-1}\frac{2}{3} - \tan^{-1}\left(-\frac{1}{5}\right)$

Hint. Call this $\alpha - \beta$ and find $\tan(\alpha - \beta)$.

42. $\tan^{-1}\frac{\sqrt{3}}{2} + \tan^{-1} 3\sqrt{3}$

43. If $\theta = \arcsin x$, show that $\frac{x}{\sqrt{1 - x^2}}$ equals $\tan\theta$.

44. If $\theta = \arctan x$, show that $\frac{\sqrt{1 + x^2}}{x^2}$ equals $\csc\theta \cot\theta$.

45. Show that the expression $\frac{x^2}{3\sqrt{x^2 + 9}}$ changes to $\sec\theta - \cos\theta$ when the substitution $\theta = \tan^{-1}(x/3)$ is made.

46. Show that $\frac{\sqrt{x^2 - 4}}{x}$ equals $\sin\theta$ under the substitution $\theta = \sec^{-1}(x/2)$.

In Problems 47–49 use a calculator or tables to find the value to four significant figures.

47. $\sin[2\cos^{-1}(-0.6187)]$

48. $\cos\left[\frac{1}{2}\tan^{-1}(-1.062)\right]$

49. $\tan(\alpha - 2\beta)$, where $\alpha = \arcsin(-0.4975)$ and $\beta = \arccos(-0.2831)$.

4.5 Review Exercise Set

A In Problems 1–16 draw the graphs.

1. $y = 2\sin 3x$

2. $y = 3\cos\frac{x}{2}$

3. $y = \frac{1}{2}\tan\pi x$

4. $y = \cot\frac{\pi x}{2}$

5. $y = 2\sec\frac{\pi x}{3}$

6. $y = 3\csc 2x$

7. $y = 1 + 2\sin x$

8. $y = 2\cos\frac{\pi x}{2} + 3$

9. $y = \cos\left(\pi x - \frac{\pi}{3}\right)$

10. $y = \sin\left(3x + \frac{\pi}{2}\right)$

11. $y = 1 + \tan\left(x + \frac{\pi}{4}\right)$

12. $y = \sec\left(2x - \frac{\pi}{3}\right)$

13. $y = \cot\left(\pi x - \frac{\pi}{6}\right)$

14. $y = -2\csc\left(3x - \frac{\pi}{4}\right)$

15. $y = -\sin\left(2x - \frac{\pi}{3}\right) + 1$

16. $y = \frac{1}{2}\cos\left(-2x + \frac{\pi}{2}\right)$

In Problems 17–22 sketch the graph using addition of ordinates.

17. $y = 2\cos x + 3\sin x$

18. $y = \sin 3x + \cos 2x$

19. $y = 2\sin\frac{1}{2}x - 3\cos x$

20. $y = \cos x + \cos 2x$

21. $y = \frac{1}{2}\sin 2x + \frac{3}{2}\cos 3x$

22. $y = 2x - \sin x$

In Problems 23–26 write each function as a sine function, give the amplitude, period, phase shift, and draw the graph.

23. $y = \sin x + \sqrt{3}\cos x$

24. $y = \sin 2x - \cos 2x$

25. $y = 4\sin x + 3\cos x$

26. $y = -2\sin 3x + \cos 3x$

In Problems 27–29 use multiplication of ordinates to sketch the graph.

27. $y = \frac{x^2}{4}\cos x$

28. $y = \frac{x}{2}\sin 2x, \quad x \geq 0$

29. $y = \frac{1}{x}\sin x, \quad x > 0$

In Problems 30–39 evaluate the given expression.

30. **a.** $\arccos\left(-\frac{1}{2}\right)$ **b.** $\tan^{-1}(-\sqrt{3})$
c. $\arcsin\left(-\frac{\sqrt{3}}{2}\right)$ **d.** $\cos^{-1}(-1)$
e. $\tan^{-1}(-1)$

31. **a.** $\sec^{-1} 2$ **b.** $\cot^{-1}(-1)$
c. $\operatorname{arcsec}\left(-\frac{2}{\sqrt{3}}\right)$ **d.** $\cot^{-1} 0$
e. $\sec^{-1}(-1)$

32. **a.** $\cos\left[\arcsin\left(-\frac{3}{5}\right)\right]$ **b.** $\sin\left[2\sin^{-1}\left(-\frac{5}{13}\right)\right]$

33. **a.** $\tan\left[2\cos^{-1}\left(-\frac{4}{5}\right)\right]$ **b.** $\cos\left(\frac{1}{2}\tan^{-1} 4\sqrt{3}\right)$

34. **a.** $\csc\left[\sec^{-1}\left(-\frac{5}{3}\right)\right]$ **b.** $\tan\left[\frac{1}{2}\tan^{-1}\left(-\frac{5}{12}\right)\right]$

35. **a.** $\tan^{-1}\left(\tan\frac{3\pi}{4}\right)$ **b.** $\sin^{-1}\left(\cos\frac{2\pi}{3}\right)$

36. **a.** $\cos^{-1}\left[\cos\left(-\frac{\pi}{6}\right)\right]$ **b.** $\sin^{-1}\left(\sin\frac{7\pi}{8}\right)$

37. **a.** $\sin(\alpha + \beta)$ if $\alpha = \sin^{-1}(-\frac{3}{5})$ and $\beta = \cos^{-1}(-\frac{5}{13})$
b. $\cos(\alpha - \beta)$ for α and β of part **a**.

38. **a.** $\tan(\alpha - \beta)$ if $\alpha = \tan^{-1}(-\frac{4}{3})$ and $\beta = \cos^{-1}(-\frac{8}{17})$

b. $\cos(\alpha + \beta)$ for α and β of part **a**.

39. **a.** $\sin\left[\sec^{-1}\left(-\frac{5}{3}\right) - \tan^{-1}\left(-\frac{12}{5}\right)\right]$

b. $\tan\left[\tan^{-1}\frac{1}{3} - \tan^{-1}\left(-\frac{2}{3}\right)\right]$

In Problems 40 and 41 use a calculator or tables to evaluate correct to four significant figures.

40. **a.** $\cos^{-1}(-0.2137)$ **b.** $\sin^{-1}(0.6022)$

41. **a.** $\arctan(1.238)$ **b.** $\arcsin(-0.3147)$

B Draw the graphs in Problems 42–46.

42. $y - 2 = 3\sin\left(\frac{x}{2} + 1\right)$

43. $2(y + 1) = 3\cos(2 - 3x)$

44. $y = \frac{1}{2}\sin^{-1} 2x + \frac{\pi}{3}$

45. $y = \cos x + \cot x$ (Use addition of ordinates.)

46. $y = \frac{4}{x^2}\sin x, \quad x \neq 0$ (Use multiplication of ordinates.)

Evaluate the expressions in Problems 47 and 48 without using tables or a calculator.

47. $\tan^{-1}\left(-\frac{1}{2}\right) - \tan^{-1}\frac{1}{3}$

48. $\cos\left[2\sin^{-1}\left(-\frac{1}{3}\right) + \tan^{-1}(-2)\right]$

5 Logarithms

Logarithms continue to play an important role in a variety of applications, even though the computational aspect of the subject is now largely accomplished by hand calculators. The following excerpt from a chemistry textbook shows an application in determining what is called hydrogen ion concentration.*

> It is rather cumbersome to express the hydrogen ion or hydroxide ion concentration in terms of some number times 10 raised to a negative exponent. A widely used and more convenient method of stating the hydrogen ion concentration of dilute acids, bases, and neutral solutions is in terms of pH. The pH of a solution is defined as follows:
>
> $$\text{pH} = \log \frac{1}{[\text{H}^+]} \quad \text{or} \quad \text{pH} = -\log\,[\text{H}^+]$$
>
> The logarithm of a number is the power to which the number 10 must be raised to give that number. . . .
>
> [Example]
>
> A solution of hydrochloric acid is 0.0063 N. If we assume complete ionization, the H^+ concentration follows from the definition of a normal solution as 0.0063 mole/liter, or 6.3×10^{-3} M. The pH of the solution is
>
> $$\begin{aligned} \text{pH} &= -\log\,[\text{H}^+] \\ &= -\log(6.3 \times 10^{-3}) \\ &= -\log 6.3 - \log 10^{-3} \end{aligned}$$

---

*From pp. 349–350 in *General College Chemistry,* 5th Edition by Charles W. Keenan, Jesse H. Wood and Donald C. Kleinfelter.

$$= -\log 6.3 - (-3)$$
$$= 3 - \log 6.3$$
$$= 3 - 0.80$$
$$= 2.20$$

The type of logarithm employed here, based on 10, is a special case of the general logarithm function which we will define and study in this chapter. Logarithms with base 10, called **common logarithms,** have typically been used for computational purposes.

5.1 Definition and Properties

The equation $y = a^x$, where $a > 0$ and $a \neq 1$, defines the **exponential function with base *a*.** We require $a > 0$ in order that a^x be defined for all real numbers x, and we restrict a to be different from 1 so that a^x will not be the constant function $1^x = 1$. The graph of $y = a^x$ is increasing if $a > 1$ and decreasing if $0 < a < 1$. We illustrate this in Figure 5.1. In either case, the function is 1-1 and so has an inverse.

Figure 5.1

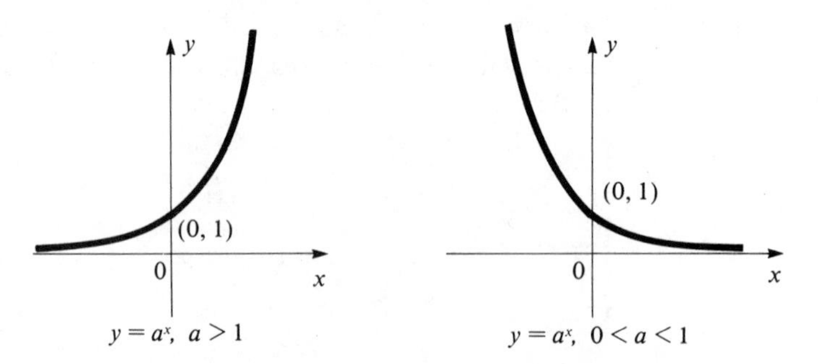

For historical reasons we name the inverse the **logarithm function with base *a*** and write $y = \log_a x$. Thus, we have the important relationship

$$y = \log_a x \quad \text{if and only if} \quad x = a^y$$

We refer to the first of these two equations as the **logarithmic form** and the second as the **exponential form.** For example, the equivalent exponential form of the equation $\log_2 8 = 3$ is $8 = 2^3$. Similarly, the logarithmic form of $10^{-2} = 0.01$ is $\log_{10} 0.01 = -2$.

Since the domain of $y = a^x$ is the set $\mathbb{R}$ of all real numbers and its range is the set $\mathbb{R}^+$ of all positive real numbers, it follows that for the inverse function, $y = \log_a x$, the domain is $\mathbb{R}^+$ and the range is $\mathbb{R}$. In particular, this says that $\log_a x$

is defined only when $x > 0$. In Figure 5.2 we show the graph of $y = \log_a x$ for $a > 1$, which is the case of primary interest to us. This graph is the reflection of the graph of $y = a^x$ in the 45° line $y = x$.

Figure 5.2

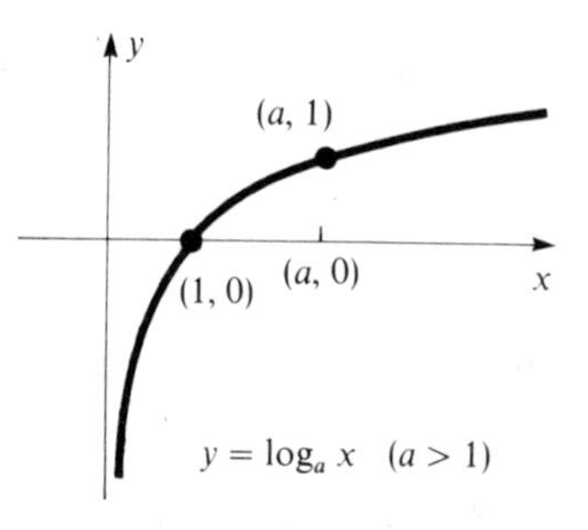

The values of the logarithm function are called **logarithms.** Just as is the case with exponential functions, most logarithms can only be approximated, using a calculator or tables, or by methods studied in calculus. We will discuss uses of tables of logarithms in Sections 5.2 and 5.3. There are, however, certain logarithms which can be obtained using the definition. For example, $\log_2 8$ is 3, since $2^3 = 8$, and $\log_{10} 100 = 2$, since $10^2 = 100$. Because of the inverse relationship between $\log_a x$ and a^x, we always have

$$\log_a a^x = x \quad \text{and} \quad a^{\log_a x} = x$$

The two most useful bases for logarithms are the number 10 and the irrational number designated by e, whose approximate value is 2.71828. This latter base, both for exponential and logarithm functions, is studied in calculus and has wide applicability in many branches of science, especially in studying exponential growth and decay. Logarithms to the base 10 are called **common logarithms,** and it is customary to omit the base in this case, writing $\log x$ to mean $\log_{10} x$.* Logarithms to the base e are called **natural logarithms** and are usually designated $\ln x$. So $\ln x$ means $\log_e x$. Most scientific hand calculators have keys for $\log x$ and $\ln x$.

The great utility of the logarithm function is a consequence of the following three properties.

Properties of Logarithms

$$\log_a xy = \log_a x + \log_a y \tag{5.1}$$

$$\log_a \frac{x}{y} = \log_a x - \log_a y \tag{5.2}$$

$$\log_a x^p = p \log_a x \tag{5.3}$$

---

*In more advanced work, however, $\log x$ frequently is understood to mean $\log_e x$.

The proof depends on the familiar laws of exponents:

$$a^x \cdot a^y = a^{x+y} \qquad \frac{a^x}{a^y} = a^{x-y} \qquad (a^x)^p = a^{px}$$

We will prove properties (5.1) and (5.3) and leave (5.2) as an exercise.

To prove (5.1), let $u = \log_a x$ and $v = \log_a y$. We write each of these in the equivalent exponential form:

$$x = a^u \quad \text{and} \quad y = a^v$$

Thus,

$$xy = a^u a^v = a^{u+v}$$

by the first law of exponents. We rewrite the result $xy = a^{u+v}$ in logarithmic form

$$\log_a xy = u + v$$

But $u = \log_a x$ and $v = \log_a y$, so (5.1) is proved.

To prove (5.3), consider

$$x^p = (a^u)^p = a^{pu}$$

by the third law of exponents. Hence,

$$\log_a x^p = pu = p \log_a x$$

In the next two sections we will see how to use these properties to simplify computations involving products, quotients, powers, and roots. This computational aspect of logarithms was first developed by a Scottish mathematician named John Napier (1550–1617), and it was a discovery of profound importance. Although the scientific hand calculator has largely replaced computation by logarithms, there is still merit in learning how to use logarithms in this way. Aside from the fact that logarithms are useful as an aid in computation, the logarithmic function is of great importance in many scientific applications.

EXAMPLE 5.1 Solve each of the following for x.

a. $\log_2 x = 4$ **b.** $\log_x 27 = \dfrac{3}{2}$

Solution In each case we first write the equivalent exponential form.

a. $2^4 = x$. So $x = 16$.

b. $x^{3/2} = 27$. Now raise both sides to the $\frac{2}{3}$ power to get

$$x = (27)^{2/3} = 3^2 = 9$$

EXAMPLE 5.2 Use the properties of logarithms to simplify the following.

$$\log \frac{x\sqrt{1 + x^2}}{(x + 2)^3}$$

Solution

$$\log \frac{x\sqrt{1 + x^2}}{(x + 2)^3} = \log(x\sqrt{1 + x^2}) - \log(x + 2)^3 \qquad \text{[Property 5.2]}$$

$$= \log x + \log(1 + x^2)^{1/2} - \log(x + 2)^3 \qquad \text{[Property 5.1]}$$

$$= \log x + \frac{1}{2}\log(1 + x^2) - 3\log(x + 2) \qquad \text{[Property 5.3]}$$

Note of Caution. The properties of logarithms enable us to simplify logarithms of products, quotients, and powers (also roots, since these can be expressed as powers), but there is no way to simplify logarithms of sums or differences. Thus $\log(x + y)$ *cannot* be changed to $\log x + \log y$, however tempting this may be. In fact we know that $\log x + \log y = \log xy$, not $\log(x + y)$.

EXAMPLE 5.3 Use the properties of logarithms to solve each of the following equations for x. Check all answers.

a. $\log 3 + \log(x - 1) = \log 2 + \log(x + 2)$
b. $\log_2 x + \log_2(x - 2) = 3$

Solution **a.** Using Property (5.1), we get

$$\log 3(x - 1) = \log 2(x + 2)$$

Since the logarithm function is 1-1, it follows that

$$3(x - 1) = 2(x + 2)$$

or

$$3x - 3 = 2x + 4$$
$$x = 7$$

Check.

$$\log 3 + \log 6 = \log 3 \cdot 6 = \log 18$$

$$\log 2 + \log 9 = \log 2 \cdot 9 = \log 18$$

So $x = 7$ is the solution.

b. Again using Property (5.1), we get

$$\log_2 x(x - 2) = 3$$

Now we write this in exponential form and solve the resulting quadratic equation.

$$x(x - 2) = 2^3$$

$$x^2 - 2x - 8 = 0$$

$$(x - 4)(x + 2) = 0$$

$$x - 4 = 0 \quad \Big| \quad x + 2 = 0$$

$$x = 4 \quad \Big| \quad x = -2$$

Check. $x = 4$.

$$\log_2 4 + \log_2 2 = \log_2 2^2 + \log_2 2 = 2 + 1 = 3$$

Check. $x = -2$.
Since $\log_2 (-3)$ is not defined, we reject $x = -2$. Thus, $x = 4$ is the only solution.

EXERCISE SET 5.1

A In Problems 1–5 find x.

1. a. $\log_4 x = 2$ **b.** $\log_x 16 = 4$

2. a. $\log_2 8 = x$ **b.** $\log_{10} x = -2$

3. a. $\ln 1 = x$ **b.** $\log_x \frac{1}{e} = -1$

4. a. $\log_{10} x = 3$ **b.** $\log_{10} 0.001 = x$

5. a. $\log_3 (x - 1) = 2$ **b.** $\log_x 3 = \frac{1}{2}$

In Problems 6–8 write in logarithmic form.

6. a. $4^3 = 64$ **b.** $3^{-2} = \frac{1}{9}$

7. a. $10^4 = 10{,}000$ **b.** $10^{-3} = 0.001$

8. a. $r^s = t$ **b.** $z^x = m$

In Problems 9–11 write in exponential form.

9. a. $\log_2 16 = 4$ **b.** $\log_{10} 100 = 2$

10. a. $\ln 1 = 0$ **b.** $\log_5 0.2 = -1$

11. a. $\log_a x = y$ **b.** $\log_x z = t$

Evaluate the expressions in Problems 12–16.

12. **a.** $\log_3 81$ **b.** $\log_{10} 0.1$

13. **a.** $\ln \frac{1}{e^2}$ **b.** $\log_3 \frac{1}{9}$

14. **a.** $\log_2 2^3$ **b.** $2^{\log_2 3}$

15. **a.** $e^{\ln 4}$ **b.** $\ln e^4$

16. **a.** $10^{\log 7}$ **b.** $e^{2 \ln x}$

In Problems 17–20 simplify the expression using the properties of logarithms. Express answers in a form which does not involve logarithms of products, quotients, powers, or roots.

17. **a.** $\log(x + 1)(x + 2)$ **b.** $\log \frac{x + 1}{x - 2}$

18. **a.** $\log(x + 4)^3$ **b.** $\log \sqrt{2x + 3}$

19. **a.** $\log \frac{x^3}{3x + 1}$ **b.** $\log \frac{x(x + 2)}{(x + 3)^2}$

20. **a.** $\log \frac{2x^2}{(x - 1)(2x + 3)}$ **b.** $\log \frac{\sqrt[3]{(x + 1)^2}}{3(x + 2)}$

In Problems 21–23 combine into a single term.

21. **a.** $\log x - \log y + \log 2$ **b.** $2 \log x + 3 \log(x + 1)$

22. $\log 3 + \frac{2}{3} \log(x^2 + 4) - 2 \log(2x - 3)$

23. $\frac{1}{2} \log(5x + 3) - [2 \log(x + 2) + \log(3x + 1)]$

B In Problems 24–32 solve for x and check your answers.

24. $\log 2 + \log x = \log 3 + \log(x - 1)$

25. $\log 4 - \log x = \log 2 + \log 3 - \log(x + 1)$

26. $\log(x - 1) - \log(x + 6) = \log(x - 2) - \log(x + 3)$

27. $2 \log x = \log 2 + \log(x + 4)$

28. $\log x + \log(2x - 5) = \log 3$

29. $2 \log_3 (x - 2) - \log_3 (x + 4) = 1$

30. $\ln(2x - 1) + \ln(x - 1) = 0$

31. $\log_{10} x + \log_{10} 2 = 1 - \log_{10} (x + 4)$

32. $\log_2 (x + 3) - \log_2 3 = \log_2 (3x + 4) - 3$

In Problems 33–35 solve for x by taking the natural logarithm of both sides. Leave answers in terms of natural logarithms.

33. **a.** $4^x = 3$ **b.** $3^x = 5^{x-1}$

34. **a.** $2(3^{-x}) = 6^{2x}$ **b.** $3^{x+1} = e^{x/2}$

35. **a.** $e^{-x} = 2^{x-3}$ **b.** $3(4^{2x}) = 2(3^{-x})$

36. In a chemical reaction a substance is converted according to the formula

$$y = 32(2^{-0.3t})$$

where y is the amount in grams of unconverted substance t seconds after the reaction began. Show that there were 32 grams of the substance originally, and find how long it will take for only 4 grams to remain.
Hint. Take logarithms to base 2 of both sides.

5.2 Computations with Common Logarithms

From the definition of common logarithms we can deduce the following:

$$\log 1 = 0 \quad \text{since} \quad 10^0 = 1$$

$$\log 10 = 1 \quad \text{since} \quad 10^1 = 10$$

$$\log 100 = 2 \quad \text{since} \quad 10^2 = 100$$

$$\log 1{,}000 = 3 \quad \text{since} \quad 10^3 = 1{,}000$$

$$\log 0.01 = -2 \quad \text{since} \quad 10^{-2} = \frac{1}{100} = 0.01$$

and in general for any integer k,

$$\log 10^k = k \tag{5.4}$$

To find the logarithm of a number N which is not a perfect power of 10, we first write the number in scientific notation to get a result of the form

$$N = n \times 10^k$$

where n is a number between 1 and 10 and k is an integer. For example, if $N = 326$, we write $N = 3.26 \times 10^2$. Similarly, if $N = 0.0104$, we write $N = 1.04 \times 10^{-2}$. If we apply Property (5.1) to Equation (5.4) we obtain $\log N = \log n + \log 10^k$, or since $\log 10^k = k$,

$$\log N = \log n + k \tag{5.5}$$

This is called the **standard form** of $\log N$. The term $\log n$ is called the **mantissa** of $\log N$ and k is called the **characteristic.** Since n is between 1 and 10, it follows that $\log n$ is between $\log 1 = 0$ and $\log 10 = 1$. The mantissa is therefore a nonnegative number usually written in decimal form. The characteristic k is an integer which may be positive, negative, or zero.

Mantissas of logarithms (that is, logarithms of numbers between 1 and 10) have been extensively tabulated. Table III is one such tabulation, giving four-decimal-place accuracy. Other tables exist giving five or more places of accuracy. The following example illustrates how to find logarithms of numbers using the foregoing discussion, along with Table III.

EXAMPLE 5.4 Find: **a.** log 326 **b.** log 0.0104

Solution **a.** $\log 326 = \log (3.26 \times 10^2) = \log 3.26 + 2$
From Table III, we read $\log 3.26 = 0.5132$.
Thus, $\log 326 = 0.5132 + 2$, or

$$\log 326 = 2.5132$$

b. $\log 0.0104 = \log (1.04 \times 10^{-2}) = \log 1.04 - 2$
$= 0.0170 - 2$ (from Table III)

Note. It is possible to combine the positive mantissa and the negative characteristic in this case to get -1.9830, but it is seldom advantageous to do this. (We might observe, however, that this single negative answer is the form we would get on a calculator.) The form $0.0170 - 2$ is an acceptable form and we will leave it and other logarithms of numbers between 0 and 1 in a form of this type unless there is a reason for doing otherwise. There are, in fact, times when it would be better to write the answer in one of the following equivalent ways:

$$1.0170 - 3$$
$$2.0170 - 4$$
$$\vdots \quad \vdots$$
$$8.0170 - 10$$

and so on. The form $8.0170 - 10$ is often used. Which form to use will be determined by what subsequent operations are to be performed on it.

The next example shows how to find a number when we know its logarithm. This process is known as finding the **antilogarithm**, or more briefly, finding the **antilog**.

EXAMPLE 5.5 Find N if $\log N = 3.8762$.

Solution We observe that the characteristic is 3 and the mantissa is 0.8762. In the body of Table III we locate the mantissa and find that it corresponds to the number 7.52. So

$$N = 7.52 \times 10^3 = 7{,}520$$

When we want the logarithm of a number with more than three significant digits, or when we want to find the antilog where the mantissa falls between tabular values, we use linear interpolation. The process is similar to that described in Section 2.2 for interpolation to find values of trigonometric functions. We illustrate this with the next two examples.

EXAMPLE 5.6 Find log 376.4.

Solution Since $376.4 = 3.764 \times 10^2$, the characteristic is 2, and the mantissa is log 3.764. Since Table III gives logarithms of numbers having at most three significant digits, we will interpolate. The number 3.764 falls between 3.760 and 3.770. We proceed as follows.

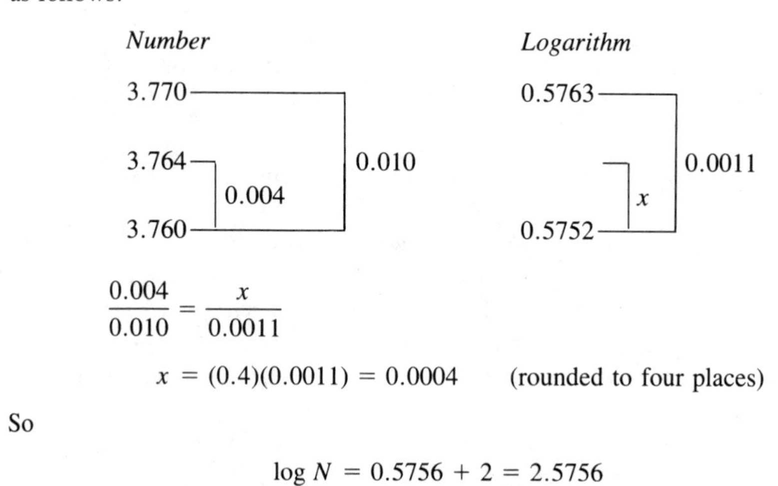

$$\frac{0.004}{0.010} = \frac{x}{0.0011}$$

$$x = (0.4)(0.0011) = 0.0004 \qquad \text{(rounded to four places)}$$

So

$$\log N = 0.5756 + 2 = 2.5756$$

EXAMPLE 5.7 Find t if $\log t = 8.9413 - 10$.

Solution The characteristic is $8 - 10 = -2$, and the mantissa is 0.9413. This mantissa falls between the tabular values 0.9410 and 0.9415, whose antilogs are 8.73 and 8.74, respectively.

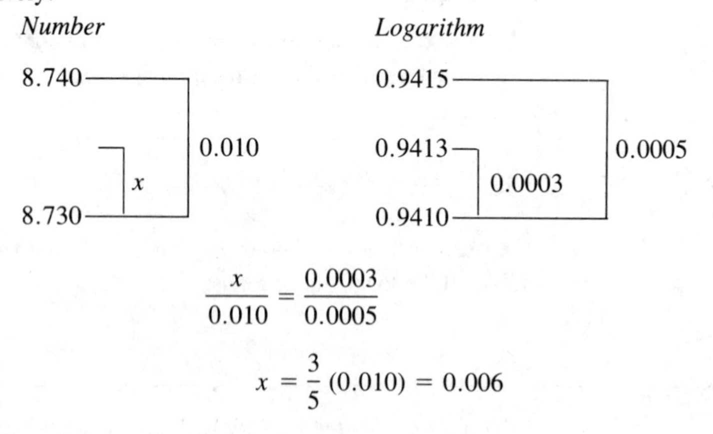

$$\frac{x}{0.010} = \frac{0.0003}{0.0005}$$

$$x = \frac{3}{5}(0.010) = 0.006$$

The antilog of the mantissa is therefore approximately 8.736. Since the characteristic is -2, we have

$$t = 8.736 \times 10^{-2} = 0.08736$$

Remark. With practice most interpolation can be done mentally. In Example 5.6, since 3.764 is 0.4 of the way between 3.760 and 3.770, the logarithm is approximately 0.4 of the way between the corresponding tabular values 0.5752 and 0.5763. Taking 0.4 of the difference and adding it to the smaller yields the result. Similarly, in Example 5.7, since the mantissa 0.9413 is $\frac{3}{5} = 0.6$ of the way between the two tabular values surrounding it, the antilog is approximately 0.6 of the way between 8.730 and 8.740, namely 8.736.

The next example shows how we can exploit the properties of logarithms to carry out computations involving products, quotients, powers, and roots.

EXAMPLE 5.8 Use logarithms to find the value of each of the following:

a. $(87{,}600)(0.0348)$ **b.** $\dfrac{13.72}{203.4}$

c. $(528.6)^4$ **d.** $\sqrt[3]{0.07254}$

Solution **a.** Let $N = (87{,}600)(0.0348)$. By Property (5.1),

$\log N = \log 87{,}600 + \log 0.0348$

$$\begin{aligned} \log 87{,}600 &= 4.9425 \\ \log 0.0348 &= \underline{0.5416 - 2} \ (+) \\ & 5.4841 - 2 = 3.4841 \end{aligned}$$

$$N = \text{antilog } 3.4841 \approx 3{,}050$$

Note that since we need only three significant figure accuracy in the answer, it is not necessary to interpolate. We find the nearest tabular value to 0.4841.

b. Let $N = \dfrac{13.72}{203.4}$. By Property (5.2),

$\log N = \log 13.72 - \log 203.4$
By interpolation we obtain the following:

$$\begin{aligned} \log 13.72 &= 1.1374 \\ \log 203.4 &= \underline{2.3084} \ (-) \end{aligned}$$

Since the second logarithm is larger than the first, this subtraction would result in a negative logarithm, and so it would not be in standard form. We remedy this by adding and subtracting 2 from the first logarithm.

$$\begin{aligned} \log 13.72 &= 3.1374 - 2 \\ \log 203.4 &= \underline{2.3084 } \ (-) \\ & 0.8290 - 2 \end{aligned}$$

$$N = \text{antilog } (0.8290 - 2) = 0.06745$$

c. Let $N = (528.6)^4$. Then by Property (5.3),

$$\log N = 4 \log 528.6 = 4\ (2.7231) = 10.8924$$

$$N = \text{antilog } 10.8924 = 7.805 \times 10^{10}$$

d. Let $N = \sqrt[3]{0.07254}$. By Property (5.3),

$$\log N = \frac{1}{3} \log 0.07254 = \frac{1}{3}\ (0.8605 - 2)$$

Dividing by 3 as the logarithm is written would complicate matters, since the characteristic is not evenly divisible by 3. We overcome this difficulty by writing the logarithm as $1.8605 - 3$. Then we get

$$\log N = \frac{1}{3}\ (1.8605 - 3) = 0.6202 - 1$$

$$N = \text{antilog } (0.6202 - 1) = 0.4171$$

Remark. The advantages of using logarithms for such calculations are that multiplication is replaced by addition, division by subtraction, raising to a power by multiplication, and extraction of a root by division (or equivalently, by multiplication by a fraction).

The next example combines all four operations in one problem.

EXAMPLE 5.9 Use logarithms to evaluate

$$N = \frac{(0.00268)\ \sqrt[3]{59{,}300}}{(7.26)^2}$$

Solution

$$\log N = \log[(0.00268)\ \sqrt[3]{59{,}300}\,] - \log(7.26)^2 \qquad \text{[Property 5.2]}$$

$$= \log 0.00268 + \log(59{,}300)^{1/3} - \log(7.26)^2 \qquad \text{[Property 5.1]}$$

$$= \log 0.00268 + \frac{1}{3} \log 59{,}300 - 2 \log 7.26 \qquad \text{[Property 5.3]}$$

$$\log 0.00268 = 0.4281 - 3$$

$$\tfrac{1}{3} \log 59{,}300 = \tfrac{1}{3}\ (4.7731) = \underline{1.5910} \quad (+)$$

$$2.0191 - 3$$

$$2 \log 7.26 = 2(0.8609) = \underline{1.7218} \quad (-)$$

$$0.2973 - 3$$

$$N = \text{antilog } (0.2973 - 3) = 0.001983$$

EXERCISE SET 5.2

A In Problems 1–8 find the logarithms using Table III.

1. log 32.7
2. log 0.0234
3. log 0.765
4. log 358
5. log 256,000
6. log 9.86
7. log (3.40×10^7)
8. log (8.02×10^{-5})

In Problems 9–16 find N, using Table III.

9. $\log N = 3.6561$
10. $\log N = 2.9624$
11. $\log N = 0.8927$
12. $\log N = 0.7582 - 2$
13. $\log N = 0.3404 - 3$
14. $\log N = 9.5224 - 10$
15. $\log N = 7.6493 - 10$
16. $\log N = 8.8831 - 10$

In Problems 17–24 use linear interpolation to find the logarithm.

17. log 5.673
18. log 247.5
19. log 0.003786
20. log 1.028
21. log 476,200
22. log 0.05932
23. log (6.757×10^6)
24. log (9.314×10^{-4})

In Problems 25–32 use linear interpolation to find N.

25. $\log N = 2.8366$
26. $\log N = 5.7585$
27. $\log N = 0.6280 - 2$
28. $\log N = 0.7163 - 4$
29. $\log N = 8.4415 - 10$
30. $\log N = 7.5813 - 10$
31. $\log N = 0.8147$
32. $\log N = 0.8906 - 1$

In Problems 33–44 use logarithms to approximate the numbers to three significant digits.

33. $(32.5)(7.24)$
34. $(254)(0.0634)$
35. $\dfrac{172}{53.6}$
36. $\dfrac{1.34}{32.1}$
37. $\dfrac{3.04}{0.0256}$
38. $\dfrac{0.00849}{0.00138}$
39. $(2.75)^{10}$
40. $(0.998)^{20}$
41. $\sqrt[3]{53.4}$
42. $\sqrt[5]{18.7}$
43. $(33.6)(142)^{2/3}$
44. $\dfrac{(9.28)^3}{\sqrt{4,780}}$

In Problems 45–50 use linear interpolation to approximate the numbers to four significant digits.

45. $\dfrac{(2.367)(18.43)}{54.72}$
46. $\dfrac{3,285}{(94.62)(3.102)}$

47. $(0.02543)^2 \sqrt{543.7}$

48. $\dfrac{\sqrt{0.06254}}{(1.325)^3}$

49. $\dfrac{\sqrt{(32.67)^3}}{(1.563)\,(0.5628)^2}$

B **50.** Use logarithms to find N correct to four significant digits if

$$N = \frac{x^{1/3}\, y^5}{z^{3/4}}$$

where $x = 0.05683$, $y = 0.2570$, and $z = 0.003896$.

51. Use logarithms to find:

a. the volume

b. the lateral surface area

of a right circular cylinder of base radius 2.037 cm and altitude 8.954 cm.

Hint. The volume is the area of the base times the altitude, and the lateral surface area is the circumference of the base times the altitude.

52. A right circular cylinder of height 4.367 cm is inscribed in a sphere of radius 3.045 cm. Use logarithms to find the volume of the cylinder. (See hint to Problem 51.)

53. A right circular cone of base radius 3.258 cm is inscribed in a sphere of diameter 12.58 cm. Use logarithms to find the volume of the cone.

Hint. The volume is one-third the area of the base times the altitude.

54. It can be shown that the length of the radius r of a circle inscribed in a triangle whose sides are of lengths a, b, and c, respectively, is given by

$$r = \sqrt{\frac{(s - a)\,(s - b)\,(s - c)}{s}}$$

where $s = (a + b + c)/2$. Using this and logarithms, find the area of the inscribed circle in a triangle of sides $a = 28.42$, $b = 16.75$, and $c = 25.68$.

55. Use logarithms to find N if

$$N = (2.681)^{1.324}$$

Hint. Find $\log(\log N)$.

56. Using logarithms to perform the division, find the value of

$$N = \frac{\log 3.75}{\log 0.0268}$$

57. By using $\log(\log N)$, find the value of

$$N = (33.46)^{\log 2.567}$$

58. Evaluate, using logarithms

$$(\log 4.562)^{5.137}$$

59. Use logarithms to evaluate

$$\frac{-(2.132)\log 7.145}{\log 54.61}$$

In Problems 60 and 61 solve for x using common logarithms.

60. $2 \cdot 3^{1-x} = 4^{2x}$

61. $(2.761)^x = (37.25)^{2-3x}$

5.3 Logarithms of Trigonometric Functions

Table IV gives values of the logarithms of trigonometric functions. This eliminates the necessity of using two tables. For example, suppose we wish to find the logarithm of sin 37°, written log sin 37°. One way is to look up sin 37° in Table I and then look up the logarithm of this number in Table III. But Table IV gives log sin 37° directly. In that table we read 9.7795, but it is understood that for all entries in this table we must attach −10 to the value given. Hence,

$$\log \sin 37° = 9.7795 - 10$$

The next two examples illustrate how to solve a right triangle making use of logarithms.

EXAMPLE 5.10 Solve the right triangle ABC in which $A = 41°\ 30'$ and $c = 27.35$ (Figure 5.3).

Figure 5.3

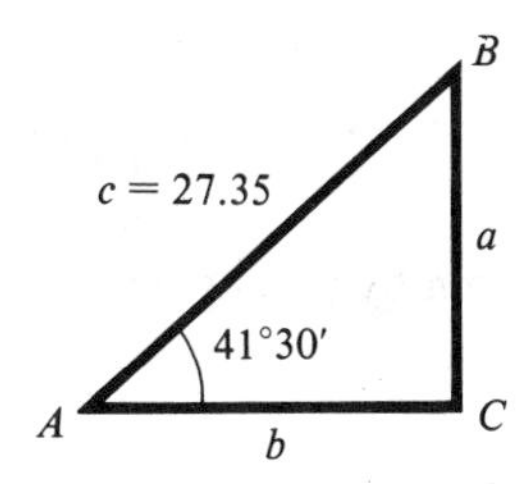

Solution Since A and B are complementary angles, we have

$$B = 90° - A = 90° - 41°\ 30' = 48°\ 30'$$

Next we use $\sin A = a/c$ to find side a.

$$\sin 41°\ 30' = \frac{a}{27.35}$$

$$a = 27.35 \sin 41°\ 30'$$

$$\log a = \log 27.35 + \log \sin 41°\ 30'$$

$$= 1.4370 + 9.8213 - 10$$

$$= 11.2583 - 10$$

$$= 1.2583$$

So

$$a = 18.12$$

To find b, we proceed in a similar way, using $\cos A = b/c$.

$$\cos 41° \, 30' = \frac{b}{27.35}$$

$$b = 27.35 \cos 41° \, 31'$$

$$\log b = \log 27.35 + \log \cos 41° \, 30'$$

$$= 1.4370 + 9.8745 - 10$$

$$= 11.3115 - 10$$

$$= 1.3115$$

So

$$b = 20.49$$

The triangle is now solved.

EXAMPLE 5.11 Solve the right triangle ABC in which $a = 2.347$ and $b = 5.102$ (Figure 5.4).

Figure 5.4

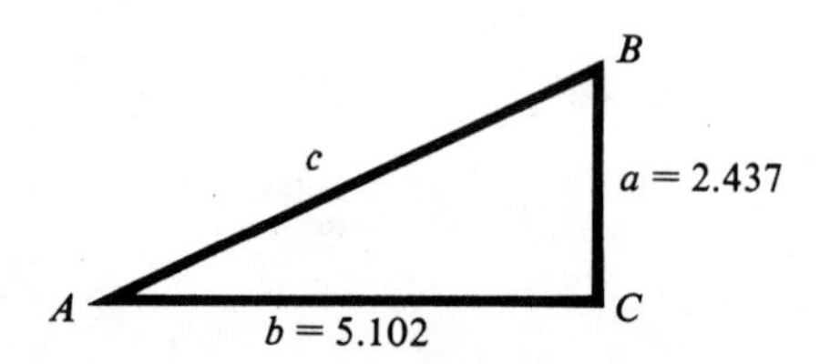

Solution We begin by finding angle A, using the tangent function.

$$\tan A = \frac{a}{b} = \frac{2.347}{5.102}$$

$$\log \tan A = \log 2.347 - \log 5.102$$

$$= 0.3705 - 0.7078$$

To perform this subtraction we write the first number in the form $10.3705 - 10$, in order to retain a positive mantissa.

$$\begin{array}{rl} & 10.3705 - 10 \\ & \underline{0.7078 \qquad\;\;} \; (-) \\ \log \tan A = & 9.6627 - 10 \end{array}$$

Now we find A from Table IV. We will show the interpolation this time.

Angle		*Log tan*	
24° 50′		9.6654 − 10	
	x ; 10	9.6627 − 10	7 ; 34
24° 40′		9.6620 − 10	

$$\frac{x}{10} = \frac{7}{34}$$

$$x = \frac{70}{34} \approx 1 \qquad \text{(to nearest integer)}$$

So

$$A = 24°21'$$

Therefore, $B = 90° - 24° 42' = 65° 18'$.

Finally, we find c using $\sin A = a/c$.

$$\sin 24° 42' = \frac{2.347}{c}$$

$$c = \frac{2.347}{\sin 24° 42'}$$

$$\log c = \log 2.347 - \log \sin 24° 42'$$

$$= 0.3705 - (9.6210 - 10)$$

Again, we write 0.3705 in the form $10.3705 - 10$ to maintain a positive mantissa.

$$\begin{array}{rl} & 10.3705 - 10 \\ & \underline{9.6210 - 10} \ (-) \\ \log c = & 0.7495 \end{array}$$

Thus, from Table III we find (after interpolating)

$$c = 5.617$$

EXERCISE SET 5.3

A In Problems 1–6 find the given logarithm using Table IV.

1. $\log \sin 31° 20'$

2. $\log \tan 14° 10'$

3. $\log \cos 67° 40'$

4. $\log \sec 48° 30'$

5. $\log \cot 12° 50'$

6. $\log \sin 73° 40'$

In Problems 7–12 use Table IV to find angle A, given that A is a positive acute angle.

7. $\log \cos A = 9.9590 - 10$

8. $\log \tan A = 8.9701 - 10$

9. $\log \sin A = 9.9268 - 10$

10. $\log \cot A = 9.6553 - 10$

11. $\log \sec A = 1.0597$

12. $\log \csc A = 0.2290$

In Problems 13–18 use Table IV and linear interpolation to find the given logarithm.

13. $\log \tan 15° 32'$

14. $\log \sin 47° 28'$

15. $\log \cos 42° 53'$

16. $\log \cot 63° 17'$

17. $\log \sin 21.4°$

18. $\log \cos 46.8°$

In Problems 19–24 use Table IV and linear interpolation to find angle A, where $0° \leq A \leq 90°$.

19. $\log \sin A = 9.7898 - 10$

20. $\log \cot A = 0.3674$

21. $\log \tan A = 9.7115 - 10$

22. $\log \sec A = 0.0212$

23. $\log \cos A = 9.4920 - 10$

24. $\log \csc A = 0.2048$

In Problems 25–34 use logarithms to solve the right triangle ABC, having right angle C.

25. $a = 25.7$, $A = 42° 21'$

26. $a = 4.56$, $B = 37° 10'$

27. $b = 35.67$, $A = 21° 36'$

28. $b = 13.15$, $c = 24.29$

29. $a = 250.1$, $b = 302.5$

30. $a = 306$, $c = 452$

31. $A = 54.62°$, $c = 132.4$

32. $B = 35.87°$, $b = 52.14$

33. $a = 41.27$, $A = 63.75°$

34. $a = 14.98$, $b = 10.22$

B **35.** Use logarithms to find the value of the expression

$$\frac{a \sin \alpha \sin \beta}{\sqrt{\sin (\alpha + \beta) \sin (\alpha - \beta)}}$$

where $a = 21.35$, $\alpha = 72° 34'$, and $\beta = 43° 52'$.

36. It can be shown (see Problem 16, Exercise Set 6.4) that the area of the triangle ABC (not necessarily a right triangle) is given by the formula

$$\text{Area} = \frac{b^2 \sin A \sin C}{2 \sin B}$$

Use this formula and logarithms to find the area of the triangle in which $b = 52.14$, $A = 27° 33'$, and $B = 41° 13'$

37. From a tower h meters high the angles of depression of points A and B are α and β, respectively. If A and B are on the same side of the tower and in line with it, show that the distance d between A and B is given by the formula

$$d = \frac{h \sin(\beta - \alpha)}{\sin \alpha \sin \beta}$$

If $h = 120$ meters, $\alpha = 18° 30'$, and $\beta = 23° 45'$, find d, using logarithms.

38. **a.** From a lighthouse the angle of depression of ship A due north of the lighthouse is $10.5°$, and the angle of depression of ship B due east of the lighthouse is $13.7°$. Use logarithms to find the bearing of ship B from ship A.

b. If the lighthouse is 135 feet high, find the distance between ships A and B.

39. From a point on the ground d feet from the base of a building, the angles of elevation of the top and bottom of a flagpole on top of the building are α and β, respectively. Show that the height h of the flagpole is given by the formula

$$h = \frac{d \sin(\alpha - \beta)}{\cos \alpha \cos \beta}$$

If $d = 110$ feet, $\alpha = 39° 10'$, and $\beta = 31°40'$, find h, using logarithms.

40. The accompanying figure is a trapezoid with parallel sides of lengths a and $b (b > a)$, and base angles α and β. Derive the following formulas for the lengths x and y of the non-parallel sides.

$$x = \frac{(b - a) \sin \alpha}{\sin(\alpha + \beta)}, \qquad y = \frac{(b - a) \sin \beta}{\sin(\alpha + \beta)}$$

Using logarithms, find x and y if $a = 7.214$, $b = 12.15$, $\alpha = 57° 54'$, and $\beta = 41° 13'$.

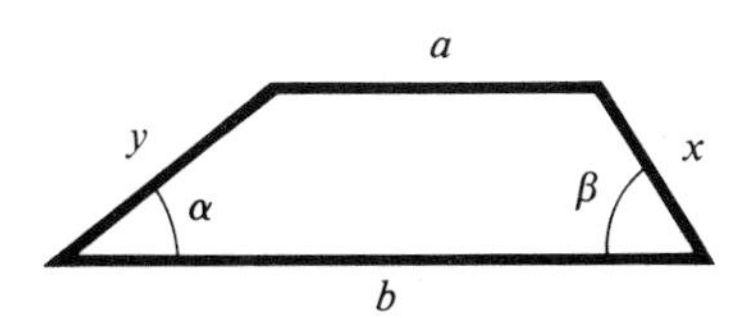

5.4 Review Exercise Set

A

1. Write in exponential form.

a. $\log_5 25 = 2$ **b.** $\log_4 256 = 4$ **c.** $\log_2 0.25 = -2$ **d.** $z = \log_r s$
e. $v = \log_k y$

2. Write in logarithmic form.

a. $4^{3/2} = 8$ **b.** $27^{-2/3} = \frac{1}{9}$ **c.** $10^3 = 1{,}000$ **d.** $p = q^s$
e. $n = a^t$

In Problems 3–5 find the value of x.

3. **a.** $\log_2 x = 4$ **b.** $\log_x 16 = -2$ **c.** $\log_3 81 = x$
4. **a.** $x = \log_3 27$ **b.** $\log_x 0.0001 = -4$ **c.** $\log_4 x = -3$
5. **a.** $x = e^{3 \ln 2}$ **b.** $x = \log_5 5^2$ **c.** $\log_2(1 - x) = 3$

In Problems 6 and 7 write each expression in a form free of logarithms of powers, roots, products, and quotients.

6. **a.** $\log \dfrac{2x^3}{\sqrt{x - 1}}$ **b.** $\log \dfrac{x(x - 2)}{2(x^2 + 1)^3}$

7. **a.** $\log \dfrac{x - 1}{(x + 2)^2 (x - 3)}$ **b.** $\log \sqrt{\dfrac{2x - 1}{3x}}$

In Problems 8 and 9 combine into a single term.

8. **a.** $4 \log x - \dfrac{1}{2} \log(x + 2) - \log(x - 1)$

b. $\log(3x + 4) - 2 \log x - \log(x + 2)$

9. **a.** $\log 2 + 3 \log x - \dfrac{1}{3} [\log(x - 1) + \log(x + 2)]$

b. $3 \log(x + 2) + \dfrac{1}{2} \log(2x - 1) - \dfrac{3}{2} \log(2 - x)$

In Problems 10–13 find the given logarithm using Table III or Table IV. Using interpolation where necessary.

10. **a.** log 2,150 **b.** log 0.0205
11. **a.** log 0.1325 **b.** log 7.534

12. **a.** $\log \sin 35° 40'$ **b.** $\log \tan 55° 10'$

13. **a.** $\log \cos 15° 32'$ **b.** $\log \sec 72° 56'$

In Problems 14–17 find x. Interpolate where necessary.

14. **a.** $\log x = 4.9474$ **b.** $\log x = 0.7931 - 1$

15. **a.** $\log x = 8.3154 - 10$ **b.** $\log x = 0.7641 - 3$

16. **a.** $\log \cos x = 9.9794 - 10$ **b.** $\log \tan x = 0.6542$

17. **a.** $\log \sin x = 9.8815 - 10$ **b.** $\log \cot x = 0.2403$

In Problems 18–21 evaluate using logarithms. Interpolate where necessary.

18. $\dfrac{(47.5)\,(8.92)}{298}$ **19.** $\dfrac{\sqrt{0.259}}{(0.0374)\,(5.89)}$

20. $\sqrt[3]{(83.75)^2}$ **21.** $\dfrac{(7.324)^2}{\sqrt{(0.04690)\,(3.215)}}$

In Problems 22–28 use logarithms to solve the right triangle ABC.

22. $A = 58° 30'$, $c = 24.7$ **23.** $B = 13° 50'$, $a = 7.54$

24. $a = 540.2$, $b = 367.5$ **25.** $A = 19° 37'$, $a = 45.72$

26. $b = 73.14$, $c = 97.42$ **27.** $B = 63.75°$, $c = 4.283$

28. $A = 21.82°$, $b = 120.5$

B In Problems 29–32 solve for x and check your answers.

29. $\log(3x + 2) - \log(2x - 1) = \log 2$

30. $\log x + \log(2x - 1) = \log 3$

31. $\log_5 (3x + 1) + \log_5 (x + 1) = 1$

32. $\log_2(2x + 1) - \log_2(x - 1) = 2$

In Problems 33 and 34 solve for x in terms of natural logarithms.

33. **a.** $3^x = 7$ **b.** $2^{x+1} = 3^{2x}$

34. **a.** $4(5^x) = 3^{x-2}$ **b.** $4^{x-1} = e^{2x+3}$

In Problems 35 and 36 find x correct to four significant figures, using common logarithms.

35. **a.** $5 \cdot 2^{3-x} = 10^{2x}$ **b.** $6^{2x+1} = \dfrac{3}{7^x}$

36. **a.** $x = (0.2047)^{-0.4256}$ **b.** $x = \dfrac{\log 752.4}{\log 123.6}$

37. From point A on the ground the angle of elevation of the top of a mountain is α, and from point B, which is on the same level as A and x meters closer to the mountain, the angle of elevation of the top is β. Derive the following formula for the height h of the mountain.

$$h = \frac{x \sin \alpha \sin \beta}{\sin(\beta - \alpha)}$$

Using logarithms, find h if $x = 420.5$ meters, $\alpha = 23° 13'$ and $\beta = 33° 42'$.

38. A ship is due north of a lighthouse at 10:00 A.M. The ship travels at an average speed of 18 miles per hour due east until 11:15 A.M. and then heads due north until 11:35 A.M. at the same speed. At that time the bearing of the lighthouse is S 32° 14′ W. Using logarithms, find the final distance and the original distance of the ship from the lighthouse.

39. In the accompanying figure, triangle ABC is a right triangle in the horizontal plane with right angle at C, and CD is a vertical line. Find h in terms of the distance c and the angles α and ϕ as shown. If $c = 3.014$, $\alpha = 24° 15'$ and $\phi = 37° 21'$, use logarithms to find h.

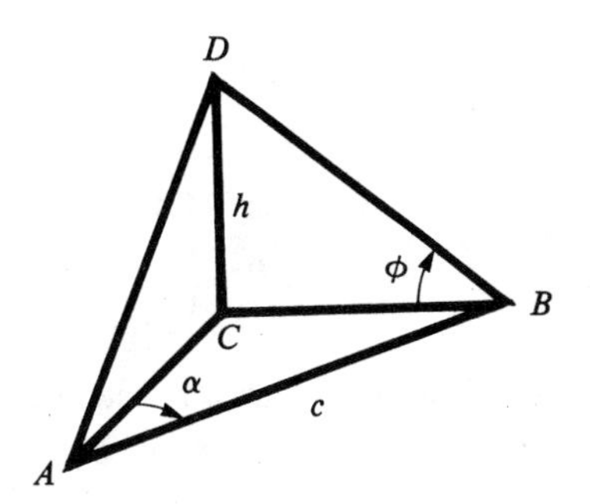

6 Oblique Triangles

The solution of oblique triangles occurs in many places in physics and engineering, as well as in the study of calculus. The following excerpt is part of a sample problem in a vector mechanics textbook.*

A hydraulic lift table is used to raise a 1,000-kg crate. It consists of a platform and two identical linkages on which hydraulic cylinders exert equal forces. . . .

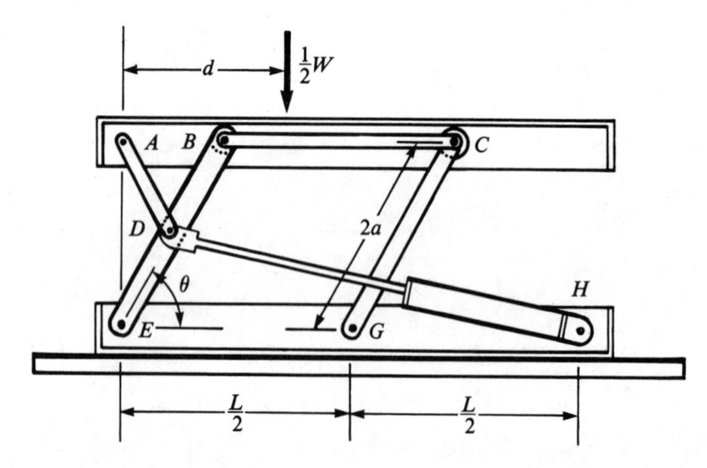

Members *EDB* and *CG* are each of length $2a$, and member *AD* is pinned to the midpoint of *EDB*. If the crate is placed on the table, so that half of its weight is supported by the system shown, determine the force exerted by each cylinder in raising the crate for $\theta = 60°$, $a = 0.70$ m, and L = 3.20 m. . . . Solution: . . .

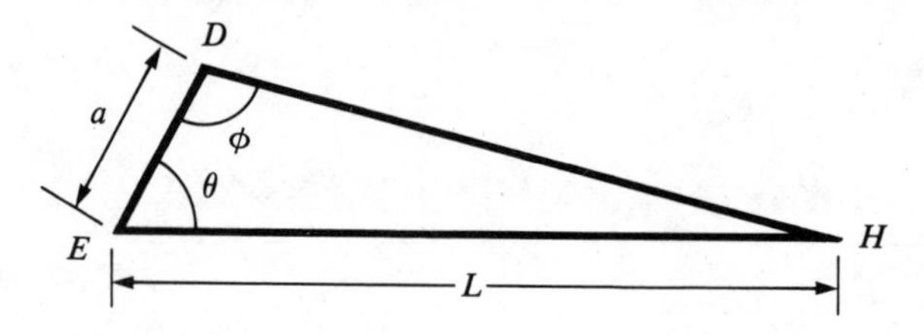

*Ferdinand P. Beer and E. Russel Johnston, Jr., *Vector Mechanics for Engineers: Statics,* 3rd ed. (New York: McGraw-Hill Book Company, 1977), pp. 254–255. Reprinted by permisssion.

Applying first the law of sines to triangles *EDH*, we write

$$\frac{\sin \phi}{EH} = \frac{\sin \theta}{DH}$$

$$\sin \phi = \frac{EH}{DH} \sin \theta$$

Using now the law of cosines, we have

$$(DH)^2 = a^2 + L^2 - 2aL \cos \theta$$

$$= (0.70)^2 + (3.20)^2 - 2(0.70)(3.20) \cos 60°$$

$$(DH)^2 = 8.49 \qquad DH = 2.91 \text{ m}$$

. . .

Substituting for $\sin \phi$ . . . and using the numerical data, we write

$$F_{DH} = W \frac{DH}{EH} \cot \theta = (9.81 \ kN) \frac{2.91 \text{ m}}{3.20 \text{ m}} \cot 60°$$

$$F_{DH} = 5.15 \ kN$$

We have omitted several parts of the solution in order to emphasize the use of the law of sines and the law of cosines, which are the primary tools for solving oblique triangles. We will study both in this chapter.

6.1 Introduction

Our purpose in this chapter is to present methods for solving oblique triangles and to use the techniques in a variety of applications. An oblique triangle, you will recall, is one which does not have a right angle. The basic approach will be to divide the triangle up in such a way that right triangle methods can be used, but we will do this in a general situation in order to derive formulas, thereby eliminating the need to go back to basic constructions in each case in the future. We will need more than one formula, depending on the nature of the given information. We distinguish four cases according to what is given: (1) two angles and one side, (2) two sides and an angle not between these sides, (3) two sides and the included angle, and (4) three sides. We consider the first two cases in Section 6.2 and the last two in Section 6.3.

6.2 The Law of Sines

Consider an oblique triangle *ABC*, as shown in Figure 6.1. The altitude from *C* divides the triangle into two right triangles *ADC* and *BDC*, as shown. From triangle *ADC* we obtain

$$\sin A = \frac{h}{b}$$

Figure 6.1

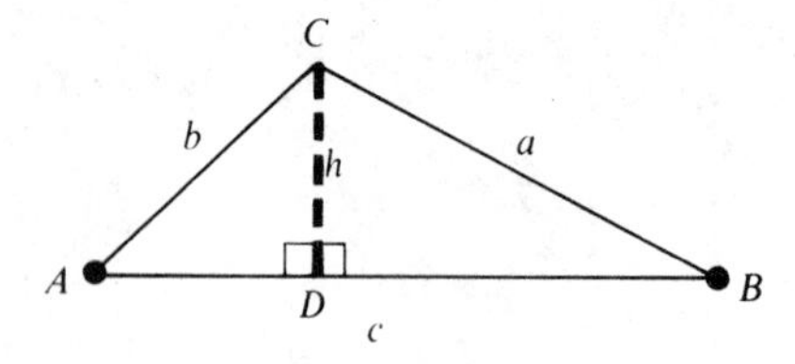

so that $h = b \sin A$. Similarly, from triangle BDC,

$$\sin B = \frac{h}{a}$$

so that $h = a \sin B$. The two expressions for h must be equal, giving $a \sin B = b \sin A$, or equivalently,

$$\frac{a}{\sin A} = \frac{b}{\sin B} \tag{6.1}$$

Now draw the altitude AD' from A to side BC extended, as shown in Figure 6.2.

Figure 6.2

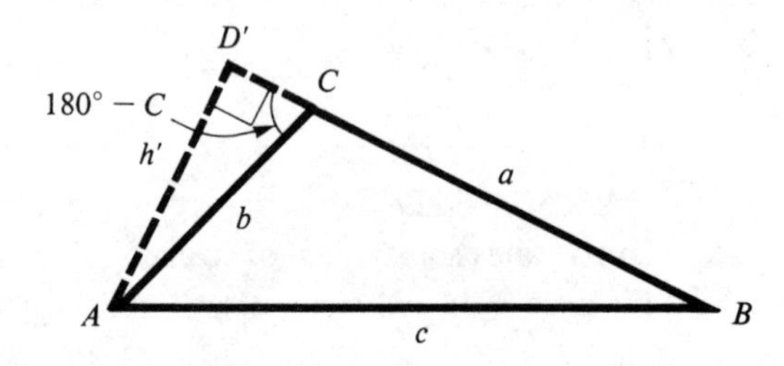

Denote the length of AD' by h'. Then from triangle $AD'C$ we have

$$\sin(180° - C) = \frac{h'}{b}$$

or $h' = b \sin(180° - C) = b \sin C$. Also, from triangle $AD'B$,

$$\sin B = \frac{h'}{c}$$

so that $h' = c \sin B$. Equating these two values of h', we get $b \sin C = c \sin B$, or equivalently,

$$\frac{b}{\sin B} = \frac{c}{\sin C} \tag{6.2}$$

When we combine Equations (6.1) and (6.2) we get the **law of sines**.

The Law of Sines

$$\frac{a}{\sin A} = \frac{b}{\sin B} = \frac{c}{\sin C}$$

Remark. The triangle we used in Figures 6.1 and 6.2 had an obtuse angle at C. An oblique triangle may not contain any obtuse angle, in which case in the second part of the proof above, the point D' will be between B and C and the proof of (6.2) follows then exactly as the proof of (6.1).

The law of sines can be used to solve oblique triangles in which the given information falls under either of the first two cases described in the previous section, namely,

Case 1. Two angles and one side are given.
Case 2. Two sides and the angle opposite one of them are given.

The next example shows how to solve Case 1 problems.

EXAMPLE 6.1 In triangle ABC, $A = 22°$, $B = 110°$, and $c = 13.4$. Solve the triangle.

Solution Since the sum of the angles must be 180°, we find that $C = 180° - (22° + 110°) = 48°$. From the law of sines we select the equation $a/\sin A = c/\sin C$. Substituting known values, we obtain

$$\frac{a}{\sin 22°} = \frac{13.4}{\sin 48°}$$

so that

$$a = \frac{13.4 \sin 22°}{\sin 48°} = 6.75$$

By calculator:

13.4 [×] 22 [SIN] [÷] 48 [SIN] [=] 6.7547108

To find b, we employ $b/\sin B = c/\sin C$.

$$\frac{b}{\sin 110°} = \frac{13.4}{\sin 48°}$$

$$b = \frac{13.4 \sin 110°}{\sin 48°} = 16.9$$

By calculator:

13.4 [×] 110 [SIN] [÷] 48 [SIN] [=] 16.944047

The triangle is now solved.

Remark. In this example we used a calculator to find sides a and b. Without a calculator it is probably easiest to use logarithms. We will illustrate the use of logarithms in Example 6.5.

Before illustrating a Case 2 problem, let us analyze the various possibilities in that situation. Suppose, for example, we are given a, b, and A. First, draw angle A with side b adjacent, thus determining vertex C, as shown in Figure 6.3. Now side a extends from C until it strikes the base. But it is quite evident that several possibilities exist, depending on the length of side a. A compass set at C with radius a clearly shows what can happen. If a exceeds the altitude from C but is shorter than b, then there are two distinct triangles and thus two distinct solutions, as Figure 6.4 shows. If a exceeds the altitude from C and also exceeds b, only one solution exists,

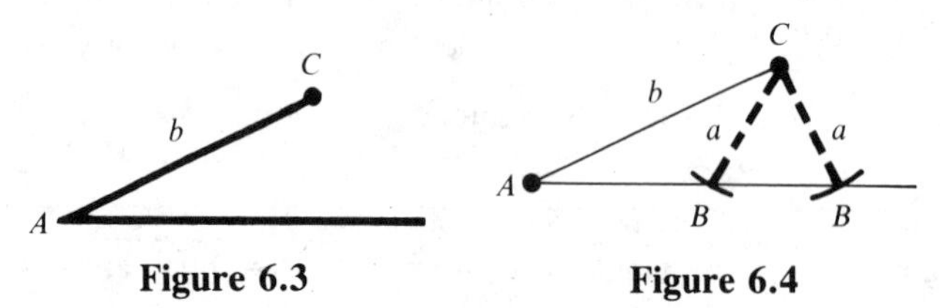

Figure 6.3 **Figure 6.4**

as shown in Figure 6.5. If a equals the altitude from C, the triangle is a right triangle, and there is again just one solution (Figure 6.6). It should be noted that while this is a theoretical possibility, in actual practice when working with data that are only approximations, it is highly unlikely that this case will occur. Finally, if a is less than the altitude from C, there is no triangle at all (Figure 6.7).

Figure 6.5

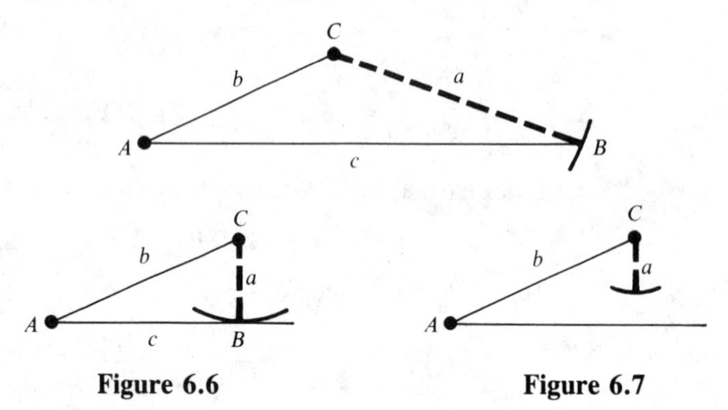

Figure 6.6 **Figure 6.7**

For obvious reasons, Case 2 is known as the **ambiguous case**. Fortunately, it is not necessary to determine in advance which situation exists; after the law of sines is employed, the situation will become clear. It is necessary, however, to be alert to the possibility of two solutions and to test this possibility. The following examples illustrate how this is done.

EXAMPLE 6.2 Solve the triangle ABC in which $a = 125$, $b = 150$, and $A = 54.0°$.

Solution From $a/\sin A = b/\sin B$, we get

$$\frac{125}{\sin 54°} = \frac{150}{\sin B}$$

so that

$$\sin B = \frac{(150)(\sin 54°)}{125} = 0.9708$$

By calculator:

150 [×] 54 [SIN] [÷] 125 [=] [INV] [SIN] 76.124805

From a calculator or tables we find that B is approximately 76.1°. But the supplement of this angle is also a possibility, since $\sin(180° - \theta) = \sin \theta$. So we must consider the alternative value of B, $180° - 76.1° = 103.9°$. In order to see if this is a feasible solution, we calculate angle C.

$$C = 180° - (A + B) = 180° - (54° + 103.9°)$$
$$= 22.1°$$

So two distinct solutions do exist. Let us designate by B_1 the value 76.1° and by B_2 the value 103.9°. Corresponding subscripts will be used for angle C and side c. So

$$C_1 = 180° - (A + B_1) = 180° - (54° + 76.1°) = 49.9°$$

and $C_2 = 22.1°$, as we have already calculated.

We use $a/\sin A = c/\sin C$ to calculate c_1 and c_2.

$$\frac{125}{\sin 54°} = \frac{c_1}{\sin 49.9°}$$

$$c_1 = \frac{(125) \sin 49.9°}{\sin 54°} = 118$$

By calculator:

125 [×] 49.9 [SIN] [÷] 54 [SIN] [=] 118.18686

and

$$\frac{125}{\sin 54°} = \frac{c_2}{\sin 22.1°}$$

$$c_2 = \frac{(125)(\sin 22.1°)}{\sin 54°} = 58.1$$

By calculator:

125 [×] 22.1 [SIN] [÷] 54 [SIN] [=] 58.129845

The two solutions are shown in Figure 6.8.

Figure 6.8

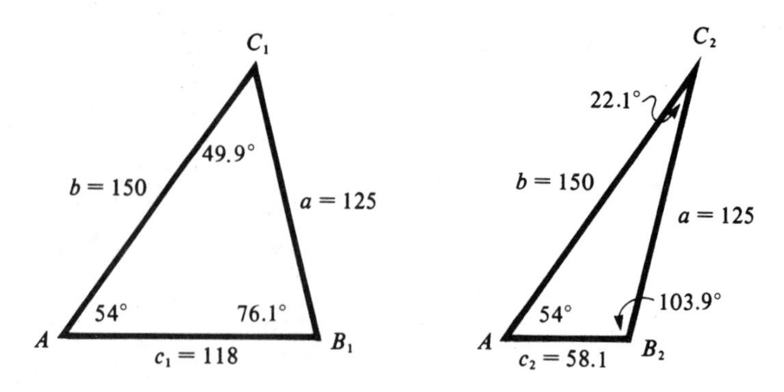

EXAMPLE 6.3 Solve the triangle ABC in which $b = 13.2$, $c = 10.5$, and $B = 42.0°$.

Solution From $b/\sin B = c/\sin C$, we find that

$$\sin C = \frac{c \sin B}{b} = \frac{(10.5)(\sin 42°)}{13.2} = 0.5323$$

so that C is either 32.2° or its supplement, 147.8°. But if $C = 147.8°$, then $B + C = 189.8°$, which is impossible since the sum of two angles of a triangle must be less than 180°. Thus, only one solution exists. Using $C = 32.2°$, we get

$$A = 180° - (B + C) = 180° - (42° + 32.2°) = 105.8°$$

Finally, we use $a/\sin A = b/\sin B$ to find side a.

$$a = \frac{b \sin A}{\sin B} = \frac{(13.2)(\sin 105.8°)}{\sin 42°} = 19.0$$

The solution is shown in Figure 6.9.

Figure 6.9

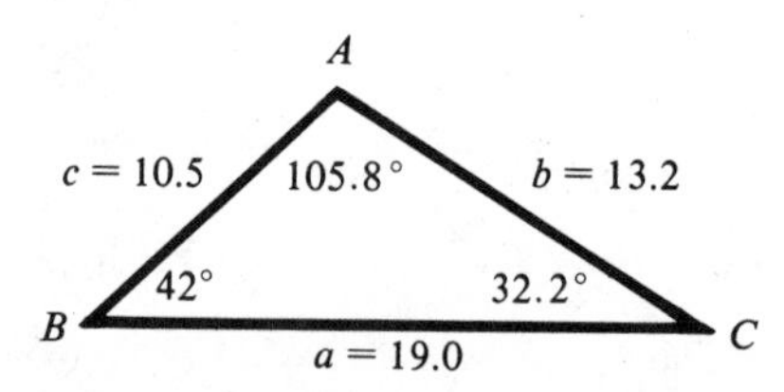

EXAMPLE 6.4 Solve the triangle ABC for which $a = 23.5$, $c = 12.0$, and $C = 35°$.

Solution From $a/\sin A = c/\sin C$, we get

$$\sin A = \frac{a \sin C}{c} = \frac{(23.5)(\sin 35°)}{12.0} = 1.123$$

But this is impossible, since it is clear from the definition that the sine never exceeds 1. So there is no solution; that is, there is no triangle having a, c, and C as given.

A Word of Caution. If the sine of the angle found in the first step of solving a Case 2 problem is less than 1, you *must* consider as possible solutions both the acute angle found from tables or a calculator *and* its supplement. The supplement will be a solution provided that it, together with the given angle, add up to an angle less than 180° (as in Example 6.2).

EXAMPLE 6.5 Use logarithms to solve the triangle ABC in which $A = 37°\ 10'$, $a = 15.80$, $b = 10.20$. (Angle A to nearest minute.)

Solution By the law of sines,

$$\frac{a}{\sin A} = \frac{b}{\sin B}$$

$$\sin B = \frac{b \sin A}{a} = \frac{10.20 \sin 37°\ 10'}{15.8}$$

$$\log \sin B = \log 10.20 + \log \sin 37°\ 10' - \log 15.80$$

$$\begin{aligned} \log 10.20 &= 1.0086 & &\text{[Table III]} \\ \log \sin 37°\ 10' &= \underline{9.7811 - 10} \ (+) & &\text{[Table IV]} \\ & 10.7897 - 10 & & \\ \log 15.80 &= \underline{1.1987} \ (-) & &\text{[Table III]} \\ \log \sin B &= 9.5910 - 10 & & \\ B &= 22°\ 57' & &\text{[Table IV, interpolating]} \end{aligned}$$

We must also check to see if the supplement of B is a feasible solution:

$$180° - 22°\ 57' = 157°\ 03'$$

But when this is added to angle A, the sum exceeds 180°. So $B = 22°\ 57'$, and there is only one solution. Next we find angle C.

$$\begin{aligned} C = 180° - (A + B) &= 180° - (37°\ 10' + 22°\ 57') \\ &= 180° - 60°\ 07' \\ &= 119°\ 53' \end{aligned}$$

To find side c we again use the law of sines.

$$\frac{a}{\sin A} = \frac{c}{\sin C}$$

$$c = \frac{a \sin C}{\sin A} = \frac{15.80 \sin 119°\ 53'}{\sin 37°\ 10'} = \frac{15.80 \sin 60°\ 07'}{\sin 37°\ 10'} \qquad \text{[Why?]}$$

$$\log c = \log 15.80 + \log \sin 60°\ 07' - \log \sin 37°\ 10'$$

$$\begin{aligned}
\log 15.80 &= 1.1987 \\
\log \sin 60^\circ\ 07' &= \underline{9.9381 - 10}\ (+) \\
& 11.1368 - 10 \\
\log \sin 37^\circ\ 10' &= \underline{9.7811 - 10}\ (-) \\
\log c &= 1.3557
\end{aligned}$$

$$c = \text{antilog } 1.3557 = 22.68$$

This completes the solution.

EXERCISE SET 6.2

A In Problems 1–15 use the law of sines to solve the triangle.

1. $a = 5.0$, $A = 40°$, $B = 20°$
2. $b = 37.4$, $A = 64.2°$, $C = 37.6°$
3. $b = 125$, $c = 85.0$, $C = 36°\ 30'$
4. $a = 10.5$, $c = 12.8$, $A = 68.0°$
5. $A = 28.5°$, $a = 21.8$, $b = 10.2$
6. $C = 30°$, $b = 15.2$, $c = 7.60$
7. $B = 47.2°$, $C = 25.8°$, $c = 7.34$
8. $A = 63.9°$, $C = 72.4°$, $a = 102$
9. $A = 37°\ 25'$, $B = 64°\ 37'$, $b = 14.72$
10. $c = 3.504$, $B = 102°\ 24'$, $C = 23°\ 42'$
11. $B = 43.25°$, $b = 17.52$, $c = 10.27$
12. $C = 23.48°$, $a = 21.83$, $c = 15.27$
13. $B = 108°\ 32'$, $b = 112.6$, $c = 73.43$
14. $A = 65.26°$, $a = 33.24$, $b = 68.73$
15. $C = 22°\ 30'$, $a = 6.432$, $c = 3.871$

16. The angle between two vectors **V** and **W** is 37.0° and the angle between **V** and **V** + **W** is 15.0°. If $|\mathbf{V}| = 12.8$, find $|\mathbf{W}|$ and $|\mathbf{V} + \mathbf{W}|$.

17. The magnitude of the resultant **R** of two forces $\mathbf{F}_1$ and $\mathbf{F}_2$ acting on an object is $|\mathbf{R}| = 223.4$ pounds, and $|\mathbf{F}_2| = 111.5$ pounds. If the angle between $\mathbf{F}_1$ and **R** is 27° 34′, and the angle between $\mathbf{F}_1$ and $\mathbf{F}_2$ is acute, find $|\mathbf{F}_1|$ and the angle between $\mathbf{F}_1$ and $\mathbf{F}_2$.

18. Points A and B are on opposite sides of a river, and it is desired to find the distance between them. Point C, on the same side of the river as A, is 35.0 feet from A. In the triangle ABC, the angle at C is found to be 42° 30′ and the angle at A to be 103° 24′. How far is it from A to B?

19. The distance x across a gorge is to be determined in order that a cable car traversing it can be constructed. One side of the gorge has only a moderate slope so that a distance

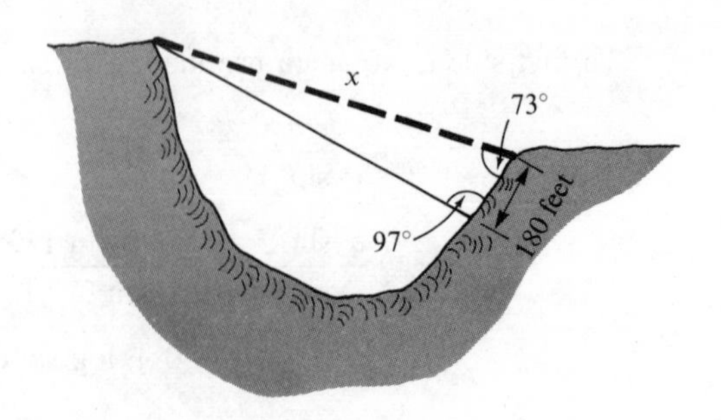

along it can be measured, as shown in the figure. The angles shown are also determined. Find the distance x.

20. Points A and B are on opposite sides of a swampy area. To find the distance between them, a point C is located 254 feet from B and 197 feet from A. The angle at A from AB to AC is found to be 47.2°. Find the distance from A to B.

B **21.** A boat leaves point A and travels in the direction S 48° E at an average speed of 15 knots (nautical miles per hour). A coast guard cutter at point B, 21 nautical miles due east of A, wishes to intercept the boat. If the cutter can average 25 knots, what should its direction be (assuming the cutter leaves at the same time as the boat)? When will the interception be made?

22. The pilot of a light airplane plans to fly from town A to town B, a distance of 650 miles. The bearing of town B from town A is S 48° 32′ E. The pilot will fly at an average airspeed of 175 miles per hour. If a constant wind is blowing from the northeast at 35 miles per hour, what should the pilot's heading be? How long will it take to make the trip?

23. Point B is on the coast 5 miles due south of A. From A the bearing of a ship is S 32.4° E, and from B the bearing of the ship is N 42.7° E. How far is it from each of the points A and B to the ship? How far is it from the ship to the nearest point on shore?

24. A pilot is planning to make a trip from A to B, 350 miles due north of A, and then back to A. There is a wind blowing from 310° at 55 miles per hour. If the average airspeed of the plane is 210 miles per hour, find the heading the pilot should take on each part of the trip. What will be the total flying time?

6.3 The Law of Cosines

The law of sines cannot be used for Case 3 (two sides and the included angle given) or for Case 4 (three sides given), since in any of the equations there will always be two unknowns. In this section we derive a formula which is suitable for these situations.

Consider first a triangle as shown in Figure 6.10, with C acute. Construct the altitude AD and designate its length by h. Then from triangle ADB and the Pythagorean theorem,

$$c^2 = h^2 + \overline{BD}^2 \tag{6.3}$$

From triangle ADC we have

$$b^2 = h^2 + \overline{CD}^2 \tag{6.4}$$

Figure 6.10

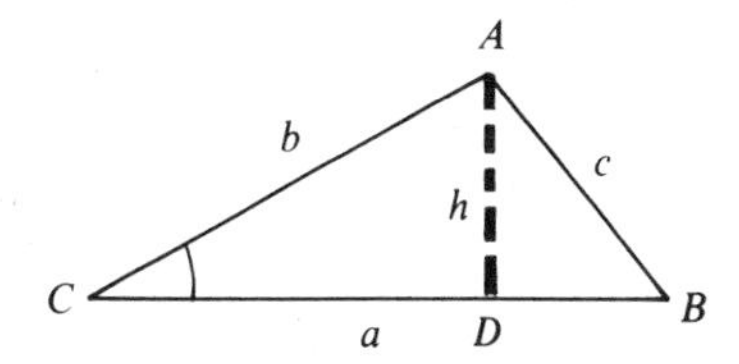

and since

$$\cos C = \frac{\overline{CD}}{b}$$

we obtain $\overline{CD} = b \cos C$. Substituting this in (6.4) gives $b^2 = h^2 + (b \cos C)^2$, or $h^2 = b^2 - (b \cos C)^2$. Also, $\overline{BD} = a - \overline{CD} = a - b \cos C$. Making these substitutions in (6.3) we obtain

$$\begin{aligned} c^2 &= b^2 - b^2 \cos^2 C + (a - b \cos C)^2 \\ &= b^2 - b^2 \cos^2 C + a^2 - 2ab \cos C + b^2 \cos^2 C \\ &= a^2 + b^2 - 2ab \cos C \end{aligned}$$

If C is obtuse, we draw the altitude AD to side BC extended, as in Figure 6.11.

Figure 6.11

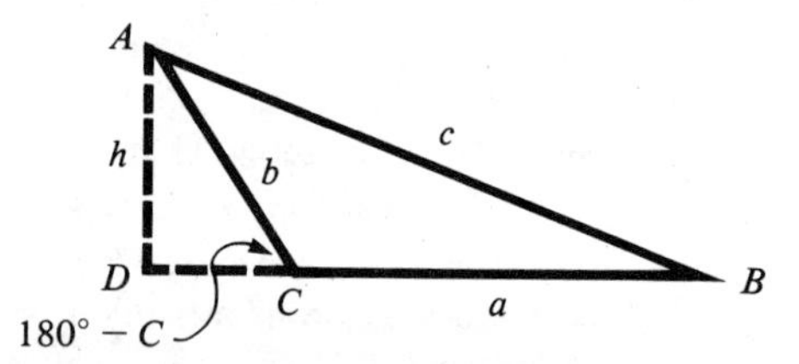

Then Equations (6.3) and (6.4) remain valid. From triangle ADC,

$$\cos(180° - C) = \frac{\overline{CD}}{b}$$

or $\overline{CD} = b \cos(180° - C) = -b \cos C$. Also, since $\overline{BD} = a + \overline{CD}$, we have $\overline{BD} = a - b \cos C$. The remainder of the derivation is identical to the previous case. So the same equation

$$c^2 = a^2 + b^2 - 2ab \cos C$$

is obtained. By starting from the other vertices, analogous formulas are obtained. We have, then, the following three formulas expressing what is known as the **law of cosines:**

$$\mathbf{a^2 = b^2 + c^2 - 2bc \cos A}$$

$$\mathbf{b^2 = a^2 + c^2 - 2ac \cos B}$$

$$\mathbf{c^2 = a^2 + b^2 - 2ab \cos C}$$

These can all be summarized in the following words:

The Law of Cosines

The square of any side of a triangle equals the sum of the squares of the other two sides, minus twice the product of these two sides and the cosine of the angle between them.

If two sides and the angle between them are given, we select the form of the law of cosines that involves the given angle. The side opposite the angle can then be calculated. To complete the solution the law of sines is used. For example, if a, c, and angle B are given, we select the formula

$$b^2 = a^2 + c^2 - 2ac \cos B$$

and calculate b. Then either angle A or angle C can be found using the law of sines.

If all three sides of a triangle are given, any one of the forms of the law of cosines can be used to calculate one of the angles. A second angle can then be found either by another application of the law of cosines or by the law of sines. For example, if a, b, and c are given, we can use

$$a^2 = b^2 + c^2 - 2bc \cos A$$

and solve for $\cos A$:

$$\cos A = \frac{b^2 + c^2 - a^2}{2bc}$$

Angle A can then be found by using a calculator or tables.

EXAMPLE 6.6 Solve the triangle ABC in which $a = 3$, $b = 5$, and $C = 120°$.

Solution To find side c we use

$$\begin{aligned} c^2 = a^2 + b^2 - 2ab \cos C &= (3)^2 + (5)^2 - 2(3)(5) \cos 120° \\ &= 9 + 25 - 30(-\tfrac{1}{2}) \\ &= 49 \end{aligned}$$

So $c = 7$.

Now we use the law of sines in the form of the equation $a/\sin A = c/\sin C$. Solving for $\sin A$ gives

$$\sin A = \frac{a \sin C}{c} = \frac{3\left(\frac{\sqrt{3}}{2}\right)}{7} = \frac{3\sqrt{3}}{14} = 0.3712$$

from which $A = 21.8°$. Finally,

$$B = 180° - (A + C) = 180° - 141.8° = 38.2°$$

EXAMPLE 6.7 Find all angles in the triangle ABC in which $a = 13$, $b = 21$, $c = 15$.

Solution By the law of cosines,

$$\cos A = \frac{b^2 + c^2 - a^2}{2bc} = \frac{441 + 225 - 169}{2(21)(15)} = 0.7889$$

from which $A = 37.9°$.

By calculator:

21 $\boxed{x^2}$ $\boxed{+}$ 15 $\boxed{x^2}$ $\boxed{-}$ 13 $\boxed{x^2}$ $\boxed{=}$
$\boxed{\div}$ $\boxed{(}$ 2 $\boxed{\times}$ 21 $\boxed{\times}$ 15 $\boxed{)}$ $\boxed{=}$ $\boxed{\text{INV}}$ $\boxed{\text{COS}}$ 37.918203

Similarly,

$$\cos B = \frac{a^2 + c^2 - b^2}{2ac} = \frac{169 + 225 - 441}{2(13)(15)} = -0.1205$$

from which $B = 96.9°$. Finally,

$$C = 180° - (A + B) = 180° - 134.8° = 45.2°$$

EXAMPLE 6.8 Find the magnitude and direction of the resultant of the two forces $\mathbf{F}_1$ and $\mathbf{F}_2$ shown in Figure 6.12 if $|\mathbf{F}_1| = 87.2$ pounds and $|\mathbf{F}_2| = 53.8$ pounds.

Solution We complete the parallelogram, as shown in Figure 6.13. The angle θ is the supplement of 39.7°,

$$\theta = 180° - 39.7° = 140.3°$$

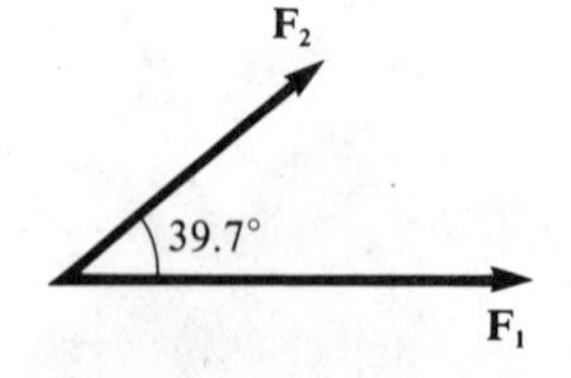

Figure 6.12

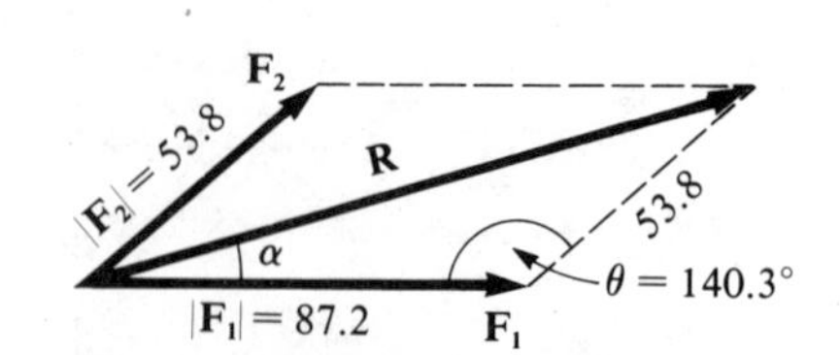

Figure 6.13

By the law of cosines,

$$\begin{aligned} |\mathbf{R}|^2 &= |\mathbf{F}_1|^2 + |\mathbf{F}_2|^2 - 2|\mathbf{F}_1|\,|\mathbf{F}_2| \cos \theta \\ &= (87.2)^2 + (53.8)^2 - 2(87.2)(53.8) \cos 140.3° \\ &= 17{,}717.34 \end{aligned}$$

so $|\mathbf{R}| = 133.1$ pounds. We indicate the direction of $\mathbf{R}$ by finding the angle α by the law of sines.

$$\frac{|\mathbf{F}_2|}{\sin \alpha} = \frac{|\mathbf{R}|}{\sin \theta}$$

$$\sin \alpha = \frac{|\mathbf{F}_2| \sin \theta}{|\mathbf{R}|} = \frac{(53.8) \sin 140.3°}{57.3} = 0.2582$$

Thus, $\alpha = 15.0°$.

EXERCISE SET 6.3

A Solve the triangles in Problems 1–12.

1. $A = 60°, b = 15, c = 10$

2. $B = 150°, a = 2.3, c = 1.2$

3. $C = 97° 20', a = 5, b = 11$

4. $a = 3, b = 4, c = 2$

5. $a = 12, b = 8, c = 15$

6. $A = 32.4°, b = 12.3, c = 15.8$

7. $B = 103.5°, a = 234, c = 160$

8. $C = 67° 42', a = 35.2, b = 42.3$

9. $a = 3.61, b = 4.72, c = 2.85$

10. $a = 21.5, b = 32.6, c = 50.7$

11. $b = 30.75, c = 25.02, A = 84° 23'$

12. $a = 0.231, c = 0.415, B = 102.4°$

13. Two sides of a parallelogram are 8 centimeters and 10 centimeters, respectively, and the shorter diagonal is 12 centimeters. Find the length of the longer diagonal.

14. Two cars leave simultaneously from the same point, one going east at an average speed of 60 kilometers per hour and the other going southwest at an average speed of 80 kilometers per hour. How far apart are they after two hours?

15. Find the magnitude and direction of the resultant of two forces of 25 pounds and 30 pounds, respectively, acting on an object, if the angle between the forces is 23°.

16. If the angle between the vectors **V** and **W** is 47.2°, and $|\mathbf{V}| = 21.5$, $|\mathbf{W}| = 37.2$, find $|\mathbf{V} + \mathbf{W}|$ and the angle from **V** to **V** + **W**.

17. A river that flows north to south has a current of 8 miles per hour. A motorboat leaves the west bank and heads in the direction S 75° E at an indicated speed of 20 miles per hour. Find the actual speed and direction of the boat.

18. A light airplane is heading in the direction 175° at an indicated airspeed of 160 miles per hour. If a wind of 43 miles per hour is blowing from 200°, find the true course and speed of the airplane.

19. Let $\mathbf{V}_1$ and $\mathbf{V}_2$ be two vectors in standard position. If the terminal points of $\mathbf{V}_1$ and $\mathbf{V}_2$ are (5, 3) and (−2, 7), respectively, find the angle between $\mathbf{V}_1$ and $\mathbf{V}_2$.

20. Repeat Problem 19 if the terminal points of $\mathbf{V}_1$ and $\mathbf{V}_2$ are (−3, 5) and (−1, −4), respectively.

21. A surveyor wishes to find the distance from point A to point B but cannot do so directly because there is a swamp between them. He sets up a transit at C and makes the following measurements: Angle $C = 27° 32'$, $\overline{CA} = 125.3$ feet, $\overline{CB} = 117.5$ feet. Find $\overline{AB}$.

B **22.** A regular pentagon is inscribed in a circle of radius 4 centimeters. Make use of the law of cosines to find the perimeter of the pentagon.

23. Airplane A leaves Chicago at 1:00 P.M. and flies at a heading of 203° at an average speed of 350 miles per hour. Airplane B leaves the same airport at 1:30 P.M. and flies at a heading of 85° at an average speed of 400 miles per hour. How far apart are the planes at 2:00 P.M.? (Neglect wind velocity.)

24. Let $\mathbf{V}_1$ and $\mathbf{V}_2$ be vectors in standard position having terminal points (x_1, y_1) and (x_2, y_2), respectively. Recall that the dot product $\mathbf{V}_1 \cdot \mathbf{V}_2$ equals the quantity $x_1 x_2 + y_1 y_2$. Use the law of cosines to prove that if θ is the angle between $\mathbf{V}_1$ and $\mathbf{V}_2$, then

$$\cos \theta = \frac{\mathbf{V}_1 \cdot \mathbf{V}_2}{|\mathbf{V}_1|\,|\mathbf{V}_2|}$$

25. Three forces of magnitude $|\mathbf{F}_1| = 30$ pounds, $|\mathbf{F}_2| = 45$ pounds, and $|\mathbf{F}_3| = 56$ pounds are acting on an object as shown. Find the magnitude of the resultant, and give its angle from $\mathbf{F}_1$.

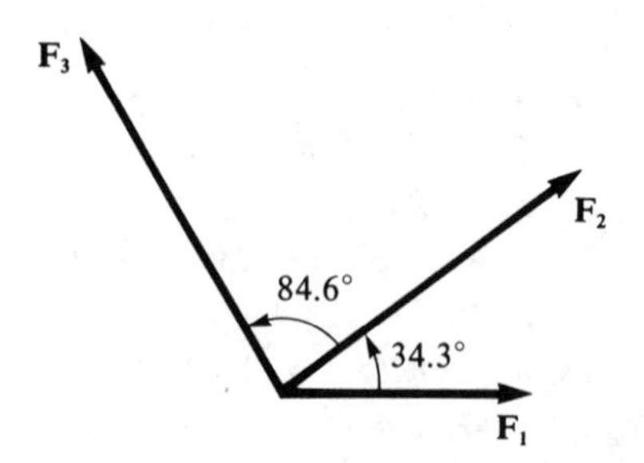

26. The heading of a jetliner flying at an indicated airspeed of 465 miles per hour is 220°. If a wind of 90 miles per hour is blowing from 300°, find the ground speed of the plane and its actual course.

27. A boat traveling at 23 miles per hour heads in the direction N 42.5° W for 45 munutes and then changes to the direction N 36.8° E and goes on this course for 1 hour and 15 minutes. How far is the boat from its starting point?

28. In order to determine the distance between two inaccessible points A and B on the opposite side of a river from an observer, two points C and D are established 50 feet apart, and angles are determined as shown. Find the distance between A and B.

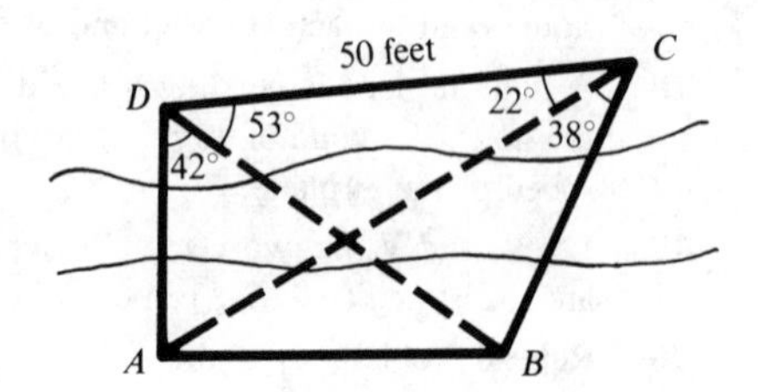

6.4 Areas of Triangles

The basic formula for the area of a triangle is well known: Area $= \frac{1}{2}bh$, where b is the length of the base, and h the length of the altitude from that base. For a right triangle either leg can be taken as the base, and the other leg is then the altitude. For oblique triangles the complication arises of how to find the altitude. The method of deriving the law of sines gives a clue as to how this can be done. Suppose an oblique triangle is oriented as shown in Figure 6.14. Since $\sin A = h/b$, we have $h = b \sin A$. The base is c, so we have

$$\textbf{Area} = \frac{1}{2} bc \sin A \tag{6.5}$$

Notice that the formula involves two sides and the angle between them. By considering altitudes drawn to the other two sides we could in a similar way obtain the formulas

$$\textbf{Area} = \frac{1}{2}\, ab \sin C = \frac{1}{2}\, ac \sin B$$

If enough information to solve the triangle is given, then the area can always be found by one of these formulas.

Figure 6.14

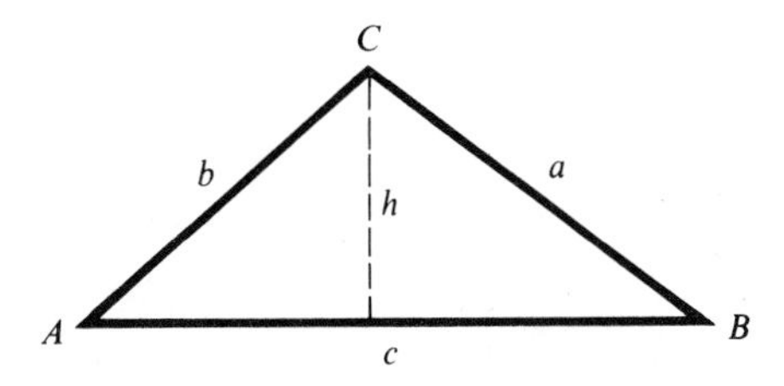

EXAMPLE 6.9 Find the area of the triangle ABC for which $A = 60°$, $b = 10$, and $c = 7$.

Solution By Equation (6.5), we have

$$\text{Area} = \frac{1}{2}\, bc \sin A = \frac{1}{2}\,(10)(7) \cdot \frac{\sqrt{3}}{2} = \frac{35\sqrt{3}}{2} \approx 30.31$$

EXAMPLE 6.10 Find the area of the triangle having sides, 3, 5, and 7.

Solution Let $a = 3$, $b = 5$, and $c = 7$. We will first use the law of cosines to find an angle. We could work with any angle, but a little scratch work will show there is an advantage in choosing angle C.

$$\cos C = \frac{a^2 + b^2 - c^2}{2ab} = \frac{9 + 25 - 49}{2(3)(5)} = \frac{-15}{30} = -\frac{1}{2}$$

So $C = 120°$. The area can now be found.

$$\text{Area} = \frac{1}{2}\, ab \sin C = \frac{1}{2}\,(3)(5)\left(\frac{\sqrt{3}}{2}\right) = \frac{15\sqrt{3}}{4} \approx 6.50$$

By using (6.5) or one of the analogous forms, together with the law of cosines, we can derive an interesting formula for the area of a triangle which involves only the sides of the triangle. We begin by squaring both sides of (6.5).

$$(\text{Area})^2 = \frac{1}{4}\, b^2c^2 \sin^2 A = \frac{b^2c^2}{4}\,(1 - \cos^2 A)$$

$$= \frac{b^2c^2}{4}\,(1 + \cos A)(1 - \cos A)$$

Now by the law of cosines,

$$\cos A = \frac{b^2 + c^2 - a^2}{2bc}$$

So

$$1 + \cos A = 1 + \frac{b^2 + c^2 - a^2}{2bc} = \frac{2bc + b^2 + c^2 - a^2}{2bc}$$

$$= \frac{(b + c)^2 - a^2}{2bc}$$

and

$$1 - \cos A = 1 - \frac{b^2 + c^2 - a^2}{2bc} = \frac{2bc - b^2 - c^2 + a^2}{2bc}$$

$$= \frac{a^2 - (b - c)^2}{2bc}$$

This gives

$$(\text{Area})^2 = \frac{b^2c^2}{4}\left[\frac{(b + c)^2 - a^2}{2bc}\right]\left[\frac{a^2 - (b - c)^2}{2bc}\right]$$

We now factor each of the numerators in brackets as the difference of two squares, and we divide out the factor b^2c^2 from numerator and denominator.

$$(\text{Area})^2 = \frac{1}{16}[(b + c) + a][(b + c) - a][a + (b - c)][a - (b - c)]$$

$$= \left(\frac{a + b + c}{2}\right)\left(\frac{b + c - a}{2}\right)\left(\frac{a + b - c}{2}\right)\left(\frac{a + c - b}{2}\right)$$

This result can be put in a simpler form by introducing a notation for the **semi-perimeter**. It is customary to use the letter s for this:

$$s = \frac{a + b + c}{2}$$

Then we note that

$$s - a = \frac{a + b + c}{2} - a = \frac{a + b + c - 2a}{2} = \frac{b + c - a}{2}$$

and in a similar way we can see that

$$s - b = \frac{a + c - b}{2} \quad \text{and} \quad s - c = \frac{a + b - c}{2}$$

Making these substitutions and taking the square root gives

$$\textbf{Area} = \sqrt{s(s - a)(s - b)(s - c)} \tag{6.6}$$

where $s = \frac{1}{2}(a + b + c)$.

We illustrate this by reworking Example 6.10. There we had $a = 3$, $b = 5$, and $c = 7$. So

$$s = \frac{1}{2}(3 + 5 + 7) = \frac{15}{2}$$

and

$$s - a = \frac{15}{2} - 3 = \frac{9}{2}, \qquad s - b = \frac{15}{2} - 5 = \frac{5}{2}, \qquad s - c = \frac{15}{2} - 7 = \frac{1}{2}$$

Thus, by (6.6)

$$\text{Area} = \sqrt{\frac{15}{2} \cdot \frac{9}{2} \cdot \frac{5}{2} \cdot \frac{1}{2}} = \frac{15}{4}\sqrt{3}$$

In Problem 16 of Exercise Set 6.4 we ask for the derivation of still another formula for the area, which is particularly convenient when the given information involves two angles and a side.

EXERCISE SET 6.4

A Find the area of triangle ABC in Problems 1–14.

1. $A = 30°$, $b = 36$, $c = 24$

2. $C = 23°$, $a = 7$, $b = 9$

3. $B = 112°$, $a = 14$, $c = 10$

4. $a = 1.2$, $b = 2.1$, $c = 2.8$

5. $A = 65.2°$, $b = 3.5$, $c = 2.8$

6. $C = 93° \, 20'$, $a = 12.3$, $b = 18.2$

7. $a = 10$, $b = 6$, $c = 14$

8. $C = 102.5°$, $a = 24$, $b = 13$

9. $A = 30°$, $a = 12$, $b = 10$

10. $B = 98° \, 33'$, $b = 4.03$, $c = 2.52$

11. $A = 69.5°$, $C = 41.7°$, $b = 54.6$

12. $B = 36.8°$, $C = 101.3°$, $a = 14.9$

13. $a = 1.023$, $b = 2.142$, $c = 1.597$

14. $C = 43° \, 15'$, $b = 25.1$, $c = 32.3$

B **15.** Show that Equation (6.5) continues to hold true when A is an obtuse angle.

16. Derive the following formula for the area of triangle ABC.

$$\text{Area} = \frac{b^2 \sin A \sin C}{2 \sin B}$$

Hint. Use Equation (6.5) and the law of sines.

17. Use the result of Problem 16 to find the area of the triangle ABC in which $b = 8$, $B = 32°$, $C = 54°$.

18. Rework Problem 11 using the result of Problem 16.

19. Rework Problem 12 using a formula analogous to the one derived in Problem 16.

20. Find the area of the triangle whose vertices in a rectangular coordinate system are (2, 1), (−4, −7), and (−1, 5).

21. Let the lengths of the two diagonals of an arbitrary quadrilateral be d_1 and d_2, and let α be their angle of intersection (either of the two angles). Prove that the area of the quadrilateral is given by $\frac{1}{2} d_1 d_2 \sin \alpha$.

6.5 Review Exercise Set

A In Problems 1–15 solve the triangle ABC and find its area.

1. $a = 4$, $b = 7$, $C = 110°$
2. $b = 15$, $c = 8$, $C = 32°$
3. $a = 8$, $b = 9$, $c = 13$
4. $A = 110°$, $C = 28°$, $c = 12$
5. $A = 34.3°$, $B = 26.8°$, $a = 13.4$
6. $B = 54° 30'$, $C = 48° 21'$, $a = 102.5$
7. $a = 22.3$, $b = 17.9$, $c = 10.5$
8. $A = 63.2°$, $a = 14.3$, $b = 25.2$
9. $B = 36° 20'$, $b = 34.54$, $c = 12.73$
10. $a = 2.05$, $c = 3.72$, $B = 61° 40'$
11. $A = 25.2°$, $c = 27.4$, $a = 15.1$
12. $B = 43.6°$, $a = 1.98$, $b = 1.53$
13. $C = 132° 13'$, $a = 32.52$, $b = 27.85$
14. $a = 15.1$, $b = 13.9$, $c = 12.3$
15. $C = 50.3°$, $b = 20.5$, $c = 16.8$
16. The sides of a parallelogram are 6 and 10 units, and the longer diagonal is 14 units. Find the interior angles of the parallelogram.
17. A lot is in the shape of a triangle in which two sides are 100 feet and 120 feet, and the angle between them is 55°. Find the length of the third side and the area of the lot.
18. A portion of a modern metal sculpture consists of a large triangular plate mounted vertically, with a horizontal base and with one vertex at the top. The length of the base is 12 feet, and the base angles are 30° and 50°, respectively. Find the lengths of the other two sides of the triangle. What is its area?
19. The angle between two vectors $\mathbf{V}_1$ and $\mathbf{V}_2$ is 27.5°, and the angle from $\mathbf{V}_1$ to $\mathbf{V}_1 + \mathbf{V}_2$ is 16.1°. If $|\mathbf{V}_1| = 10.7$, find $|\mathbf{V}_2|$ and $|\mathbf{V}_1 + \mathbf{V}_2|$.
20. Two forces, of magnitudes $|\mathbf{F}_1| = 8.67$ newtons and $|\mathbf{F}_2| = 6.43$ newtons, act on an object. If the angle between $\mathbf{F}_1$ and $\mathbf{F}_2$ is $72° 33'$, find the magnitude and direction of the resultant.
21. Two forces, of magnitudes $|\mathbf{F}_1| = 36.5$ pounds and $|\mathbf{F}_2| = 53.4$ pounds, are acting on an object. If the resultant $\mathbf{R}$ makes an angle of $32° 15'$ with $\mathbf{F}_1$, find $|\mathbf{R}|$. What is the angle between $\mathbf{F}_1$ and $\mathbf{F}_2$?
22. A river flows from west to east at a rate of 4.25 miles per hour. A boat pilot leaves point A on the south bank of the river and arrives at point B on the north bank. The boat was headed in the direction N 47.3° W, and the indicated speed of the boat was 12.5 miles per hour. Find the actual speed of the boat and the bearing of B from A.
23. Points A and B are on opposite sides of a pond, and it is desired to find the distance between then. Point C is located so that $\overline{AC} = 198.7$ feet and $\overline{BC} = 254.3$ feet. The angle at A from AB to AC is found to be $47° 21'$. Find the distance from A to B.
24. Two guy wires are attached on opposite sides of a pole. They make angles of 36° and 42°, respectively, with the horizontal, and the points where they meet the ground are 50 feet apart. Find the length of each wire.

B

25. An airplane pilot flies for 30 minutes at a compass heading of 140° and then changes to a heading of 36° and flies at the new heading for 2 hours and 15 minutes. His airspeed throughout the trip was 385 miles per hour, and a wind was blowing at an average of 63.5 miles per hour from 250°. What was the actual speed and direction of the airplane on each leg of the trip? At the end of this time how far was the airplane from its starting point?

26. Prove that the area of the parallelogram having the vectors **V** and **W** as adjacent sides is $|\mathbf{V}|\,|\mathbf{W}|\sin\theta$, where θ is the angle between **V** and **W**.

27. If there were no wind, a pilot would head in the direction 215° to go from city A to city B. The distance between the cities is 425 miles. If a 40 mile per hour wind is blowing from 290°, and if the average airspeed of the plane is 185 miles per hour, find the heading the pilot should take to go from A to B. If she also wishes to return to A, what should her heading be? What will the total flying time be for the round trip?

28. Find the area of the triangle whose vertices are the points $(3, -5)$, $(-1, -2)$, and $(8, 7)$.

29. A ranger in a lookout tower spots a fire in the direction N 40° W. A second ranger in a tower 10 miles from the first tower and in the direction S 75° W from it also sees the fire and finds it to be in the direction N 35° E. How far is the fire from each lookout tower?

30. A hill slopes at an angle of 23° 10′, and a monument is erected on top of the hill, as shown in the figure. From point A at the base of the hill the angle of elevation of the top of the monument is 34° 37′, and from point B, 200 feet up the hill from A, the angle of elevation of the top of the monument is 41° 06′. How high is the monument?

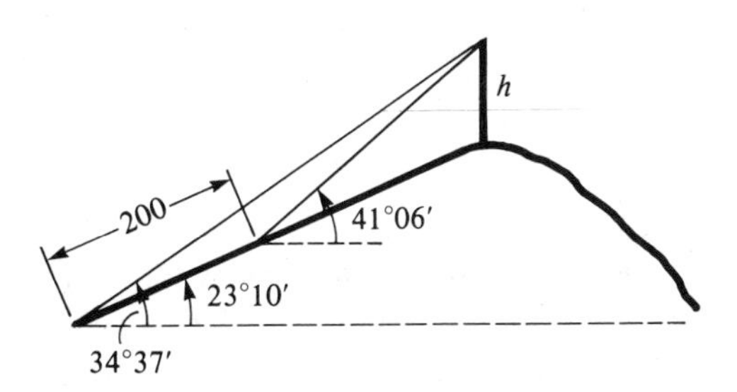

31. In the accompanying figure, triangle ABC is in the horizontal plane and line CD is vertical. If side c and angles α, β, and ϕ are known, derive the following formula for the length h of side CD:

$$h = \frac{c \sin \alpha \tan \phi}{\sin(\alpha + \beta)}$$

If $c = 21.34$, $\alpha = 18°\ 37'$, $\beta = 43°\ 16'$, and $\phi = 56°\ 42'$, find h. Also find angle θ.

32. From the top of a lighthouse 120 feet high the angle of depression of ship A is 12.8° and the angle of depression of ship B is 14.2°. The bearing of ship A from the lighthouse is N 35.7° E and the bearing of B is S 64.3° E. Find the distance between the ships and the bearing of ship B from ship A.

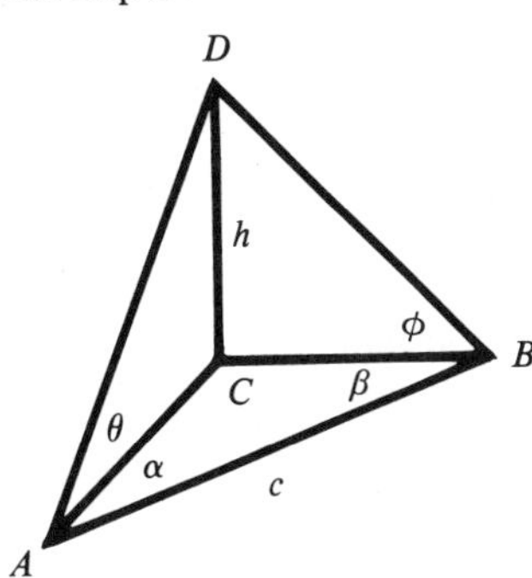

7 Complex Numbers

Complex numbers have important applications in physics and chemistry, especially in the study of vibrations. Also they are particularly useful in aeronautical and electrical engineering and in hydrodynamics. The following excerpt from a physics textbook illustrates one type of application.*

1. We start with the physical equation of motion . . . :

$$\frac{md^2x}{dt^2} + kx = F_0 \cos \omega t$$

2. We imagine the driving force $F_0 \cos \omega t$ as being the projection on the x axis of a rotating vector $F_0 \exp(j\omega t)$, as shown in . . . (a), and we imagine x as being the projection of a vector z that rotates at the same frequency ω [(b)].

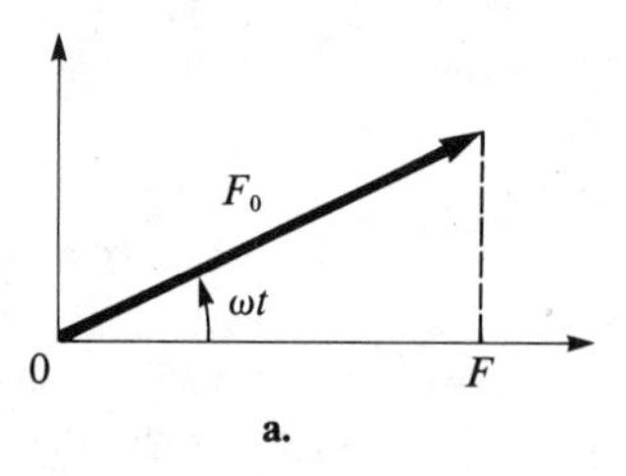

a.

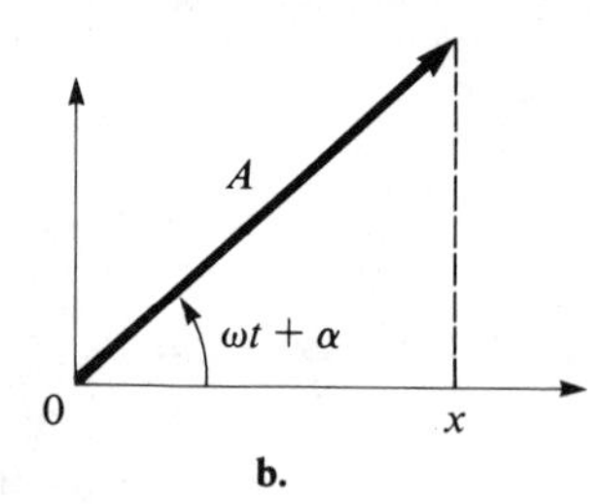

b.

3. We then write the differential equation that governs z:

$$\frac{md^2z}{dt^2} + kz = F_0e^{j\omega t}$$

4. We try the solution

$$z = Ae^{j(\omega t + \alpha)}$$

Substituting . . . , this gives us

$$(-m\omega^2A + kA)e^{j(\omega t + \alpha)} = F_0e^{j\omega t}$$

---

*Example is reprinted from *Vibrations and Waves* by A. P. French, with the permission of W. W. Norton & Company, Inc., Copyright © 1971, 1966 by the Massachusetts Institute of Technology.

which can be rewritten as follows:

$$(\omega_0^2 - \omega^2)A = \frac{F_0}{m} e^{-j\alpha}$$

$$= \frac{F_0}{m} \cos \alpha - j\frac{F_0}{m} \sin \alpha$$

This contains two conditions, corresponding to the real and imaginary parts on the two sides of the equation:

$$(\omega_0^2 - \omega)A = \frac{F_0}{m} \cos \alpha$$

$$0 = -\frac{F_0}{m} \sin \alpha$$

The letter j used here has the same meaning as the letter i we will use in this chapter, namely the so-called "imaginary unit." It is shown in calculus that $e^{j\omega t} = \cos \omega t + j \sin \omega t$, and the right-hand side of this equation is an example of the trigonometric form of a complex number, which will be our primary concern in this chapter.

7.1 Introduction

Trigonometry provides a useful means of representing the complex numbers studied in algebra. We will see this relationship to trigonometry in Section 7.3 and explore the consequences in Sections 7.4 and 7.5. We begin by recalling some basic definitions and properties.

The numbers we have dealt with thus far have all been **real**. The real numbers can be characterized as being the set of all decimal quantities. Those numbers having either a finite decimal representation, such as 1.25, or a repeating decimal representation, such as 0.272727 . . . , are called **rational**. It is always possible to express rational numbers in the form m/n, where m and n are integers, with $n \neq 0$. All other real numbers are **irrational**. Some irrational numbers are those represented by $\sqrt{2}$, π, and sin 35°, for example. It is characteristic of irrational numbers that they have infinite, non-repeating decimal representations. We can associate real numbers in a one-to-one manner with points on a line. In fact, both the x axis and the y axis in a rectangular coordinate system make use of such an association. We designate the set of all real numbers by $\mathbb{R}$.

In studying quadratic equations in algebra, it quickly becomes apparent that there are equations whose solutions cannot be found in $\mathbb{R}$. The simplest such equation is $x^2 + 1 = 0$. The complex numbers were invented to remedy this situation. It is a remarkable fact that by enlarging the number system so that all quadratic equations can be solved, we get the bonus that the solutions of *all* polynomial equations of every degree lie within the new system. (That is not to say you can actually *find* these solutions, but they can be shown to exist.) In fact, the solutions of an enormous variety of equations, both algebraic and non-algebraic, exist within the complex number system.

7.2 Definitions and Properties

To define the complex number system we begin by introducing the symbol i for $\sqrt{-1}$ and call this the **imaginary unit**. Thus, $i^2 = -1$. If a and b are any two real numbers, we call $a + bi$ a **complex number**. The real number a is called the **real part** of the complex number $a + bi$, and b is called the **imaginary part** (note that the imaginary part is the real number b, and not bi). We designate the set of all complex numbers by $\mathbb{C}$. Thus,

$$\mathbb{C} = \{a + bi \colon a, b \in \mathbb{R}, i^2 = -1\}.$$

Two complex numbers $a+bi$ and $c+di$ are equal if and only if $a=c$ and $b=d$, that is, if and only if their real parts are equal and their imaginary parts are equal. Complex numbers are added by adding their real parts and their imaginary parts, individually, as follows:

$$(a + bi) + (c + di) = (a + c) + (b + d)i \tag{7.1}$$

Multiplication is defined by the following:

$$(a + bi)(c + di) = (ac - bd) + (ad + bc)i \tag{7.2}$$

This will not appear mysterious if you multiply as in elementary algebra and then use the fact that $i^2 = -1$. Doing multiplication in this way, in fact, is probably easier than using Equation (7.2) directly. For example,

$$(2 + 3i)(4 - 7i) = 8 - 2i - 21i^2 = 8 - 2i + 21 = 29 - 2i$$

The real numbers can be considered as a subset of the complex numbers, since any real number a can be written in the form $a + 0i$. It can easily be verified that when (7.1) and (7.2) are applied to two numbers of this form, the results correspond to the ordinary sum and product of real numbers, that is, $(a + 0i) + (b + 0i) = (a + b) + 0i$ and $(a + 0i)(b + 0i) = ab + 0i$. (You should check to see that these are true.) To distinguish between complex numbers that are real and those which are not, we sometimes refer to the latter as **imaginary**. So a complex number of the form $a + bi$ with $b \neq 0$ is imaginary. Those of the form $0 + bi$ are called **pure imaginary**.

The number $0 = 0 + 0i$ is the zero element (or the **additive identity**) for $\mathbb{C}$, and $1 = 1 + 0i$ is the unity element (or **multiplicative identity**). The negative of $a + bi$ (also called the **additive inverse**) is $(-a) + (-b)i$. Subtraction of $c + di$ from $a + bi$ is defined as the sum of $a + bi$ and the additive inverse of $c + di$. So

$$\begin{aligned}(a + bi) - (c + di) &= (a + bi) + [(-c) + (-d)i] \\ &= [a + (-c)] + [b + (-d)]i \\ &= (a - c) + (b - d)i\end{aligned}$$

From the definitions given so far, you can see that the operations of addition, subtraction, and multiplication are carried out just as in elementary algebra, with the proviso that in multiplication whenever i^2 is encountered, it is replaced by -1. Division is handled differently. First we show how to find the reciprocal (also called the **multiplicative inverse**) of a non-zero complex number.

$$\begin{aligned}(a + bi)^{-1} &= \frac{1}{a + bi} = \frac{1}{a + bi} \cdot \frac{a - bi}{a - bi} = \frac{a - bi}{a^2 - b^2i^2} = \frac{a - bi}{a^2 + b^2} \\ &= \frac{a}{a^2 + b^2} + \frac{-b}{a^2 + b^2}i\end{aligned}$$

The important thing to remember here is the fact that we multiplied numerator and denominator by $a - bi$ to get the result. Memorizing the result itself is not recommended. For example, to find the multiplicative inverse of $2 + 3i$, we have

$$(2 + 3i)^{-1} = \frac{1}{2 + 3i} \cdot \frac{2 - 3i}{2 - 3i} = \frac{2 - 3i}{4 - 9i^2} = \frac{2 - 3i}{4 + 9} = \frac{2}{13} + \frac{-3}{13}i$$

This can be checked by multiplying the result by $2 + 3i$ to see that the answer is 1. The number $a - bi$ is called the **conjugate** of $a + bi$. For example, the conjugate of $2 + 3i$ is $2 - 3i$ and the conjugate of $5 - 4i$ is $5 + 4i$. What we have shown, then, is that *we find the multiplicative inverse of a complex number by multiplying numerator and denominator of its reciprocal by its conjugate.*

Observe that the product of a complex number $a + bi$ and its conjugate $a - bi$ is always real:

$$(a + bi)(a - bi) = a^2 - b^2i^2 = a^2 + b^2$$

Division of $a + bi$ by $c + di$ (where $c + di \neq 0$) is defined as the product $(a + bi)(c + di)^{-1}$. So to divide $a + bi$ by $c + di$ we write

$$\frac{a + bi}{c + di}$$

and then multiply numerator and denominator by the conjugate of $c + di$ and simplify the result. For example,

$$\frac{3 + 2i}{5 - 4i} = \frac{3 + 2i}{5 - 4i} \cdot \frac{5 + 4i}{5 + 4i} = \frac{15 + 22i + 8i^2}{25 - 16i^2} = \frac{7}{41} + \frac{22}{41}i$$

To indicate the operation of taking the conjugate, a bar is used above the number: $\overline{a + bi} = a - bi$. For example,

$$\overline{2 + 3i} = 2 - 3i$$

$$\overline{5 - 4i} = 5 + 4i$$

$$\overline{-7i} = \overline{0 - 7i} = 7i$$

$$\overline{-2} = \overline{-2 + 0i} = -2 - 0i = -2$$

Often, the single letter z is used to designate a complex number. Then $\overline{z}$ indicates its conjugate. The following facts about conjugates can be shown:

1. $\overline{z_1 + z_2} = \overline{z_1} + \overline{z_2}$ (conjugate of a sum = sum of conjugates)
2. $\overline{z_1 \cdot z_2} = \overline{z_1} \cdot \overline{z_2}$ (conjugate of a product = product of conjugates)
3. $\overline{\left(\dfrac{z_1}{z_2}\right)} = \dfrac{\overline{z_1}}{\overline{z_2}}$ (conjugate of a quotient = quotient of conjugates)
4. $z\overline{z}$ is real
5. If z is real, then $\overline{z} = z$.

Since $i^2 = -1$, it follows that $i^3 = i^2 \cdot i = -i$ and $i^4 = i^2 \cdot i^2 = (-1)(-1) = 1$. Higher powers can now be readily calculated. For example,

$$i^{22} = i^{20} \cdot i^2 = (i^4)^5 \cdot i^2 = 1^5 \cdot i^2 = 1 \cdot i^2 = -1$$

and

$$i^{35} = i^{32} \cdot i^3 = (i^4)^8 \cdot i^3 = 1^8 \cdot i^3 = i^3 = -i$$

The object is to factor out the highest power of i^4, which is 1. The remaining factor will be one of the powers: i, $i^2 = -1$, or $i^3 = -i$. More generally, if $n > 4$ and we write $n = 4k + r$, where $0 \le r \le 3$, then

$$i^n = i^{4k+r} = (i^{4k}) \cdot i^r = (i^4)^k \cdot i^r = 1^k \cdot i^r = i^r$$

If p is a positive real number, we define $\sqrt{-p}$ as follows:

$$\sqrt{-p} = \sqrt{(-1)p} = \sqrt{-1}\,\sqrt{p} = i\sqrt{p}$$

When such an expression occurs, it is best to change it immediately to the form $i\sqrt{p}$. This helps to avoid such pitfalls as the following *incorrect* multiplication:

$$\sqrt{-2}\,\sqrt{-5} = \sqrt{(-2)(-5)} = \sqrt{10}$$

The correct way is

$$\sqrt{-2}\,\sqrt{-5} = (i\sqrt{2})(i\sqrt{5}) = i^2(\sqrt{2}\,\sqrt{5}) = -\sqrt{10}$$

Recall from algebra that the solutions to the quadratic equation

$$ax^2 + bx + c = 0 \qquad (a \ne 0)$$

are given by

$$\frac{-b + \sqrt{b^2 - 4ac}}{2a} \quad \text{and} \quad \frac{-b - \sqrt{b^2 - 4ac}}{2a}$$

the result known as the *quadratic formula*. Although the quadratic formula holds true when a, b, and c are arbitrary complex numbers (with $a \neq 0$), for the present we assume them to be real. The nature of the two solutions depends upon the number $b^2 - 4ac$ under the radical. This number is called the **discriminant** of the quadratic equation. When the discriminant is positive, the square root is a positive real number, so the solutions are real and distinct. When the discriminant is zero, the square root is zero, so the solutions are real and equal (we say there is a double root). Finally, when the discriminant is negative, the square root is imaginary, so the solutions are non-real complex numbers, that is, they are imaginary. In fact, the solutions are complex conjugates. We summarize these results:

$$\text{If } b^2 - 4ac \begin{cases} > 0 \\ = 0 \\ < 0 \end{cases} \text{ the solutions are } \begin{cases} \text{real and unequal} \\ \text{real and equal} \\ \text{imaginary} \end{cases}$$

EXAMPLE 7.1 Solve the equation $x^2 - 6x + 10 = 0$.

Solution By the quadratic formula

$$x = \frac{6 \pm \sqrt{36 - 40}}{2} = \frac{6 \pm \sqrt{-4}}{2} = \frac{6 \pm i\sqrt{4}}{2}$$

$$= \frac{6 \pm 2i}{2} = 3 \pm i$$

EXAMPLE 7.2 Without solving, determine the nature of the roots of each of the following:

a. $2x^2 - 3x - 4 = 0$ **b.** $9x^2 - 24x + 16 = 0$ **c.** $3x^2 - 2x + 4 = 0$.

Solution

a. $b^2 - 4ac = (-3)^2 - 4(2)(-4) = 9 + 32 = 41 > 0$
So the solutions are real and unequal.

b. $b^2 - 4ac = (-24)^2 - 4(9)(16) = 576 - 576 = 0$
So the solutions are real and equal.

c. $b^2 - 4ac = 4 - 4(3)(4) = 4 - 48 = -44 < 0$.
So the solutions are imaginary.

EXERCISE SET 7.2

A In Problems 1–8 perform the indicated operations. Express answers in the form $a + bi$.

1. **a.** $(2 - 5i) + (7 + 3i)$ **b.** $(3 + 2i) + (-5 - 7i)$ **c.** $(5 + 8i) - (6 - i)$
d. $(3 - i) - (2i - 3)$

2. a. $(3 + 7i) + (2 + 3i)$ b. $(5 - 3i) + (2 + i)$ c. $(6 - 2i) - (3 + 4i)$
   d. $(4i + 2) - (3 - 8i)$
3. a. $(4 + 3i)(3 - 2i)$ b. $(2 - i)(2 + i)$ c. $(5i - 9)(4i + 7)$
   d. $(i - 3)(4 - 3i)$
4. a. $(5 + 4i)(6 + 7i)$ b. $(2 - 5i)(3 + i)$ c. $(3 + 4i)(3 - 4i)$
   d. $(2i - 3)(4 + 5i)$
5. a. $\dfrac{1}{1 + i}$ b. $(3 - 2i)^{-1}$ c. $\dfrac{1}{i}$ d. $\dfrac{1}{7 + 8i}$
6. a. $\dfrac{1}{2 + 3i}$ b. $(4 - i)^{-1}$ c. $\dfrac{1}{i - 1}$ d. $(5 - 4i)^{-1}$
7. a. $\dfrac{2 + 3i}{3 - 2i}$ b. $\dfrac{1 - i}{1 + i}$ c. $\dfrac{3 + 4i}{5 - 3i}$ d. $\dfrac{3i - 1}{6 + 5i}$
8. a. $\dfrac{3 - i}{2 + 3i}$ b. $\dfrac{4 + 3i}{4 - 3i}$ c. $\dfrac{i}{i + 1}$ d. $\dfrac{2i - 3}{5 - 2i}$

In Problems 9–11 solve for x and y.

9. a. $x + yi = 2 - 3i$ b. $3 + yi = x - 4i$
10. a. $2x - 3yi = 5 + 2i$ b. $x + 1 + 2yi = -2i$
11. $(x - y) + (x + 2y)i = 4 + i$

In Problems 12–15 simplify the expressions using properties of i.

12. a. $\sqrt{-4}$ b. $\sqrt{-162}$ c. $\sqrt{-40}$ d. $\sqrt{-4}\,\sqrt{-9}$
    e. $\sqrt{-2}\,\sqrt{-8}$
13. a. $\sqrt{-75}$ b. $\dfrac{1}{\sqrt{-12}}$ c. $\sqrt{-3}\,\sqrt{-6}$ d. $\dfrac{\sqrt{20}}{\sqrt{-5}}$
    e. $\sqrt{-8}\,\sqrt{2}$
14. a. i^7 b. i^9 c. i^{12} d. i^{25} e. i^{18}
15. a. $\dfrac{1}{i^2}$ b. $\dfrac{1}{i^3}$ c. $\dfrac{1}{i^9}$ d. i^{-5} e. i^{-10}

In Problems 16–23 solve the quadratic equation.

16. a. $x^2 + 4 = 0$ b. $x^2 - x + 2 = 0$
17. a. $x^2 = -9$ b. $4x^2 + 25 = 0$
18. a. $x^2 + 3x + 3 = 0$ b. $2x^2 + 5 = 6x$
19. a. $x^2 - 2x + 6 = 0$ b. $t^2 + 3t = -4$
20. a. $3m^2 - 4m + 6 = 0$ b. $5t^2 + 2 = 4t$
21. a. $2s^2 - 5s + 6 = 0$ b. $3s^2 + 2s + 1 = 0$
22. a. $2t^2 = 6t - 7$ b. $t(5t + 6) + 4 = 0$
23. a. $5 + 7x^2 = 8x$ b. $x(8 - 5x) = 4$

In Problems 24–26 determine whether the solutions are real and unequal, real and equal, or imaginary, by using the discriminant.

24. a. $3x^2 - 9x + 5 = 0$ b. $7x^2 + 6x + 3 = 0$
25. a. $2t^2 - 8 = 3t$ b. $9m^2 = 5(6m - 5)$
26. a. $y(4y - 7) + 3 = 0$ b. $8 + 6s^2 = 13s$

B In Problems 27 and 28 expand and simplify.

27. **a.** $(2 + 3i)^3$ **b.** $(1 + i)^5$

28. **a.** $(2 - i)^4$ **b.** $(3 - 4i)^3$

29. Evaluate the polynomial $x^3 - 5x^2 + 17x - 13$ when $x = 2 - 3i$.

30. Evaluate the polynomial $2x^3 - 7x^2 - 10x - 6$ when $x = 1 - i$.

31. Solve the following quadratic equations:

a. $3x^2 - 8ix - 4 = 0$ **b.** $x^2 - x + 1 + i = 0$

32. Prove the following, where $z_1 = a + bi$ and $z_2 = c + di$.

a. $\overline{z_1 z_2} = \overline{z_1}\,\overline{z_2}$ **b.** $\overline{\left(\dfrac{z_1}{z_2}\right)} = \dfrac{\overline{z_1}}{\overline{z_2}}$

33. For what values of k will the equation $k^2x^2 + (k - 1)x + 4 = 0$ have imaginary solutions?

7.3 Graphical Representation of Complex Numbers and the Trigonometric Form

If we associate with the complex number $a + bi$ the ordered pair (a, b), we can establish a 1-1 correspondence between complex numbers and points in the plane. We identify $a + bi$ with the point whose coordinates are (a, b), as shown in Figure 7.1.

Figure 7.1

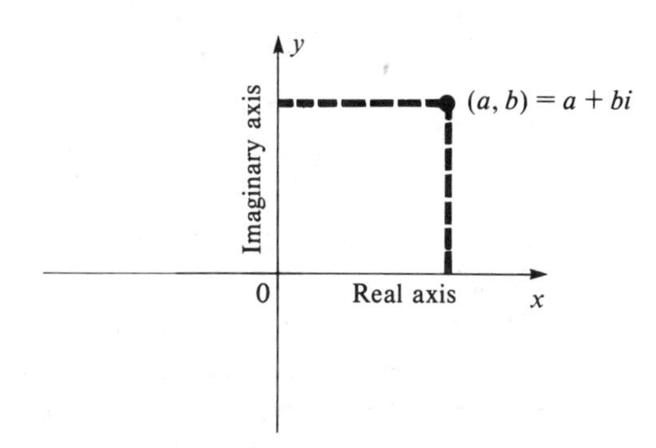

Real numbers correspond to points on the x axis, since real numbers are of the form $a + 0i$ and hence are represented by points of the form $(a, 0)$. Similarly, pure imaginary numbers, which are of the form $0 + bi$, correspond to points with coordinates of the form $(0, b)$, that is, to points on the y axis. For this reason we often refer to the x axis as the **real axis** and the y axis as the **imaginary axis** when the plane is used to plot complex numbers. When we use the plane in this way, we call it the **complex plane**.

It is useful also to identify the complex number $a + bi$ with the vector drawn from the origin to the point having coordinates (a, b). Since $a + bi$ uniquely determines a point (a, b), which in turn uniquely determines the vector from the origin to the point (a, b), we may speak of "the point $a + bi$" or "the vector $a + bi$" without ambiguity. (See Figure 7.2.)

Figure 7.2

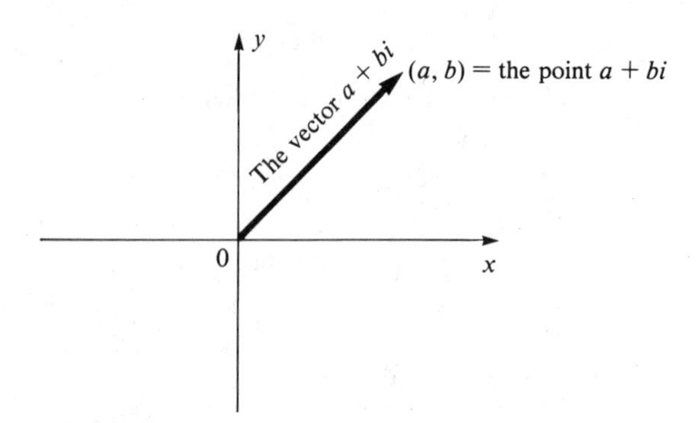

It is interesting to note that if $z_1 = a_1 + b_1 i$ and $z_2 = a_2 + b_2 i$, then the graphical addition of z_1 and z_2 by the parallelogram law for addition of vectors agrees with the algebraic definition $z_1 + z_2 = (a_1 + a_2) + (b_1 + b_2)i$. This is illustrated in Figure 7.3.

Figure 7.3

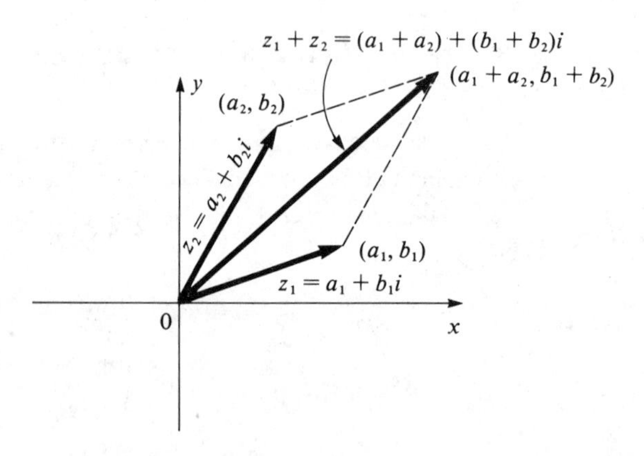

If $z = a + bi$, we call $\sqrt{a^2 + b^2}$ the **modulus** of z and denote it by $|z|$. Geometrically, $|z|$ is the distance from the origin to the point (a, b), that is, it is the length of the vector z, as shown in Figure 7.4. This is a natural extension of the concept of the absolute value of a real number, since if a is a real number $|a|$ can be interpreted geometrically as the distance from the origin to the point representing a on the number line (the real axis). The following properties hold true for the modulus of a complex number:

1. $|z| \geq 0$ and $|z| = 0$ if and only if $z = 0$
2. $|z_1 z_2| = |z_1| \, |z_2|$
3. $|z_1 + z_2| \leq |z_1| + |z_2|$ (Triangle inequality)

You will be asked to verify these in the next exercise set.

In order to see how to relate trigonometry to complex numbers, we designate by θ the smallest positive angle from the positive x axis to the vector z. This angle is

Figure 7.4

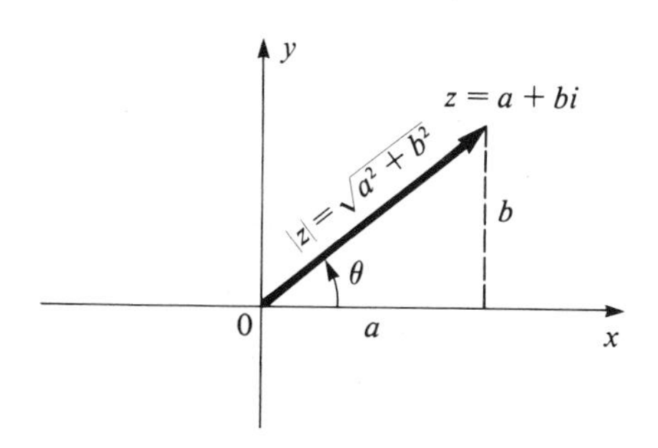

called the **primary argument** of z. If we add any integral multiple of 2π to the primary argument, the result is called an **argument** of z. Thus if θ is the primary argument, then all other arguments are of the form $\theta + 2k\pi$, where k is an integer. If $z = a + bi$, then we see from Figure 7.4 that $\tan \theta = b/a$. It may not be true, however, that $\theta = \tan^{-1}(b/a)$, since we are limiting the inverse tangent function to principal values. If we designate the modulus $|z|$ by r, we can also write

$$a = r \cos \theta$$

$$b = r \sin \theta$$

Therefore, $z = a + bi = r \cos \theta + (r \sin \theta)i$, or

$$\mathbf{z = r(\cos \theta + i \sin \theta)} \tag{7.3}$$

This is called the **trigonometric form**, or **polar form**, of the complex number z. Because of periodicity, we also have, for any integer k,

$$z = r[\cos(\theta + 2k\pi) + i \sin(\theta + 2k\pi)]$$

That is, an equivalent trigonometric form results if any argument is used in place of its primary argument. We call $a + bi$ the **rectangular form** of a complex number. The next two examples show how to change from rectangular form to trigonometric form, and vice versa.

EXAMPLE 7.3 Give the modulus and primary argument of each of the following, and write the number in trigonometric form.

a. $1 + i$ **b.** $\sqrt{3} - i$ **c.** $-4\sqrt{2} - 4i\sqrt{2}$ **d.** $-\sqrt{3} + 3i$

Solution It is helpful in each case to draw a figure.

a. In Figure 7.5, $|1 + i| = \sqrt{1^2 + 1^2} = \sqrt{2}$; and $\tan \theta = 1$, so $\theta = \pi/4$. Therefore, by Equation (7.3),

$$1 + i = \sqrt{2}\left(\cos \frac{\pi}{4} + i \sin \frac{\pi}{4}\right)$$

b. In Figure 7.6, $|\sqrt{3} - i| = \sqrt{3 + 1} = 2$; $\tan \theta = -1/\sqrt{3}$ and θ is in quadrant IV, so $\theta = 11\pi/6$. By (7.3),

$$\sqrt{3} - i = 2\left(\cos\frac{11\pi}{6} + i\sin\frac{11\pi}{6}\right)$$

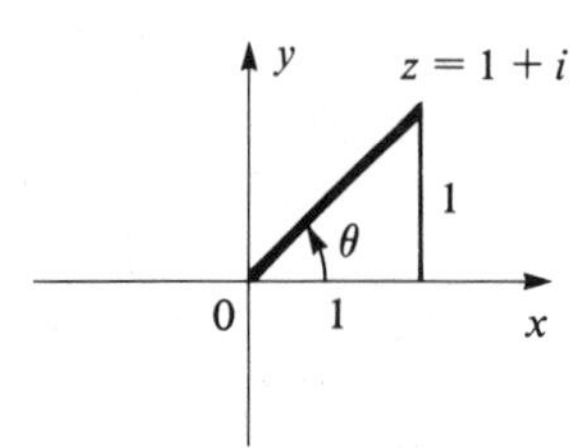

Figure 7.5

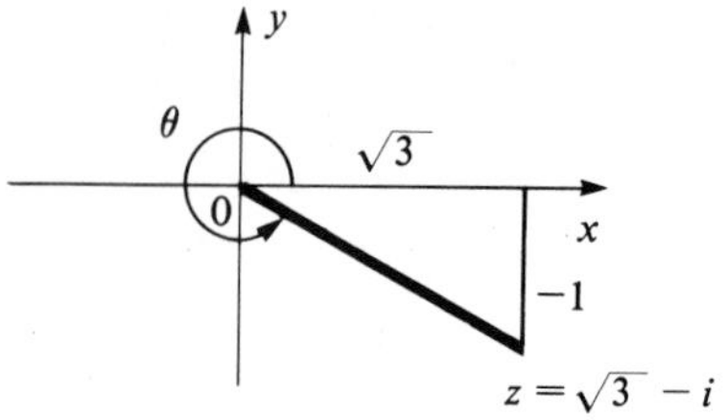

Figure 7.6

c. In Figure 7.7, $|-4\sqrt{2} - 4i\sqrt{2}| = \sqrt{32 + 32} = \sqrt{64} = 8$; $\tan\theta = 1$ and θ is in quadrant III, so $\theta = 5\pi/4$. Therefore, by Equation (7.3),

$$-4\sqrt{2} - 4i\sqrt{2} = 8\left(\cos\frac{5\pi}{4} + i\sin\frac{5\pi}{4}\right)$$

d. In Figure 7.8, $|-\sqrt{3} + 3i| = \sqrt{3 + 9} = \sqrt{12} = 2\sqrt{3}$; $\tan\theta = -3/\sqrt{3} = -\sqrt{3}$ and θ is in quadrant II, so $\theta = 2\pi/3$. By (7.3),

$$-\sqrt{3} + 3i = 2\sqrt{3}\left(\cos\frac{2\pi}{3} + i\sin\frac{2\pi}{3}\right)$$

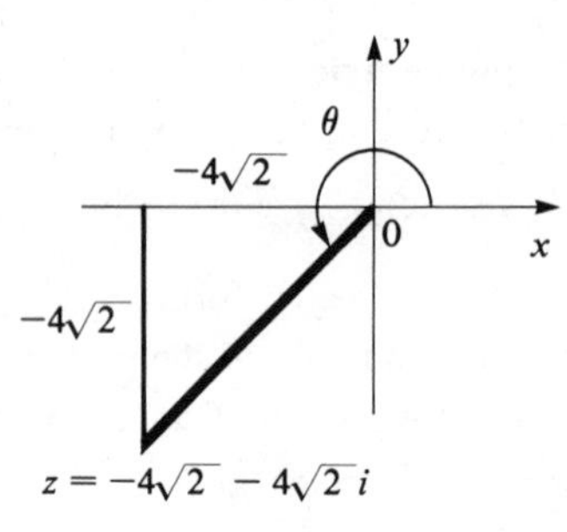

Figure 7.7

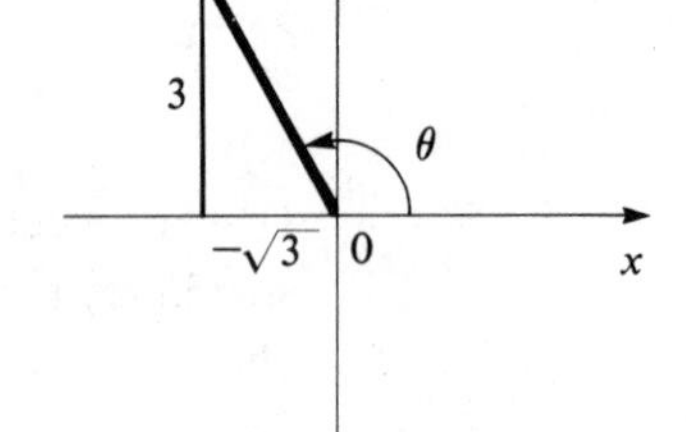

Figure 7.8

EXAMPLE 7.4 Change each of the following to rectangular form.

a. $2\left(\cos\frac{\pi}{3} + i\sin\frac{\pi}{3}\right)$ **b.** $8\left(\cos\frac{7\pi}{4} + i\sin\frac{7\pi}{4}\right)$

c. $5\left[\cos\left(-\frac{\pi}{6}\right) + i\sin\left(-\frac{\pi}{6}\right)\right]$ **d.** $16(\cos 540^\circ + i\sin 540^\circ)$

Solution In each case it is sufficient to give the values of the sine and cosine and to simplify the result. Notice that in parts **c** and **d** the primary arguments are not given, but this presents no problem since we can still evaluate the functions.

a. $2\left(\cos\frac{\pi}{3} + i\sin\frac{\pi}{3}\right) = 2\left(\frac{1}{2} + i\frac{\sqrt{3}}{2}\right) = 1 + i\sqrt{3}$

b. $8\left(\cos\frac{7\pi}{4} + i\sin\frac{7\pi}{4}\right) = 8\left[\frac{\sqrt{2}}{2} + i\left(-\frac{\sqrt{2}}{2}\right)\right] = 4\sqrt{2} - 4i\sqrt{2}$

c. $5\left[\cos\left(-\frac{\pi}{6}\right) + i\sin\left(-\frac{\pi}{6}\right)\right] = 5\left[\frac{\sqrt{3}}{2} + i\left(-\frac{1}{2}\right)\right] = \frac{5\sqrt{3}}{2} - \frac{5i}{2}$

d. $16(\cos 540° + i\sin 540°) = 16(\cos 180° + i\sin 180°)$

$= 16(-1 + i \cdot 0) = -16$

EXERCISE SET 7.3

A In Problems 1 and 2 plot the given number in the complex plane.

1. **a.** $2 + 2i$ **b.** $-1 - i\sqrt{3}$ **c.** $-8i$ **d.** -4 **e.** $-2\sqrt{3} + i$

2. **a.** $2\left(\cos\frac{3\pi}{4} + i\sin\frac{3\pi}{4}\right)$ **b.** $4(\cos 240° + i\sin 240°)$

c. $8\left(\cos\frac{11\pi}{6} + i\sin\frac{11\pi}{6}\right)$ **d.** $\cos\frac{3\pi}{2} + i\sin\frac{3\pi}{2}$

e. $6(\cos 585° + i\sin 585°)$

In Problems 3–7 find the modulus and primary argument, and write each number in trigonometric form.

3. **a.** $2\sqrt{3} - 2i$ **b.** $\sqrt{2}(1 + i)$

4. **a.** $-1 - i\sqrt{3}$ **b.** $-3i$

5. **a.** $-4 + 4i$ **b.** -16

6. **a.** $i - 1$ **b.** $\sqrt{12} + 2i$

7. **a.** $2\sqrt{2} + i\sqrt{24}$ **b.** $i\sqrt{3} - 3$

In Problems 8–12 write each number in rectangular form.

8. **a.** $4\left(\cos\frac{2\pi}{3} + i\sin\frac{2\pi}{3}\right)$ **b.** $8\left(\cos\frac{5\pi}{4} + i\sin\frac{5\pi}{4}\right)$

9. **a.** $2\left(\cos\frac{5\pi}{6} + i\sin\frac{5\pi}{6}\right)$ **b.** $12\left(\cos\frac{3\pi}{2} + i\sin\frac{3\pi}{2}\right)$

10. **a.** $\sqrt{2}\left[\cos\left(-\frac{3\pi}{4}\right) + i\sin\left(-\frac{3\pi}{4}\right)\right]$

b. $5(\cos 3\pi + i\sin 3\pi)$

11. **a.** $4(\cos 600° + i\sin 600°)$ **b.** $8\left(\cos\frac{17\pi}{6} + i\sin\frac{17\pi}{6}\right)$

12. **a.** $6\left(\cos\frac{7\pi}{4} + i\sin\frac{7\pi}{4}\right)$ **b.** $3\left[\cos\left(-\frac{\pi}{3}\right) + i\sin\left(-\frac{\pi}{3}\right)\right]$

Verify the modulus properties 2 and 3 for each pair of complex numbers in Problems 13–17.

13. **a.** $z_1 = 3 - 4i$, $z_2 = 5 + 12i$ **14.** $z_1 = -1 + 2i$, $z_2 = 3 - i$

15. **a.** $z_1 = 8 - 15i$, $z_2 = 8 + 6i$ **16.** $z_1 = -7 - 3i$, $z_2 = 5 - 2i$

17. **a.** $z_1 = 6i - 5$, $z_2 = -4 - 5i$

18. Show that if $z = r(\cos\theta + i\sin\theta)$, then $\bar{z} = r[\cos(-\theta) + i\sin(-\theta)]$.

19. **a.** Show that $|z| = |\bar{z}|$

b. Show that $|z| = \sqrt{z\bar{z}}$

20. **a.** Show that $|-z| = |z|$

b. Use part **a** and the triangle inequality to show that $|z_1 - z_2| \le |z_1| + |z_2|$. Give an example to show that $|z_1 - z_2| \not\le |z_1| - |z_2|$.

B **21.** Prove modulus properties 1 and 2.

22. Use a geometric argument to prove modulus property 3.

23. Prove that if $z \ne 0$ and $z = r(\cos\theta + i\sin\theta)$ then

$$\frac{1}{z} = \frac{1}{r}(\cos\theta - i\sin\theta)$$

24. Prove that the definition of addition given by Equation (7.1) is equivalent geometrically to the parallelogram law for the addition of vectors.

25. For each of the following use a hand calculator to find (i) the modulus of z, (ii) the argument of z, and (iii) the multiplicative inverse of z.

a. $z = 2.732 - 5.031i$ **b.** $z = -72.51 - 54.96i$ **c.** $z = 0.231i - 0.476$

26. Prove that $|z_1 - z_2|$ is the distance between the points z_1 and z_2 in the complex plane.

7.4 Products and Quotients of Complex Numbers; De Moivre's Theorem

The trigonometric form of complex numbers is especially useful in finding products and quotients. Consider first the product $z_1 z_2$, where $z_1 = r_1(\cos\theta_1 + i\sin\theta_1)$ and $z_2 = r_2(\cos\theta_2 + i\sin\theta_2)$.

$$\begin{aligned} z_1 z_2 &= r_1(\cos\theta_1 + i\sin\theta_1)\cdot r_2(\cos\theta_2 + i\sin\theta_2) \\ &= r_1 r_2[(\cos\theta_1\cos\theta_2 - \sin\theta_1\sin\theta_2) + i(\sin\theta_1\cos\theta_2 + \cos\theta_1\sin\theta_2)] \end{aligned}$$

By the addition formulas for the sine and cosine this can be written in the form

$$z_1 z_2 = r_1 r_2[\cos(\theta_1 + \theta_2) + i\sin(\theta_1+\theta_2)] \tag{7.4}$$

This tells us that **the modulus of the product is the product of the moduli** and that **the argument of the product is the sum of the arguments**. In a similar way (see Problem 36, Exercise Set 7.4), if $z_2 \ne 0$, the following formula for the quotient z_1/z_2 can be obtained.

$$\frac{z_1}{z_2} = \frac{r_1}{r_2}[\cos(\theta_1 - \theta_2) + i\sin(\theta_1 - \theta_2)] \tag{7.5}$$

So **for quotients, the modulus is the modulus of the numerator divided by the modulus of the denominator**, and **the argument is the argument of the numerator minus the argument of the denominator**.

EXAMPLE 7.5 Find the product of the complex numbers

$$z_1 = 2\left(\cos\frac{\pi}{6} + i\sin\frac{\pi}{6}\right) \quad \text{and} \quad z_2 = 3\left(\cos\frac{\pi}{3} + i\sin\frac{\pi}{3}\right)$$

Solution By Equation (7.4), the product is

$$\begin{aligned} z_1 z_2 &= 2 \cdot 3\left[\cos\left(\frac{\pi}{6} + \frac{\pi}{3}\right) + i\sin\left(\frac{\pi}{6} + \frac{\pi}{3}\right)\right] \\ &= 6\left(\cos\frac{\pi}{2} + i\sin\frac{\pi}{2}\right) \end{aligned}$$

In rectangular form the answer is therefore $z_1 z_2 = 6i$.

EXAMPLE 7.6 Find the quotient z_1/z_2 if

$$z_1 = 4\left(\cos\frac{5\pi}{6} + i\sin\frac{5\pi}{6}\right) \quad \text{and} \quad z_2 = 2\left(\cos\frac{\pi}{2} + i\sin\frac{\pi}{2}\right)$$

Solution By Equation (7.5), we have

$$\begin{aligned} \frac{z_1}{z_2} &= \frac{4}{2}\left[\cos\left(\frac{5\pi}{6} - \frac{\pi}{2}\right) + i\sin\left(\frac{5\pi}{6} - \frac{\pi}{2}\right)\right] \\ &= 2\left(\cos\frac{\pi}{3} + i\sin\frac{\pi}{3}\right) \end{aligned}$$

The answer in rectangular form is

$$\frac{z_1}{z_2} = 2\left(\frac{1}{2} + i\frac{\sqrt{3}}{2}\right) = 1 + i\sqrt{3}$$

Equation (7.4) can be used to find the square of a complex number, since if $z = r(\cos\theta + i\sin\theta)$,

$$\begin{aligned} z^2 &= [r(\cos\theta + i\sin\theta)][r(\cos\theta + i\sin\theta)] \\ &= r \cdot r[\cos(\theta + \theta) + i\sin(\theta + \theta)] \\ &= r^2(\cos 2\theta + i\sin 2\theta) \end{aligned}$$

Now, if we use this result and Equation (7.4) again, we obtain

$$\begin{aligned} z^3 = z^2 \cdot z &= [r^2(\cos 2\theta + i \sin 2\theta)][r(\cos \theta + i \sin \theta)] \\ &= r^2 \cdot r[\cos(2\theta + \theta) + i \sin(2\theta + \theta)] \\ &= r^3(\cos 3\theta + i \sin 3\theta) \end{aligned}$$

We could continue this approach to find higher powers of z. The result, stated below, is due to the French mathematician Abraham De Moivre (1667–1754). The proof of the result for all positive integers n makes use of *mathematical induction*, which you may have studied in algebra.

DE MOIVRE'S THEOREM **Let $z = r(\cos \theta + i \sin \theta)$. Then for all positive integers n,**

$$z^n = r^n(\cos n\theta + i \sin n\theta) \tag{7.6}$$

This is an important result, and it illustrates one of the main advantages of the trigonometric form of a complex number over its rectangular form.

EXAMPLE 7.7 Expand $(1 + i)^6$.

Solution With the aid of Figure 7.9 we determine that $r = \sqrt{2}$ and $\theta = \pi/4$. Writing $z = 1 + i$, we have by Equation (7.6)

$$\begin{aligned} z^6 = (1 + i)^6 &= \left[\sqrt{2}\left(\cos \frac{\pi}{4} + i \sin \frac{\pi}{4}\right)\right]^6 \\ &= (\sqrt{2})^6\left(\cos \frac{6\pi}{4} + i \sin \frac{6\pi}{4}\right) \\ &= 8\left(\cos \frac{3\pi}{2} + i \sin \frac{3\pi}{2}\right) \end{aligned}$$

Or, in rectangular form,

$$(1 + i)^6 = -8i$$

Figure 7.9

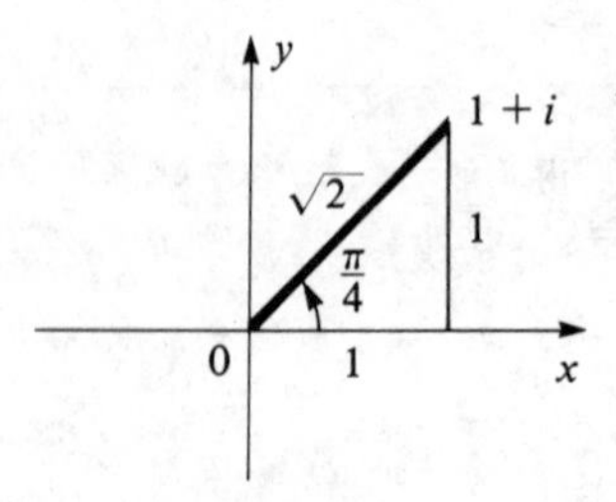

Remark. You might wish to compare the work involved in doing this problem as we did here with doing it by expanding $(1 + i)^6$ by the binomial theorem.

EXERCISE SET 7.4

A In Problems 1–10 use Equation (7.4) to find the product $z_1 z_2$. Express the answer in rectangular form.

1. $z_1 = 2\left(\cos \frac{\pi}{4} + i \sin \frac{\pi}{4}\right), \quad z_2 = 3\left(\cos \frac{5\pi}{4} + i \sin \frac{5\pi}{4}\right)$

2. $z_1 = 8(\cos \pi + i \sin \pi), \quad z_2 = 4\left(\cos \frac{2\pi}{3} + i \sin \frac{2\pi}{3}\right)$

3. $z_1 = 4\left(\cos \frac{\pi}{2} + i \sin \frac{\pi}{2}\right), \quad z_2 = 2\left(\cos \frac{3\pi}{4} + i \sin \frac{3\pi}{4}\right)$

4. $z_1 = 7\left(\cos \frac{5\pi}{6} + i \sin \frac{5\pi}{6}\right), \quad z_2 = 3\left(\cos \frac{4\pi}{3} + i \sin \frac{4\pi}{3}\right)$

5. $z_1 = 6(\cos 85^\circ + i \sin 85^\circ), \quad z_2 = 5(\cos 35^\circ + i \sin 35^\circ)$

6. $z_1 = \sqrt{2}(\cos 237^\circ + i \sin 237^\circ), \quad z_2 = \sqrt{8}(\cos 78^\circ + i \sin 78^\circ)$

7. $z_1 = 12\left(\cos \frac{5\pi}{12} + i \sin \frac{5\pi}{12}\right), \quad z_2 = 3\left(\cos \frac{\pi}{3} + i \sin \frac{\pi}{3}\right)$

8. $z_1 = \sqrt{3} - i, \quad z_2 = 1 + i\sqrt{3}$

9. $z_1 = -2\sqrt{2} - 2i\sqrt{2}, \quad z_2 = -4 + 4i$

10. $z_1 = -5i, \quad z_2 = -2 + i\sqrt{12}$

In Problems 11–20 use Equation (7.5) to find the quotient z_1/z_2. Express the quotient in rectangular form.

11. z_1 and z_2 of Problem 1.

12. z_1 and z_2 of Problem 2.

13. z_1 and z_2 of Problem 3.

14. $z_1 = 12\left(\cos \frac{11\pi}{6} + i \sin \frac{11\pi}{6}\right), \quad z_2 = 3\left(\cos \frac{2\pi}{3} + i \sin \frac{2\pi}{3}\right)$

15. $z_1 = 6(\cos 256^\circ + i \sin 256^\circ), \quad z_2 = 4(\cos 121^\circ + i \sin 121^\circ)$

16. $z_1 = 2(\cos 87^\circ + i \sin 87^\circ), \quad z_2 = 6(\cos 327^\circ + i \sin 327^\circ)$

17. $z_1 = 15\left(\cos \frac{\pi}{6} + i \sin \frac{\pi}{6}\right), \quad z_2 = 5\left(\cos \frac{5\pi}{12} + i \sin \frac{5\pi}{12}\right)$

18. z_1 and z_2 of Problem 8.

19. z_1 and z_2 of Problem 9.

20. z_1 and z_2 of Problem 10.

In Problems 21–30 use De Moivre's theorem to find the indicated powers. Express answers in rectangular form.

21. $\left[2\left(\cos \frac{\pi}{6} + i \sin \frac{\pi}{6}\right)\right]^4$

22. $\left[3\left(\cos \frac{3\pi}{4} + i \sin \frac{3\pi}{4}\right)\right]^6$

23. $[2(\cos 240^\circ + i \sin 240^\circ)]^5$

24. $(\sqrt{3} - i)^8$

25. $(-1 + i)^{10}$

26. $(2 - 2i\sqrt{3})^4$

27. $(-\sqrt{2} - i\sqrt{2})^5$

28. $(-2\sqrt{3} + 2i)^6$

29. $[2(\cos 50^\circ + i \sin 50^\circ)]^6$

30. $[\cos(-20^\circ) + i \sin(-20^\circ)]^9$

B **31.** Make use of the trigonometric form to show that $1 + i\sqrt{3}$ and $1 - i\sqrt{3}$ are solutions of the equation $x^4 - 2x^3 + 3x^2 + 2x - 4 = 0$.

32. Let $z_1 = 3 - i\sqrt{3}$, $z_2 = -2\sqrt{2} - 2i\sqrt{2}$, and $z_3 = 2 + 2i\sqrt{3}$. Make use of the trigonometric form to evaluate the following. Leave answers in trigonometric form.

a. $\dfrac{z_1 \cdot z_2}{z_3}$ **b.** $\dfrac{z_2^{\,3}}{z_1 \cdot z_2}$ **c.** $\dfrac{\overline{z}_1 \cdot z_3}{z_2^{\,4}}$

33. If $f(z) = z^3 - 8z - 32$, show that $f(z_1) = 0$ and $f(\overline{z}_1) = 0$, where $z_1 = -2 + 2i$, by making use of polar form.

34. Use De Moivre's theorem together with a calculator or table to find the following. Express answers in rectangular form.

a. $(0.6 + 0.8\,i)^{10}$ **b.** $(i - 2)^7$ **c.** $(-2 - 3i)^4$

35. **a.** State and prove a formula for the product of three complex numbers, $z_1 \cdot z_2 \cdot z_3$, in polar form. Generalize this to n complex numbers.

b. Use your result of part **a** to find $z_1 \cdot z_2 \cdot z_3$ if $z_1 = -\sqrt{3} + i$, $z_2 = -\frac{3}{2}(1 + i\sqrt{3})$, and $z_3 = -4i$.

36. Prove Equation (7.5) is ture.

37. Prove De Moivre's theorem using mathematical induction.
Hint. Two steps are involved. First, prove it is true for $n = 1$. Second, prove that its truth for $n = k$ implies its truth for $n = k + 1$.

38. Prove that De Moivre's theorem remains true for n a negative integer.
Hint. Write $n = -m$, where $m > 0$.

7.5 Roots of Complex Numbers

We can use De Moivre's theorem to find roots of complex numbers. Let $z = r(\cos\theta + i\sin\theta)$, and suppose we want to find $\sqrt[n]{z}$, where n is a positive integer. If we denote such an nth root by ζ, then we are seeking all values of ζ for which $\zeta^n = z$. Let $\zeta = \rho(\cos\phi + i\sin\phi)$ be the trigonometric form of ζ, with ρ and ϕ yet to be determined. By De Moivre's theorem,

$$\zeta^n = \rho^n(\cos n\phi + i\sin n\phi)$$

Now in order for this to equal z, its modulus must be the same as that of z, and its argument must either be the same as the argument of z or differ from it by an integral multiple of 2π. That is, $\rho^n = r$ and $n\phi - \theta = 2k\pi$ for some integer k, or equivalently

$$\rho = r^{1/n}, \qquad \phi = \frac{\theta + 2k\pi}{n} \qquad (k = 0, \pm 1, \pm 2, \ldots)$$

By $r^{1/n}$ we mean the positive real nth root of r (called the **principal *n*th root**). If we take the n different values of k given by $k = 0, 1, 2, \ldots, n - 1$, we get n distinct values of ϕ, yielding n different roots ζ. All other values of k result in angles ϕ coterminal with one of the n values just found and hence do not result in any new values for ζ. For example, if $k = n$, $\phi = (\theta + 2n\pi)/n = \theta/n + 2\pi$, which is coterminal with θ/n, the value corresponding to $k = 0$. So there are exactly n distinct values of ζ, which can be found by taking $k = 0, 1, 2, \ldots, n - 1$.

To summarize, we have found the following:

$$z^{1/n} = r^{1/n}\left[\cos\left(\frac{\theta + 2k\pi}{n}\right) + i \sin\left(\frac{\theta + 2k\pi}{n}\right)\right] \tag{7.7}$$

The n distinct values are found by taking $k = 0, 1, 2, \ldots, n - 1$.

The modulus of each of the n roots is $r^{1/n}$, the principal nth root of the modulus of the original number z. If we designate the n roots which correspond to $k = 0, 1, 2, \ldots, n - 1$ by $\zeta_0, \zeta_1, \zeta_2, \ldots, \zeta_{n-1}$, respectively, then their arguments are as follows.

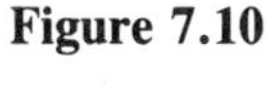

Figure 7.10

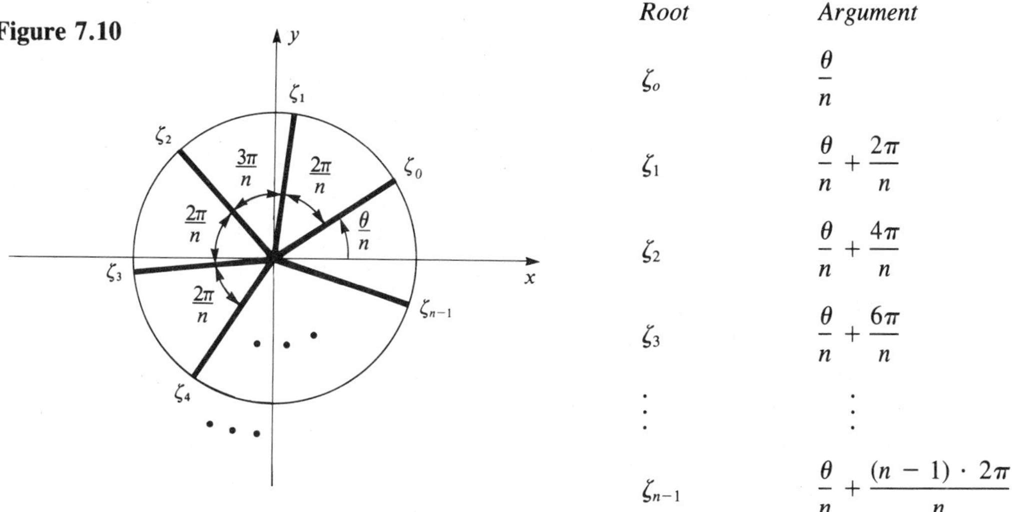

Root	*Argument*
ζ_o	$\frac{\theta}{n}$
ζ_1	$\frac{\theta}{n} + \frac{2\pi}{n}$
ζ_2	$\frac{\theta}{n} + \frac{4\pi}{n}$
ζ_3	$\frac{\theta}{n} + \frac{6\pi}{n}$
⋮	⋮
ζ_{n-1}	$\frac{\theta}{n} + \frac{(n-1) \cdot 2\pi}{n}$

Observe that the arguments of any two successive roots differ by $2\pi/n$. It follows that in the complex plane the roots are equally spaced on a circle of radius $r^{1/n}$. This is illustrated in Figure 7.10.

EXAMPLE 7.8 Find all fourth roots of $-8 - 8i\sqrt{3}$.

Solution Let $z = -8 - 8i\sqrt{3}$. From Figure 7.11 we see that $r = 16$ and $\theta = 4\pi/3$. So in polar form $z = 16[\cos(4\pi/3) + i \sin(4\pi/3)]$. By Equation (7.7), the fourth roots are given by

$$z^{1/4} = (16)^{1/4}\left[\cos\frac{\left(\frac{4\pi}{3} + 2k\pi\right)}{4} + i\sin\frac{\left(\frac{4\pi}{3} + 2k\pi\right)}{4}\right]$$

$$= 2\left[\cos\left(\frac{\pi}{3} + \frac{k\pi}{2}\right) + i\sin\left(\frac{\pi}{3} + \frac{k\pi}{2}\right)\right], \qquad k = 0, 1, 2, 3$$

Designating these roots by ζ_0, ζ_1, ζ_2, and ζ_3, we get

$$k = 0: \quad \zeta_0 = 2\left(\cos\frac{\pi}{3} + i\sin\frac{\pi}{3}\right) = 2\left(\frac{1}{2} + i\frac{\sqrt{3}}{2}\right) = 1 + i\sqrt{3}$$

$$k = 1: \quad \zeta_1 = 2\left[\cos\left(\frac{\pi}{3} + \frac{\pi}{2}\right) + i\sin\left(\frac{\pi}{3} + \frac{\pi}{2}\right)\right] = 2\left(\cos\frac{5\pi}{6} + i\sin\frac{5\pi}{6}\right)$$

$$= 2\left(-\frac{\sqrt{3}}{2} + i\frac{1}{2}\right) = -\sqrt{3} + i$$

$$k = 2: \quad \zeta_2 = 2\left[\cos\left(\frac{\pi}{3} + \pi\right) + i\sin\left(\frac{\pi}{3} + \pi\right)\right] = 2\left(\cos\frac{4\pi}{3} + i\sin\frac{4\pi}{3}\right)$$

$$= 2\left(-\frac{1}{2} - i\frac{\sqrt{3}}{2}\right) = -1 - i\sqrt{3}$$

$$k = 3: \quad \zeta_3 = 2\left[\cos\left(\frac{\pi}{3} + \frac{3\pi}{2}\right) + i\sin\left(\frac{\pi}{3} + \frac{3\pi}{2}\right)\right]$$

$$= 2\left(\cos\frac{11\pi}{6} + i\sin\frac{11\pi}{6}\right) = 2\left(\frac{\sqrt{3}}{2} - i\frac{1}{2}\right) = \sqrt{3} - i$$

The roots are shown graphically in Figure 7.12. Notice that they are equally spaced on a circle of radius 2.

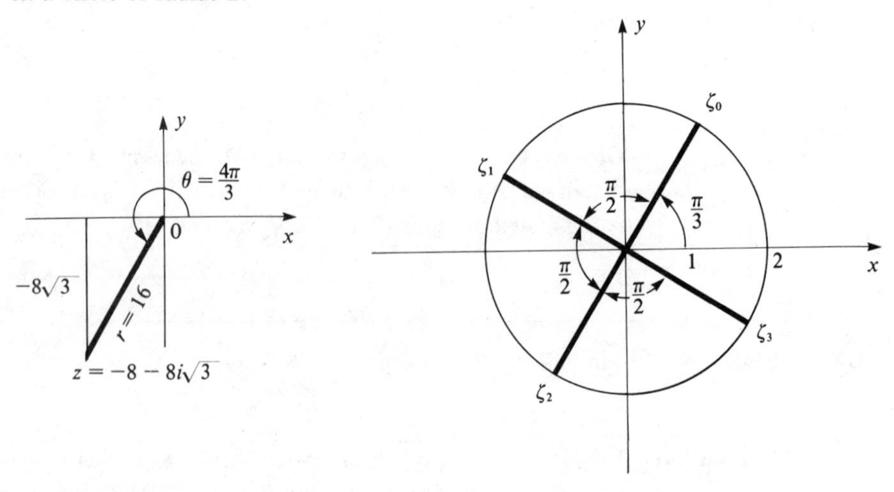

Figure 7.11

Figure 7.12

EXAMPLE 7.9 Find all roots of the equation $x^3 + 8 = 0$.

Solution The problem is equivalent to solving $x^3 = -8$, that is, to finding all cube roots of -8. From Figure 7.13, the complex number -8 can be written in the trigonometric form $-8 = 8(\cos \pi + i \sin \pi)$. So the cube roots are given by

$$8^{1/3}\left[\cos\left(\frac{\pi + 2k\pi}{3}\right) + i \sin\left(\frac{\pi + 2k\pi}{3}\right)\right], \qquad k = 0, 1, 2$$

Thus,

$$\zeta_0 = 2\left(\cos \frac{\pi}{3} + i \sin \frac{\pi}{3}\right) = 1 + i\sqrt{3}$$

$$\zeta_1 = 2(\cos \pi + i \sin \pi) = -2$$

$$\zeta_2 = 2\left(\cos \frac{5\pi}{3} + i \sin \frac{5\pi}{3}\right) = 1 - i\sqrt{3}$$

So the solution set to the equation $x^3 + 8 = 0$ is $\{1 + i\sqrt{3}, -2, 1 - i\sqrt{3}\}$.

Figure 7.13

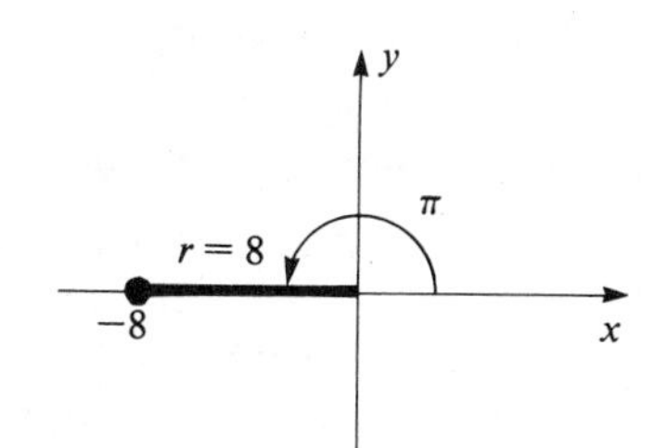

EXERCISE SET 7.5

A In Problems 1–8 find all indicated roots. Express answers in rectangular form when this can be done without using tables or a calculator.

1. Cube roots of $-1 + i$

2. Square roots of -1

3. Sixth roots of 64

4. Cube roots of $-4\sqrt{3} + 4i$

5. Fourth roots of -16

6. Sixth roots of $\frac{1}{2} - \frac{i\sqrt{3}}{2}$

7. Fourth roots of $-\frac{1}{2} - \frac{i\sqrt{3}}{2}$

8. Fifth roots of $-16\sqrt{3} + 16i$

In Problems 9–12 find the complete solution set.

9. $x^3 - 1 = 0$

10. $x^4 + 16 = 0$

11. $x^5 = 32$

12. $x^3 + 27i = 0$

13. An nth root of unity is a number ζ such that $\zeta^n = 1$. Find all sixth roots of unity and plot them on a unit circle in the complex plane.

B In Problems 14–17 find the indicated roots, using a calculator or tables where necessary. Express answers in rectangular form.

14. Fifth roots of $3 - 4i$

15. Eighth roots of $-8 - 8i\sqrt{3}$

16. Cube roots of $12(\cos 75° + i \sin 75°)$

17. Sixth roots of $27(\cos 222° + i \sin 222°)$

18. Factor $x^6 + 64$ completely

a. over the complex field

b. over the real field

Hint. Use part **a** and multiply together factors involving complex conjugates.

19. Find the complete solution set of the equation

$$x^{3/2} = 4\sqrt{2}(1 + i)$$

20. Evaluate $\left(\frac{1}{2} - \frac{i\sqrt{3}}{2}\right)^{3/4}$

7.6 Polar Coordinates

In graphing we have used the rectangular coordinate system thus far. There are alternative ways of pairing numbers with points in the plane which sometimes have advantages over the rectangular system. The most important of these is the **polar coordinate system** in which points are determined by radial distances from a fixed point, called the **pole**, and angles from a fixed ray, called the **polar axis**. The polar axis extends from the pole horizontally to the right. By choosing a unit of distance we make the polar axis into the positive half of a number line, with 0 corresponding to the pole. Now let θ be any angle with the pole as its vertex and the polar axis as its initial side, as in Figure 7.14. For any positive number r, if we measure r units from the pole along the terminal side of θ, a unique point P is determined, and we call r and θ the **polar coordinates** of P, designated by the ordered pair (r, θ). For a negative number r, we measure $|r|$ units from the pole on the ray directed opposite to the terminal side of θ. In Figure 7.15 we illustrate the points $(2, \pi/6)$ and $(-2, \pi/6)$.

In the manner described, the pair (r, θ) determines a unique point P in the plane. On the other hand, if we are given a point P there are infinitely many sets of polar coordinates corresponding to it, since the angle to the ray through P is not unique. For example, $(2, \pi/6)$ and $(2, 13\pi/6)$ correspond to the same point, as does

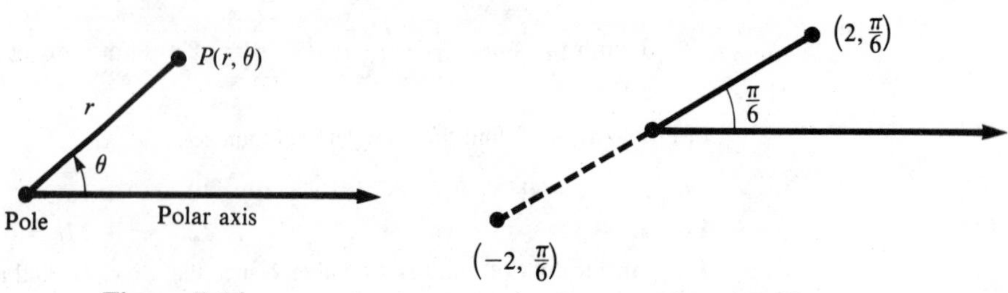

Figure 7.14

Figure 7.15

$(2, \pi/6 + 2k\pi)$ for any integer k. Less obvious is the fact that $(-2, 7\pi/6)$ is still another set of coordinates for this point. (Check this.) More generally, we can say that if P has coordinates (r, θ), then for all integers k both $(r, \theta + 2k\pi)$ and $(-r, \theta + \pi + 2k\pi)$ are valid coordinates of P. In general, we will use the coordinates for which $r > 0$ and $0 \leq \theta < 2\pi$. We note that for any value of θ, the point $(0, \theta)$ is the pole.

If we superimpose a rectangular coordinate system on a polar system so that the origin coincides with the pole and the positive x axis with the polar axis, as in Figure 7.16, we can determine relationships between rectangular and polar coordinates. Let P have coordinates (x, y) in the rectangular system and (r, θ) in the polar system; then the following relationships hold true:

$$\left.\begin{aligned} r^2 &= x^2 + y^2 \\ \sin\theta &= \frac{y}{r} \\ \cos\theta &= \frac{x}{r} \end{aligned}\right\} \tag{7.8}$$

and

$$\left.\begin{aligned} x &= r\cos\theta \\ y &= r\sin\theta \end{aligned}\right\} \tag{7.9}$$

These equations enable us to change from one coordinate system to the other, as we illustrate in the next four examples.

Figure 7.16

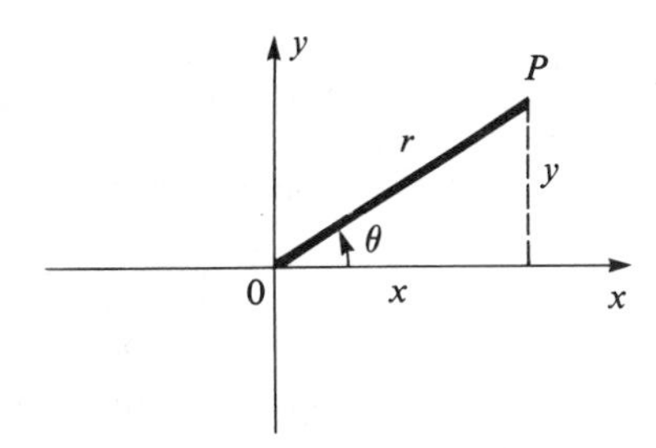

EXAMPLE 7.10 Change to polar coordinates with $r > 0$ and $0 \leq \theta < 2\pi$.

a. $(-2, 2)$ **b.** $(\sqrt{3}, -1)$

Solution We use Equations (7.8) to determine r and θ in each case.

a. $r^2 = 4 + 4$, so $r = 2\sqrt{2}$. Thus, $\sin\theta = 1/\sqrt{2}$ and $\cos\theta = -1/\sqrt{2}$, from which we conclude that $\theta = 3\pi/4$. So the polar coordinates are $(2\sqrt{2}, 3\pi/4)$.

b. $r^2 = 4$, so $r = 2$. So $\sin\theta = -\frac{1}{2}$, $\cos\theta = \sqrt{3}/2$, and $\theta = 11\pi/6$. Thus, the polar coordinates are $(2, 11\pi/6)$.

EXAMPLE 7.11 Change to rectangular coordinates.

a. $\left(2, \dfrac{4\pi}{3}\right)$ **b.** $\left(-\sqrt{2}, \dfrac{3\pi}{4}\right)$

Solution We use Equations (7.9).

a. $x = 2\cos\dfrac{4\pi}{3} = -1, \qquad y = 2\sin\dfrac{4\pi}{3} = -\sqrt{3}$

So the rectangular coordinates are $(-1, -\sqrt{3})$.

b. $x = -\sqrt{2}\cos\dfrac{3\pi}{4} = -\sqrt{2}\left(-\dfrac{1}{\sqrt{2}}\right) = 1,$

$$y = -\sqrt{2}\sin\frac{3\pi}{4} = -\sqrt{2}\left(\frac{1}{\sqrt{2}}\right) = -1$$

So the rectangular coordinates are $(1, -1)$.

EXAMPLE 7.12 Find the rectangular equation of the curve whose polar equation is $r = 2\cos\theta$.

Solution If $r \neq 0$, we replace $\cos\theta$ by x/r and then clear fractions.

$$r = 2\left(\frac{x}{r}\right)$$

$$r^2 = 2x$$

Now we use the fact that $r^2 = x^2 + y^2$ to obtain the desired equation.

$$x^2 + y^2 = 2x$$

This is a circle with center on the x axis and passing through the origin. The fact that the origin is a point on the curve assures us that no further points on the graph are obtained if $r = 0$, since the only point for which $r = 0$ is the pole, which coincides with the origin.

Note. In changing an equation from polar coordinates to rectangular, it is sometimes useful, as in this example, to leave r in the equation initially in order to see if r^2 eventually appears, since r^2 is particularly easy to express in rectangular coordinates.

EXAMPLE 7.13 Find the polar equation of the curve whose rectangular equation is $y^2 = 2x$.

Solution From Equations (7.9), we obtain

$$r^2 \sin^2 \theta = 2r \cos \theta$$

$$r(r \sin^2 \theta - 2 \cos \theta) = 0$$

$$r = 0 \quad | \quad r = 2 \cos \theta \csc^2 \theta \qquad (\theta \neq 2k\pi)$$

When $\theta = \pi/2$, the equation $r = 2 \cos \theta \csc^2 \theta$ gives $r = 0$, so that the first solution $r = 0$ is redundant. Therefore, the answer is

$$r = 2 \cos \theta \csc^2 \theta$$

For the remainder of this section we are going to concentrate on drawing graphs of polar equations. For this purpose it is helpful to use polar graph paper, available in college bookstores. Such graph paper provides a grid of concentric circles and radial lines similar to that shown in Figure 7.17. To draw the graph of an equation in polar coordinates we can either change to rectangular coordinates or plot points directly from the polar equation by substituting various values of θ and solving for r. Which procedure is easier depends on the nature of the equation. For example, the equation

$$r = \frac{3}{\cos \theta - 2 \sin \theta}$$

can be written

$$r \cos \theta - 2r \sin \theta = 3$$

which in rectangular coordinates is

$$x - 2y = 3$$

We recognize this as the equation of a line. In this case changing to rectangular coordinates is clearly the easier way to proceed. On the other hand, if we change

Figure 7.17

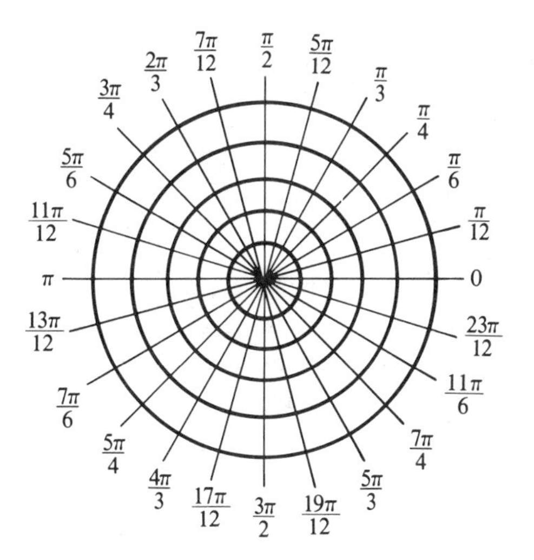

$$r^2 = 4 \cos 2\theta$$

to rectangular coordinates, we get (check this)

$$(x^2 + y^2)^2 = 4(x^2 - y^2)$$

and it would be very difficult to draw the graph of this. It is easier to use the polar equation in this case. We illustrate in the examples that follow the most important curves for which polar coordinates are particularly convenient.

EXAMPLE 7.14 Draw the graph of $r = 2$.

Solution Since no restriction is placed on θ, the graph consists of all points 2 units from the pole. This is a circle of radius 2 with center at the pole, as shown in Figure 7.18.

Figure 7.18

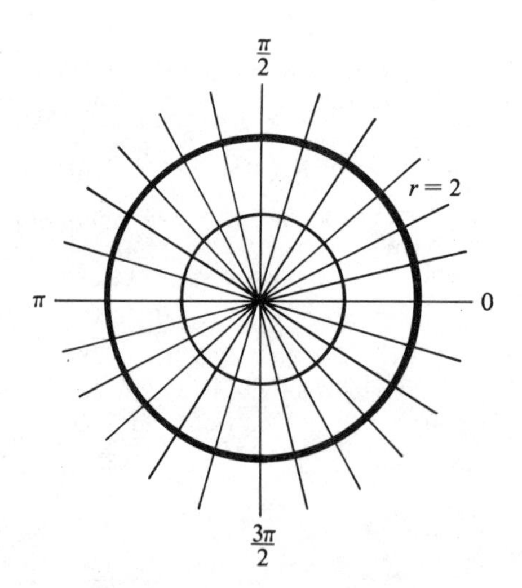

EXAMPLE 7.15 Draw the graph of $\theta = \pi/4$.

Solution The set of all points $(r, \pi/4)$ where r can have any real value constitutes the graph. Thus, the graph is the $\pi/4$ ray together with the ray directed opposite to this (since r can be negative). That is, the graph is the entire line through the pole and making a 45° angle with the polar axis (Figure 7.19). More generally, the graph of $\phi = \alpha$ is a line making an angle α with the polar axis.

Figure 7.19

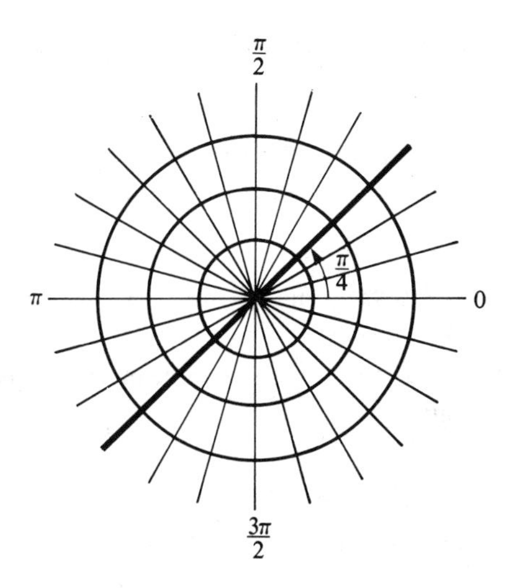

EXAMPLE 7.16 Draw the graph of the equation $r = 2 \cos \theta$.

Solution We saw in Example 7.12 that the corresponding rectangular equation is

$$x^2 + y^2 = 2x$$

By rearranging terms and completing the square on x, we can bring this to the standard form

$$(x - 1)^2 + y^2 = 1$$

and we recognize this as a circle of radius 1 with center at the point (1, 0), as shown in Figure 7.20.

Figure 7.20

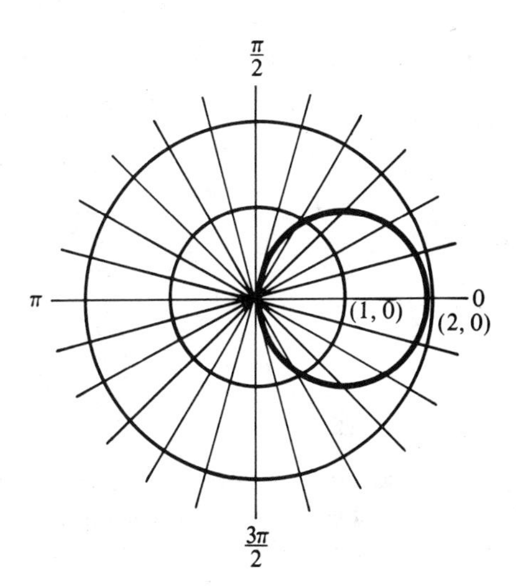

This is a special case of an equation of the form

$$y = 2a \cos \theta \qquad (a > 0)$$

In a manner similar to that of our example, we can show that the graph of this equation is a circle with radius a and center at $(a, 0)$.

EXAMPLE 7.17 Draw the graph of the equation $r = 2(1 + \cos \theta)$.

Solution This time we make a table of values, using special angles θ. The function values are given to two decimal places only.

θ	0	$\frac{\pi}{6}$	$\frac{\pi}{4}$	$\frac{\pi}{3}$	$\frac{\pi}{2}$	$\frac{2\pi}{3}$	$\frac{3\pi}{4}$	$\frac{5\pi}{6}$	π
r	4	3.73	3.41	3	2	1	0.59	0.27	0

Now, rather than extend the table to 2π, we make use of the fact that $\cos(-\theta) = \cos \theta$. This says that for the negatives of the values of θ in the table, r will be unchanged. The graph therefore is symmetric with respect to the polar axis. That is, below the polar axis the curve is the mirror image of that above the axis. This curve (Figure 7.21) is called a **cardioid** because of its heartlike shape. The general equation of a cardioid in this position is

$$r = a(1 + \cos \theta) \qquad (a > 0)$$

Figure 7.21

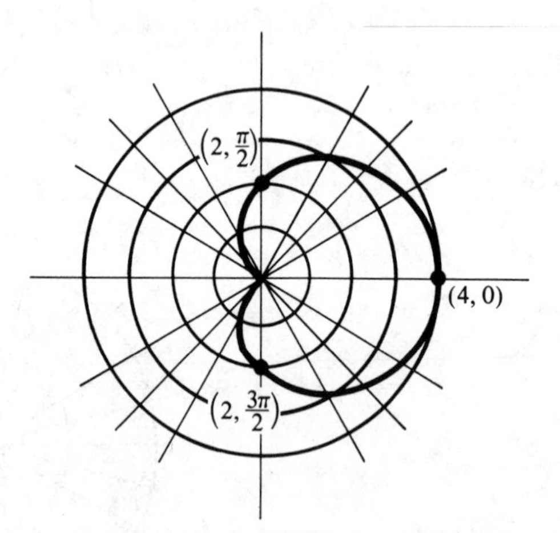

The discussion of symmetry in this example is typical. We can say in general that **if a polar equation is unchanged when θ is replaced by $-\theta$, the graph is symmetric to the polar axis**. This will always be the case if the equation contains $\cos \theta$ (or $\cos n\theta$) but no other function of θ.

EXAMPLE 7.18 Draw the graph of the equation

a. $r = 4 + 3 \cos \theta$ **b.** $r = 3 + 4 \cos \theta$

Solution **a.**

θ	0	$\frac{\pi}{6}$	$\frac{\pi}{4}$	$\frac{\pi}{3}$	$\frac{\pi}{2}$	$\frac{2\pi}{3}$	$\frac{3\pi}{4}$	$\frac{5\pi}{6}$	π
r	7	6.60	6.12	5.50	4	2.50	1.88	1.40	1

Again, we use the fact that the curve is symmetric to the polar axis, obtaining the graph shown in Figure 7.22.

Figure 7.22

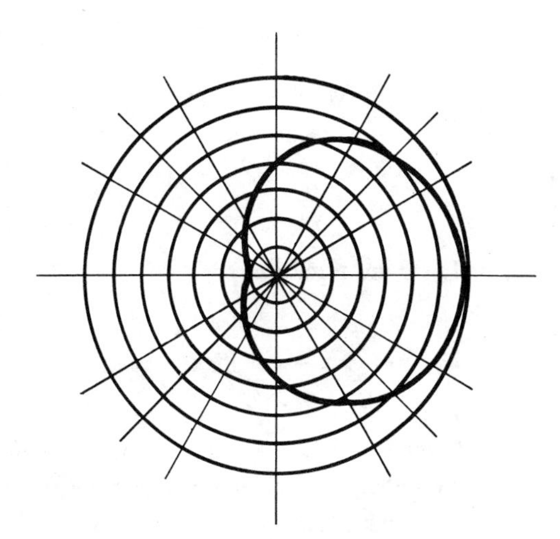

b.

θ	0	$\frac{\pi}{6}$	$\frac{\pi}{4}$	$\frac{\pi}{3}$	$\frac{\pi}{2}$	$\frac{2\pi}{3}$	$\frac{3\pi}{4}$	$\frac{5\pi}{6}$	π
r	7	6.46	5.83	5.00	3	1	0.17	−0.46	−1

Plotting the values obtained from the table, and relying once again on symmetry, we obtain the graph shown in Figure 7.23.

Note that $r > 0$ for $\theta = 3\pi/4$ and $r < 0$ for $\theta = 5\pi/6$. We might expect, therefore, that there is some value of θ for which $r = 0$. To show this is so, we set $r = 0$ and solve for θ, using a calculator.

$$3 + 4 \cos \theta = 0$$

$$\cos \theta = -\frac{3}{4}$$

$$\theta = \cos^{-1}(-0.75) \approx 2.42 \text{ radians}$$

Figure 7.23

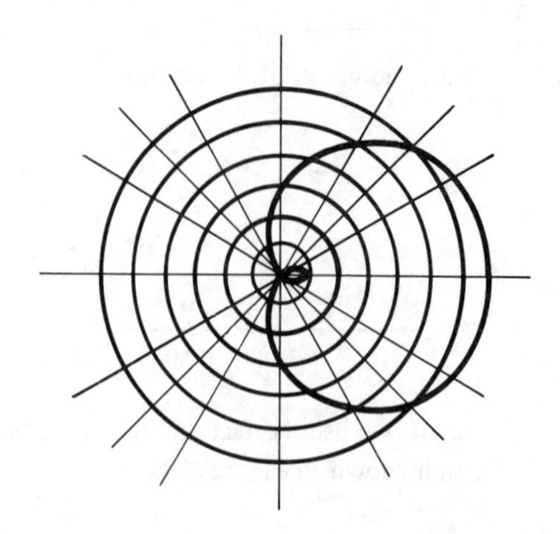

So the curve does go through the pole when $\theta = 2.42$.

The curves in this example are called **limaçons**. The general equation for this position is

$$r = a + b \cos \theta$$

There are three cases:

1. $a > b$ — without a loop
2. $a < b$ — with a loop
3. $a = b$ — cardioid

In the first case the graph is similar to part **a** of our example, although the "dimple" on the left may or may not be present, depending on the relative sizes of a and b. Note that a cardioid is a special case of a limaçon.

EXAMPLE 7.19 Draw the graph of the equation $r^2 = 4 \cos 2\theta$.

Solution We could write equivalently

$$r = \pm 2\sqrt{\cos 2\theta}$$

and we see from this that values of θ for which $\cos 2\theta < 0$ are excluded. For θ between 0 and π these excluded values are those for which $\pi/4 < \theta < 3\pi/4$.

θ	0	$\frac{\pi}{6}$	$\frac{\pi}{4}$	$\frac{3\pi}{4}$	$\frac{5\pi}{6}$	π
r	± 2	± 1.41	0	0	± 1.41	± 2

Plotting the values obtained from the table, we obtain the curve shown in Figure 7.24. This curve is called a **lemniscate** and has the general equation

$$r^2 = a^2 \cos 2\theta$$

Figure 7.24

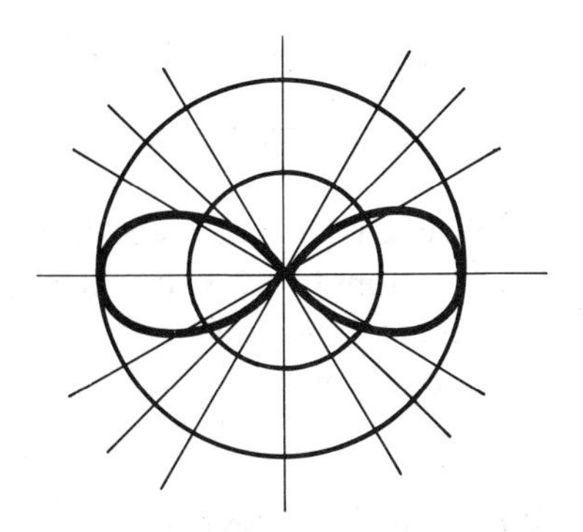

EXAMPLE 7.20 Draw the graph of the equation.

a. $r = 4 \cos 2\theta$ **b.** $r = 4 \cos 3\theta$

Solution **a.** It is instructive in this case to take values of θ all the way from 0 to 2π to observe how the curve is traced out as θ increases. From this table of values, we obtain the graph shown in Figure 7.25.

θ	0	$\frac{\pi}{6}$	$\frac{\pi}{4}$	$\frac{\pi}{3}$	$\frac{\pi}{2}$	$\frac{2\pi}{3}$	$\frac{3\pi}{4}$	$\frac{5\pi}{6}$	π	$\frac{7\pi}{6}$	$\frac{5\pi}{4}$	$\frac{4\pi}{3}$	$\frac{3\pi}{2}$	$\frac{5\pi}{3}$	$\frac{7\pi}{4}$	$\frac{11\pi}{6}$
r	4	2	0	−2	−4	−2	0	2	4	2	0	−2	−4	−2	0	2

b.

θ	0	$\frac{\pi}{6}$	$\frac{\pi}{4}$	$\frac{\pi}{3}$	$\frac{\pi}{2}$	$\frac{2\pi}{3}$	$\frac{3\pi}{4}$	$\frac{5\pi}{6}$	π	$\frac{7\pi}{6}$	$\frac{5\pi}{4}$	$\frac{4\pi}{3}$	$\frac{3\pi}{2}$	$\frac{5\pi}{3}$	$\frac{7\pi}{4}$	$\frac{11\pi}{6}$
r	4	0	−2.83	−4	0	4	2.83	0	−4	0	2.83	4	0	−4	−2.83	0

After locating these points, we observe that some intermediate values of θ would be useful. We choose these so that 3θ is a convenient angle.

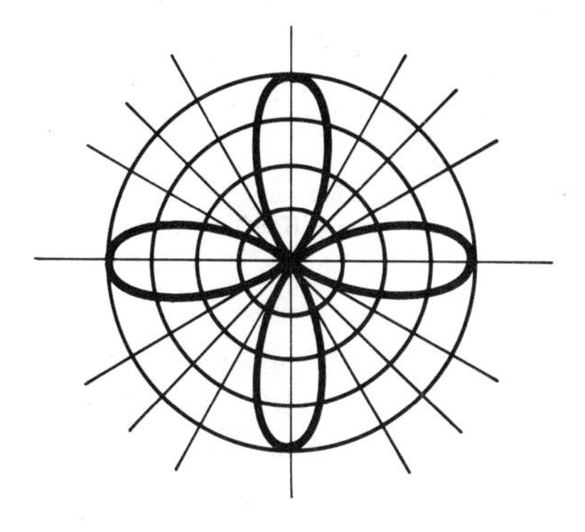

Figure 7.25

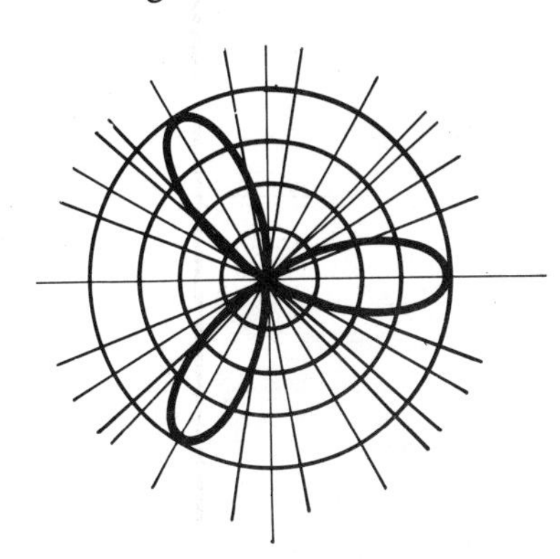

Figure 7.26

θ	$\frac{\pi}{9}$	$\frac{2\pi}{9}$	$\frac{4\pi}{9}$	$\frac{5\pi}{9}$	$\frac{7\pi}{9}$	$\frac{8\pi}{9}$
r	2	-2	-2	2	2	-2

Combining both tables, we obtain the graph shown in Figure 7.26. Notice that as θ varies from 0 to 2π, the curve is traced out twice.

These two curves are examples of what are called **rose curves**. The general form of the equation is

$$r = a \cos n\theta \qquad (a > 0)$$

If *n is even*, the rose has $2n$ *leaves*, whereas if *n is odd*, the rose has *n leaves*. In each case one leaf is along the polar axis, and the leaves are equally spaced.

While there are other standard polar curves, some of which are given in the exercises, these examples illustrate the most important types. We summarize these below.

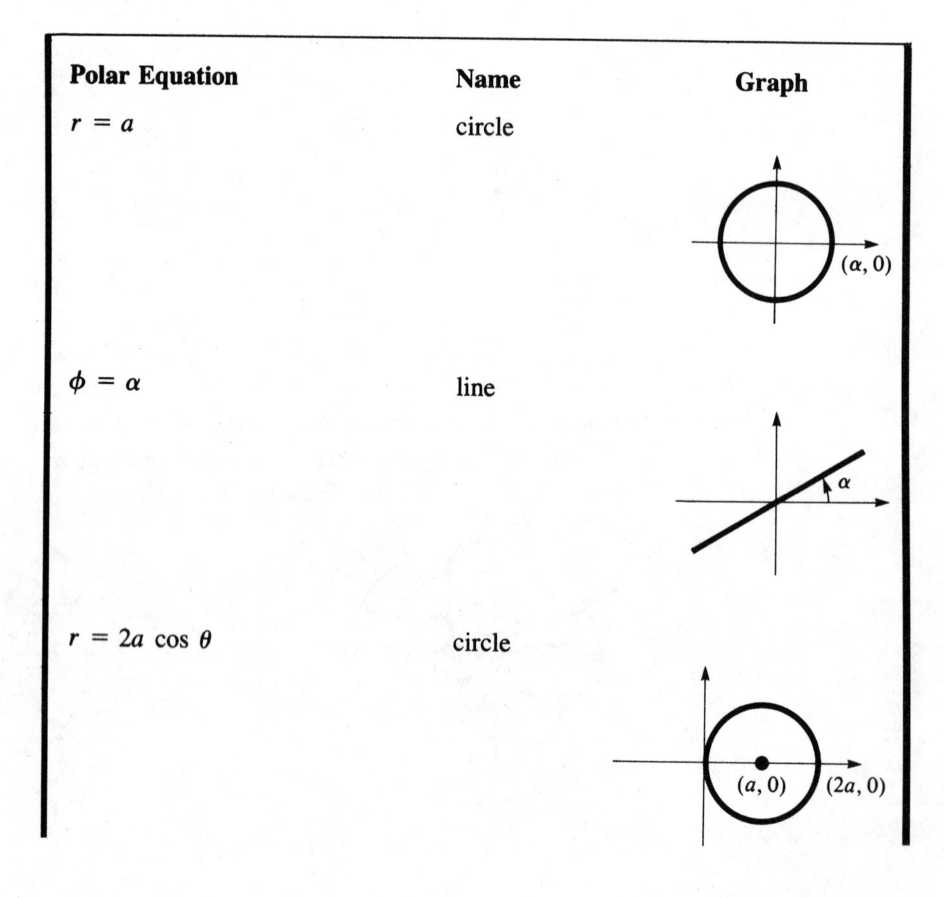

Polar Equation	Name	Graph
$r = a$	circle	
$\phi = \alpha$	line	
$r = 2a \cos \theta$	circle	

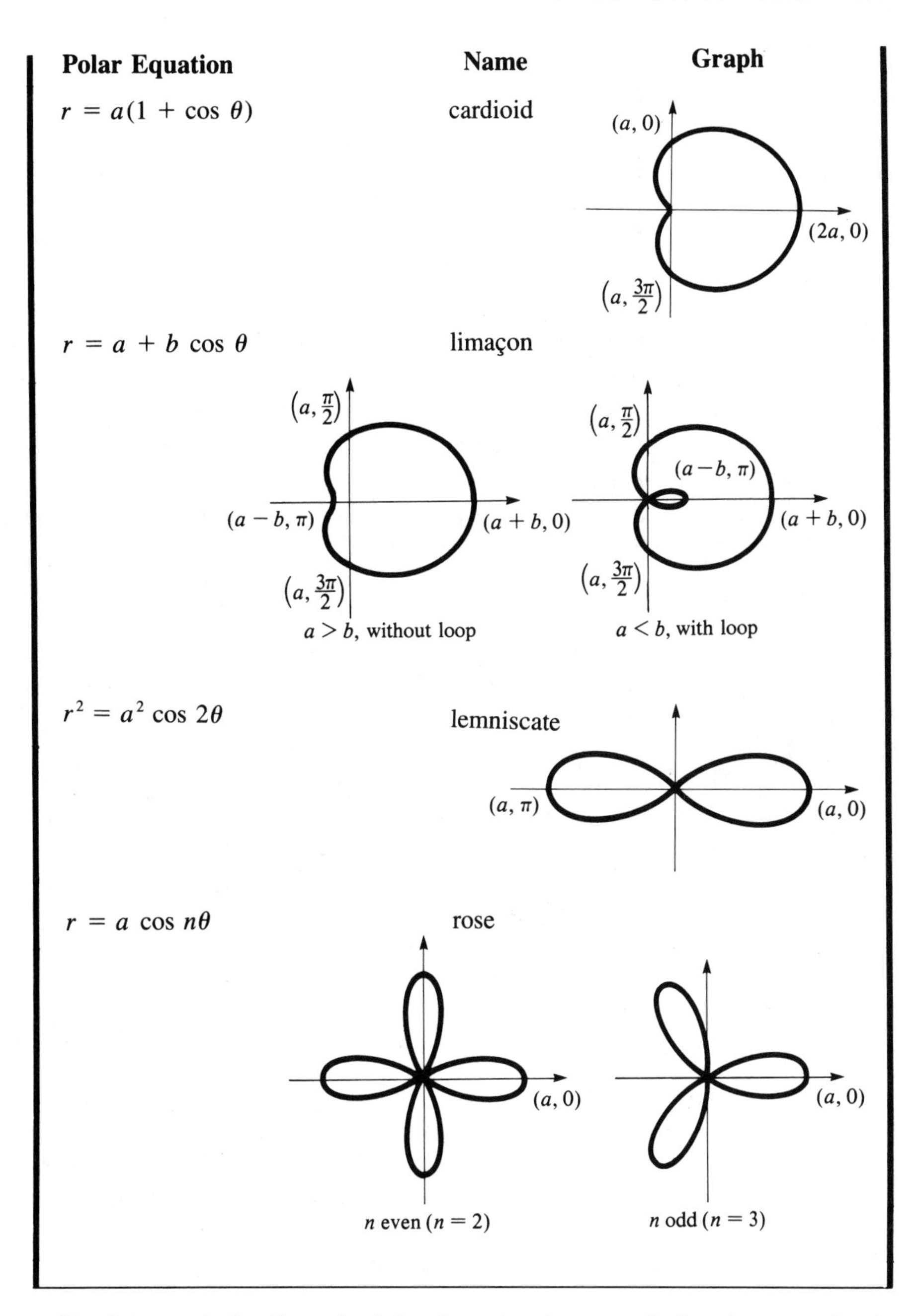

You have undoubtedly noticed that the only trigonometric function occurring in any of the equations is the cosine. We will show below that if the cosine is replaced by sine, the curve is the same but is rotated through a certain angle. The angle is 90° if $\cos\theta$ is replaced by $\sin\theta$ and is $90°/n$ if $\cos n\theta$ is replaced by $\sin n\theta$. This is a consequence of the following general result.

Rotation of Axes

If a point P in a polar coordinate system has coordinates (r, θ), and the polar axis is rotated through an angle α, then the point P has coordinates $(r, \theta - \alpha)$ with respect to the rotated polar coordinate system.

This result follows from an analysis of Figure 7.27. We see that the radial distance r from the pole to P is unchanged by the rotation. If we let θ' be the angle from the new polar axis to the ray from the pole through P, then we see that $\theta = \theta' + \alpha$, or $\theta' = \theta - \alpha$. This shows that r and $\theta - \alpha$ are the new coordinates of P.

As a consequence of the rotation property, a curve whose equation is of the form $r = f(\theta - \alpha)$ can be drawn by rotating the polar axis through an angle α and then drawing $r = f(\theta)$ with respect to the rotated system. For example, to draw

$$r = 2\cos\left(\theta - \frac{\pi}{4}\right)$$

we rotate through an angle of $\pi/4$ radians and draw $r = 2\cos\theta'$, which is the circle of Example 7.16, as shown in Figure 7.28.

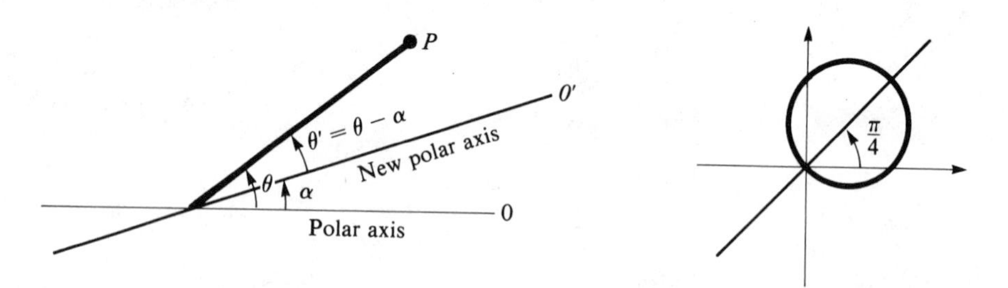

Figure 7.27

Figure 7.28

Suppose now that $\sin\theta$ replaces $\cos\theta$ in one of our standard equations. Since $\sin\theta = \cos(\theta - \pi/2)$, we see that the original curve has been rotated through $\pi/2$ radians. For example, $r = 2(1 + \sin\theta)$ is the cardioid $r = 2(1 + \cos\theta)$ rotated 90°, as shown in Figure 7.29. In a similar way, if $\sin n\theta$ replaces $\cos n\theta$, we write

$$\sin n\theta = \cos\left(n\theta - \frac{\pi}{2}\right) = \cos n\left(\theta - \frac{\pi}{2n}\right)$$

from which we conclude that the original curve is rotated through $\pi/2n$ radians. For example, $r^2 = 4\sin 2\theta$ is the lemniscate $r^2 = 4\cos 2\theta$ rotated $\pi/4$ radians, as shown in Figure 7.30. Since

$$-\cos\theta = \cos(\theta - \pi)$$

and

$$-\sin\theta = \cos\left(\theta - \frac{3\pi}{2}\right)$$

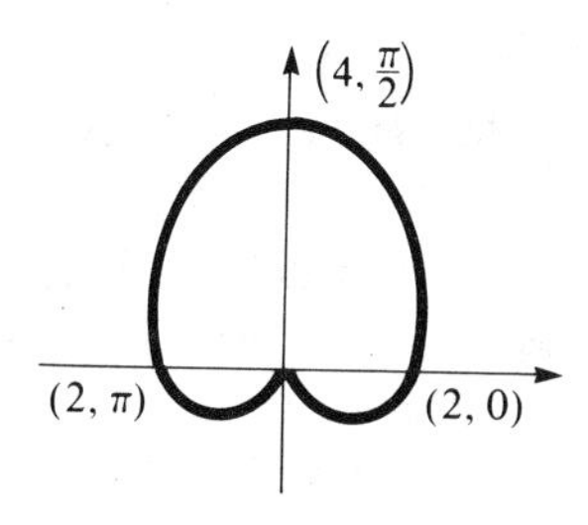

Figure 7.29

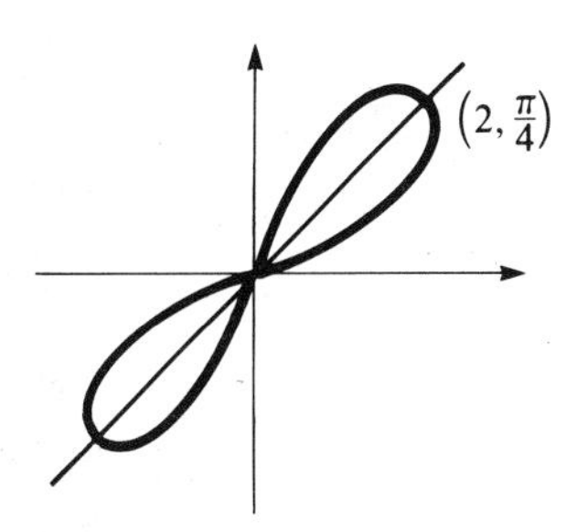

Figure 7.30

we also conclude that when $\cos\theta$ is replaced by $-\cos\theta$, the effect is to rotate π radians, and when $\cos\theta$ is replaced by $-\sin\theta$, the curve is rotated through $3\pi/2$ radians.

We conclude with an example of a type of problem that arises in calculus.

EXAMPLE 7.21 Draw the curves $r = 3\cos\theta$ and $r = 1 + \cos\theta$ on the same polar coordinate system and crosshatch the area inside the first curve which lies outside the second. Also find all points of intersection of the curves.

Solution The first curve is a circle of radius $\frac{3}{2}$ centered at $(\frac{3}{2}, 0)$, and the second is a cardioid, as shown in Figure 7.31. The crosshatched area is the one described in the problem. To find the points of intersection we set the two r values equal to one another and solve for θ.

$$3\cos\theta = 1 + \cos\theta$$

$$2\cos\theta = 1$$

$$\cos\theta = \frac{1}{2}$$

$$\theta = \frac{\pi}{3}, \frac{5\pi}{3}$$

Figure 7.31

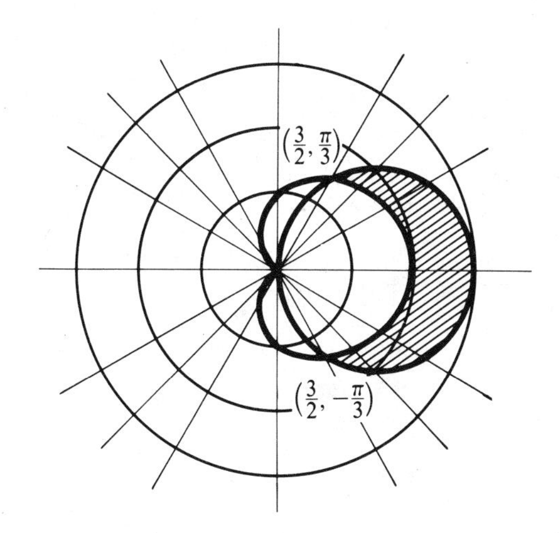

We find r to be $\frac{3}{2}$ in each case. So the curves intersect at $(\frac{3}{2}, \pi/3)$ and $(\frac{3}{2}, 5\pi/3)$. It is probably better to give the equivalent representation $(\frac{3}{2}, -\pi/3)$ for the second point.

By solving the two equations simultaneously we found two points of intersection. But from the graph we see that the curves also intersect at the pole. How is it that this did not show up in our solution? The answer lies in the fact that the curves pass through the pole for different values of θ. The circle goes through the pole when $\theta = \pi/2$, whereas the cardioid goes through the pole when $\theta = \pi$. Nevertheless, the pole is a point of intersection.

EXERCISE SET 7.6

A In Problems 1 and 2 plot the points in a polar coordinate system.

1. a. $\left(1, \frac{2\pi}{3}\right)$ **b.** $\left(2, -\frac{\pi}{2}\right)$ **c.** $\left(-2, \frac{3\pi}{2}\right)$ **d.** $\left(3, \frac{\pi}{6}\right)$
e. $\left(-1, -\frac{3\pi}{4}\right)$

2. a. $(4, \pi)$ **b.** $\left(-2, \frac{\pi}{2}\right)$ **c.** $\left(3, -\frac{5\pi}{6}\right)$ **d.** $\left(0, \frac{\pi}{5}\right)$
e. $\left(-3, -\frac{5\pi}{3}\right)$

3. Give five different sets of polar coordinates for the point $(2, 2\pi/3)$. Include at least one set for which $r < 0$ and one for which $\theta < 0$.

In Problems 4 and 5 give the corresponding rectangular coordinates for the given polar coordinates.

4. a. $\left(2, \frac{\pi}{6}\right)$ **b.** $\left(-\sqrt{2}, \frac{\pi}{4}\right)$ **c.** $\left(4, \frac{2\pi}{3}\right)$ **d.** $\left(6, \frac{3\pi}{2}\right)$
e. $\left(-8, -\frac{5\pi}{3}\right)$

5. a. $\left(2\sqrt{2}, -\frac{7\pi}{4}\right)$ **b.** $\left(6, \frac{11\pi}{6}\right)$ **c.** $\left(-10, -\frac{9\pi}{4}\right)$ **d.** $(4, 3\pi)$
e. $\left(\sqrt{8}, \frac{9\pi}{4}\right)$

In Problems 6 and 7 give the polar coordinates (for which $r > 0$ and $0 \leq \theta < 2\pi$) corresponding to the points with the given rectangular coordinates.

6. a. $(1, \sqrt{3})$ **b.** $(-2, 2)$ **c.** $(0, 5)$ **d.** $(-4, 0)$
e. $(-3, -\sqrt{3})$

7. a. $(4, -4)$ **b.** $(-2\sqrt{3}, 2)$ **c.** $(0, -2)$ **d.** $\left(\frac{1}{2}, -\frac{\sqrt{3}}{2}\right)$
e. $(-3\sqrt{2}, -\sqrt{6})$

In Problems 8–18 change to the equivalent polar equation.

8. $x^2 + y^2 = 9$

9. $x + y = 0$

10. $2x - 3y = 4$

11. $x = 3$

12. $y + 1 = 0$

13. $x^2 + y^2 - 4y = 0$

14. $x^2 - y^2 = 4$

15. $xy = 2$

16. $y^2 = 2x + 1$

17. $(x + 1)^2 + y^2 = 1$

18. $\sqrt{x^2 + y^2} - 2 = x$

In Problems 19–30 change to an equivalent rectangular equation.

19. $r = 4$

20. $\theta = \dfrac{\pi}{3}$

21. $r \cos \theta = 2$

22. $r \sin \theta = -1$

23. $2r \cos \theta - 3r \sin \theta = 5$

24. $r = 2 \sin \theta$

25. $r = 3 \sec \theta$

26. $r = 4 \csc \theta$

27. $r = -2 \cos \theta$

28. $\theta = \tan^{-1} \dfrac{2}{3}$

29. $r = \dfrac{1}{1 - \cos \theta}$

30. $r^2 \sin 2\theta = 1$

In Problems 31–52 identify the curve and draw its graph.

31. $r = 3$

32. $r = -2$

33. $\theta = \dfrac{3\pi}{4}$

34. $r \cos \theta = 2$

35. $r = -3 \csc \theta$

36. $r = 4 \cos \theta$

37. $r = 1 - \cos \theta$

38. $r = 1 + \sin \theta$

39. $r = 5 - 4 \cos \theta$

40. $r = 4 - 5 \sin \theta$

41. $r^2 = 9 \sin 2\theta$

42. $r = 2 \cos 2\theta$

43. $r = \sin 3\theta$

44. $r = -\cos 3\theta$

45. $r = 3 \sin 2\theta$

46. $r^2 = \cos^2 \theta - \sin^2 \theta$

47. $r = 2 + \sin \theta$

48. $r = 2 \cos \theta - 1$

49. $r = 2 \cos\left(\theta - \dfrac{3\pi}{4}\right)$

50. $r = 2 + 2 \cos\left(\theta + \dfrac{\pi}{6}\right)$

51. $r = 3(\sin \theta - 1)$

52. $r^2 = 8 \sin \theta \cos \theta$

B **53.** Write in polar form.

a. $x^2y^2 = 4(x^2 + y^2)^{3/2}$ **b.** $x^2 - y^2 - 4x = 0$

54. Write in rectangular form.

a. $r = \dfrac{4}{1 - 2 \sin \theta}$ **b.** $r \sec^2 \theta = \sin \theta - 3 \cos \theta$

Draw the graphs in Problems 55–61.

55. $r = 2 \sin^2 \dfrac{\theta}{2}$

56. $r = \dfrac{4}{2 + \cos \theta}$

57. $r = \cos \theta + \sin \theta$

Hint. Write in the form $r = a \cos(\theta - \alpha)$

58. $r = 3 \cos \theta - 4 \sin \theta$ (See hint to Problem 57.)

59. $r = 2 \cos \dfrac{\theta}{2}$

60. $r = \sqrt{4 \sin \theta}$

61. $r = 3 \cos 5\theta$

62. By changing to rectangular coordinates, show that the graph of $r = 4 \sec^2 (\theta/2)$ is a parabola. Draw the graph.

Problems 63–65 involve three spiral curves. Draw them for the indicated values of a.

63. **Spiral of Archimedes,** $r = a\theta$ $(a = 1)$

64. **Hyperbolic spiral,** $r = a/\theta$ $(a = 2)$

65. **Logarithmic spiral,** $r = e^{a\theta}$ $(a = 1)$

66. Graph the **bifolium** $r = a \sin 2\theta \cos \theta$ for $a = 1$.

In Problems 67–71 show the two curves on the same coordinate system, indicate the specified area by crosshatching, and find all points of intersection.

67. Outside $r = 2$ and inside $r = 2(1 + \cos \theta)$.

68. Outside $r = 1 + \sin \theta$ and inside $r = 3 \sin \theta$.

69. Inside $r = 3 \cos \theta$ and inside $r = 2 - \cos \theta$.

70. Inside $r = -4 \cos \theta$ and outside $r = 4 + 4 \cos \theta$.

71. Inside $r = 5 \cos \theta$ and outside $r = 2 + \cos \theta$.

7.7 Review Exercise Set

A In Problems 1–4 perform the indicated operations. Express answers in the form $a + bi$.

1. **a.** $(3 + 2i) + (5 - i)$ **b.** $(2 - 5i) - (-3 + 2i)$

2. **a.** $(4 + i)(3 - 2i)$ **b.** $(i - 1)(2 + i)$

3. **a.** $(2 - 3i)^{-1}$ **b.** $\dfrac{i + 1}{i - 1}$

4. **a.** $\dfrac{3 + i}{4 - 5i}$ **b.** $(2 - 3i)(5 + 7i) - \dfrac{4 - 3i}{1 + i}$

5. Solve for x and y.

a. $2ix - 3y = 4 + 6i$ **b.** $x + 2y - 3i = 1 + (x - y)i$

Simplify the expressions in Problems 6 and 7.

6. **a.** $\sqrt{-6}\,\sqrt{-24}$ **b.** $\dfrac{\sqrt{-2}}{\sqrt{-8}}$ **c.** $\sqrt{-12}\,\sqrt{27}$

7. **a.** i^{15} **b.** i^{-21} **c.** $\dfrac{1}{i^7}$

Solve the quadratic equations in Problems 8–13.

8. $x^2 - 2x + 4 = 0$

9. $2x^2 + x + 1 = 0$

10. $3t^2 - 5t + 6 = 0$

11. $y(3 - y) = 5$

12. $z(2z - 4) + 3 = 0$

13. $x^2 + 1 = -4(x + 1)$

14. Without solving, determine the nature of the roots of each of the following equations.

a. $6x^2 + 8x - 5 = 0$ **b.** $5y^2 = 3(y - 1)$ **c.** $t(t - 1) + 4 = 0$

d. $3x^2 + 2 = 5x$ **e.** $3x(3x + 8) = -16$

15. Plot each of the following in the complex plane, determine the modulus and argument, and write in trigonometric form.

a. $2\sqrt{3} - 2i$ **b.** $4(1 - i)$ **c.** $-\sqrt{3} - 3i$ **d.** $-8i$

e. $\sqrt{12} + \sqrt{-4}$

16. Change to rectangular form.

a. $2\sqrt{2}\left(\cos\dfrac{3\pi}{4} + i\sin\dfrac{3\pi}{4}\right)$ **b.** $6\left(\cos\dfrac{7\pi}{6} + i\sin\dfrac{7\pi}{6}\right)$

c. $5\left(\cos\dfrac{5\pi}{2} + i\sin\dfrac{5\pi}{2}\right)$ **d.** $4\left(\cos\dfrac{5\pi}{3} + i\sin\dfrac{5\pi}{3}\right)$

e. $9(\cos 585° + i\sin 585°)$

In Problems 17–21 use the polar form to perform the indicated operations. Express answers in rectangular form.

17. **a.** $3\left(\cos\dfrac{3\pi}{4} + i\sin\dfrac{3\pi}{4}\right) \cdot 4\left(\cos\dfrac{\pi}{2} + i\sin\dfrac{\pi}{2}\right)$

b. $\dfrac{6\left(\cos\dfrac{5\pi}{6} + i\sin\dfrac{5\pi}{6}\right)}{2\left(\cos\dfrac{\pi}{3} + i\sin\dfrac{\pi}{3}\right)}$

18. **a.** $(1 - i\sqrt{3})(-\sqrt{3} - i)$ **b.** $\dfrac{-2i}{1 + i}$

19. **a.** $\dfrac{2(\cos 350° + i\sin 350°)}{16(\cos 200° + i\sin 200°)}$ **b.** $\dfrac{(-2 + 2i)(\sqrt{3} - 3i)}{2\sqrt{3} + 2i}$

20. **a.** $\left[3\left(\cos\dfrac{2\pi}{3} + i\sin\dfrac{2\pi}{3}\right)\right]^4$ **b.** $\left[2\left(\cos\dfrac{5\pi}{9} + i\sin\dfrac{5\pi}{9}\right)\right]^3$

21. **a.** $(2 - 2i)^5$ **b.** $\left(\dfrac{-\sqrt{3} - i}{2}\right)^4$

In Problems 22–25 find the indicated roots by De Moivre's theorem. Express answers in rectangular form.

22. Square roots of $1 - i\sqrt{3}$.

23. Fourth roots of 81.

24. Cube roots of $-64i$.

25. Cube roots of i.

26. Solve completely, making use of Equation (7.7).

a. $x^4 + 1 = 0$ **b.** $8x^3 - 27 = 0$

27. Plot the following points in a polar coordinate system. Give the equivalent rectangular coordinates of each point.

a. $\left(2\sqrt{2}, \frac{3\pi}{4}\right)$ **b.** $\left(-4, \frac{5\pi}{6}\right)$ **c.** $\left(6, -\frac{\pi}{3}\right)$ **d.** $\left(5, \frac{3\pi}{2}\right)$

e. $\left(-\sqrt{8}, -\frac{3\pi}{4}\right)$

28. Change the given rectangular coordinates to polar coordinates with $r > 0$ and such that $|\theta|$ is as small as possible. Plot the points in a polar coordinate system.

a. $(2, -2)$ **b.** $(-2, \sqrt{12})$ **c.** $(-\sqrt{3}, -1)$ **d.** $(0, -2)$

e. $(\sqrt{6}, \sqrt{18})$

In Problems 29–31 write an equivalent polar equation.

29. **a.** $x^2 + (y - 2)^2 = 4$ **b.** $\sqrt{3}\,x + y = 0$

30. **a.** $x^2 = 16 - y^2$ **b.** $y = 3x - 4$

31. **a.** $x^2 + 8y = 16$ **b.** $x^2 - y^2 = \dfrac{2xy}{x^2 + y^2}$

In Problems 32–34 write an equivalent rectangular equation.

32. **a.** $r = -3$ **b.** $r = \sec\theta$

33. **a.** $r = -2\sin\theta$ **b.** $\theta = \sin^{-1}\left(-\frac{1}{3}\right)$

34. **a.** $r = \dfrac{4}{2 - 3\cos\theta}$ **b.** $r^2\cos 2\theta = 4$

In Problems 35–44 draw the graph.

35. $r = 5$

36. $\theta = \tan^{-1}\left(-\frac{1}{2}\right)$

37. $r = -3\cos\theta$

38. $r = 3 - \sin\theta$

39. $r = 2 - 3\cos\theta$

40. $r^2 = -4\cos 2\theta$

41. $r = 2 - 2\cos\theta$

42. $r + 2\cos 2\theta = 0$

43. $r = 1 - \cos\left(\theta + \frac{\pi}{3}\right)$

44. $r = \cos\left(3\theta - \frac{3\pi}{4}\right)$

B **45.** Evaluate

$$\frac{z_1 \cdot z_2^{\,4}}{\sqrt{z_3}}$$

where $z_1 = -4 - 4i$, $z_2 = 2\sqrt{3} - 2i$, and $z_3 = 4i$.

46. Prove that

$$\frac{z_1}{z_2} = \frac{z_1\bar{z}_2}{|z_2|^2}$$

47. Let $z = -1 + i\sqrt{3}$. Verify by use of De Moivre's theorem that both z and $\bar{z}$ are solutions of the equation

$$2x^4 + 4x^3 + 5x^2 - 6x - 12 = 0$$

48. Factor $x^6 - 729$ completely (a) over the complex field, and (b) over the real field.

49. Make use of the trigonometric form, together with a calculator or tables, to evaluate each of the following. Write answers in rectangular form.

a. $\left(\frac{5}{13} + \frac{12i}{13}\right)^7$ **b.** $(3 - 4i)^{1/5}$

50. Solve the equation

$$x^2 - 2ix - 1 - i = 0$$

and express your answers in the form $a + bi$.

In Problems 51–55 draw the graph of the given polar equation.

51. $r = \sqrt{3} \cos \theta + \sin \theta$ **52.** $r = \dfrac{2}{1 - 2 \sin \theta}$

53. $r = -4 \cos^2 \dfrac{\theta}{2}$ **54.** $r = -2 \sin \dfrac{\theta}{2}$

55. $r = 2 \sin \theta \tan \theta$ (This is a special case of the **cissoid**.)

In Problems 56 and 57 draw both curves on the same polar coordinate system, crosshatch the specified area, and find all points of intersection.

56. Outside $r = 2$ and inside $r^2 = 8 \sin 2\theta$.

57. Outside $r = 3 - 4 \cos \theta$ and inside $r = 2\sqrt{3}(1 - \cos \theta)$.

Answers to Selected Problems

Exercise Set 1.2

1. **a.** $\pi/6$ **b.** $\pi/4$ **3.** **a.** $2\pi/3$ **b.** $3\pi/4$ **5.** **a.** $4\pi/3$ **b.** $3\pi/2$ **7.** **a.** $11\pi/6$ **b.** $\pi/12$ **9.** **a.** $10\pi/3$ **b.** $-4\pi/5$ **11.** **a.** $135°$ **b.** $300°$ **13.** **a.** $150°$ **b.** $-240°$
15. **a.** $100°$ **b.** $(360/\pi)° \approx 114.59°$ **17.** **a.** $(540/\pi)° \approx 171.89°$ **b.** $15°$
19. **a.** $-(720/\pi)° \approx -229.18°$ **b.** $330°$ **21.** **a.** $32.85°$ **b.** $102.58°$ **23.** **a.** $27.22°$ **b.** $-56.32°$
25. **a.** $-13° 03'$ **b.** $153° 25'$ **27.** **a.** $171° 53'$ **b.** $406° 21'$ **29.** **a.** 0.233 **b.** 5.065
31. **a.** $s = 6$ **b.** $r = \frac{4}{3}$ **33.** **a.** $r = 32/\pi \approx 10.19$ **b.** $\theta = \frac{2}{3}$ **35.** $s = 22\pi$ mi. ≈ 69.12 mi.
37. $s = 5\pi$ in. ≈ 15.7 in. **39.** 60π rad./sec. **41.** $\frac{3}{4}$ rad, $(135/\pi)° \approx 43.0°$ **43.** 66,705 mph
45. $r_B = 7.78$ in., $r_C = 9.78$ in.

Exercise Set 1.3

1. $a = \frac{5}{2}$, $b = 5\sqrt{3}/2$ **3.** $a = 10\sqrt{3}$, $c = 20$ **5.** $a = 2\sqrt{3}$, $c = 4\sqrt{3}$ **7.** $a = 4$, $b = 4$
9. $a = 5\sqrt{3}/2$, $b = \frac{5}{2}$ **11.** $b = 3$, $c = 3\sqrt{2}$ **13.** $c = \sqrt{13}$ **15.** $b = 24$ **17.** $b = 48$ **19.** $c = 3$
21. $81° 09'$ **23.** $23.82°$ **25.** $5\pi/12$ **29.** **a.** $10\sqrt{6}$ **b.** $\frac{75}{2}(\sqrt{3} + 4)$
31. length $AB = 2a$, length $BD = a$ $\therefore$ $\triangle CBD$ is isosceles. Since $B = 60°$ and $\angle BDC = \angle BCD$, it follows that these two angles are both 60°. So $\triangle CBD$ is equilateral. $\therefore$ $CD = a$.

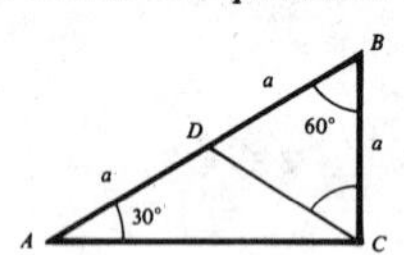

33. $\pi a^2/3$

Exercise Set 1.4

1. **a.** $\sin 60° = \sqrt{3}/2$, $\cos 60° = \frac{1}{2}$, $\tan 60° = \sqrt{3}$, $\cot 60° = 1/\sqrt{3}$, $\sec 60° = 2$, $\csc 60° = 2/\sqrt{3}$
b. $\sin 45° = 1/\sqrt{2}$, $\cos 45° = 1/\sqrt{2}$, $\tan 45° = 1$, $\cot 45° = 1$, $\sec 45° = \sqrt{2}$, $\csc 45° = \sqrt{2}$
3. $\sin B = \frac{8}{17}$, $\cos B = \frac{15}{17}$, $\tan B = \frac{8}{15}$, $\cot B = \frac{15}{8}$, $\sec B = \frac{17}{15}$, $\csc B = \frac{17}{8}$
5. $\sin A = 1/\sqrt{5}$, $\cos A = 2/\sqrt{5}$, $\tan A = \frac{1}{2}$, $\cot A = 2$, $\sec A = \sqrt{5}/2$, $\csc A = \sqrt{5}$
7. $\sin B = \sqrt{5}/3$, $\cos B = \frac{2}{3}$, $\tan B = \sqrt{5}/2$, $\cot B = 2/\sqrt{5}$, $\sec B = \frac{3}{2}$, $\csc B = 3/\sqrt{5}$
9. $\sin U = z/x$, $\cos U = y/x$, $\tan U = z/y$, $\cot U = y/z$, $\sec U = x/y$, $\csc U = x/z$
11. $\sin M = f/e$, $\cos M = d/e$, $\tan M = f/d$, $\cot M = d/f$, $\sec M = e/d$, $\csc M = e/f$
13. $(1 + 2\sqrt{2})/2$ **15.** $2\sqrt{3}$ **17.** 3 **19.** 1 **21.** $(1 + \sqrt{3})/2\sqrt{2}$
31. $\sin\theta = 2\sqrt{2}/3$, $\tan\theta = 2\sqrt{2}$, $\cot\theta = 1/2\sqrt{2}$, $\sec\theta = 3$, $\csc\theta = 3/2\sqrt{2}$
33. $\sin\theta = 2/\sqrt{5}$, $\cos\theta = 1/\sqrt{5}$, $\cot\theta = \frac{1}{2}$, $\sec\theta = \sqrt{5}$, $\csc\theta = \sqrt{5}/2$
35. $\sin\theta = \sqrt{5}/3$, $\cos\theta = \frac{2}{3}$, $\tan\theta = \sqrt{5}/2$, $\cot\theta = 2/\sqrt{5}$, $\csc\theta = 3/\sqrt{5}$
37. $\cos\theta = \sqrt{1 - x^2}$, $\tan\theta = x/\sqrt{1 - x^2}$, $\cot\theta = \sqrt{1 - x^2}/x$, $\sec\theta = 1/\sqrt{1 - x^2}$, $\csc\theta = 1/x$
39. $\sin\theta = x/\sqrt{1 + x^2}$, $\cos\theta = 1/\sqrt{1 + x^2}$, $\cot\theta = 1/x$, $\sec\theta = \sqrt{1 + x^2}$, $\csc\theta = \sqrt{1 + x^2}/x$

Exercise Set 1.5

1. **a.** $5\sqrt{2}$ **b.** $3\sqrt{10}$ **c.** 13
3. $\sin 330° = -\frac{1}{2}$, $\cos 330° = \sqrt{3}/2$, $\tan 330° = -1/\sqrt{3}$, $\cot 330° = -\sqrt{3}$, $\sec 330° = 2/\sqrt{3}$, $\csc 330° = -2$

5. $\sin 120° = \sqrt{3}/2$, $\cos 120° = -\frac{1}{2}$, $\tan 120° = -\sqrt{3}$, $\cot 120° = -1/\sqrt{3}$, $\sec 120° = -2$, $\csc 120° = 2/\sqrt{3}$

7. $\sin 270° = -1$, $\cos 270° = 0$, $\tan 270°$ undefined, $\cot 270° = 0$, $\sec 270°$ undefined, $\csc 270° = -1$

9. $\sin(7\pi/4) = -1/\sqrt{2}$, $\cos(7\pi/4) = 1/\sqrt{2}$, $\tan(7\pi/4) = -1$, $\cot(7\pi/4) = -1$, $\sec(7\pi/4) = \sqrt{2}$, $\csc(7\pi/4) = -\sqrt{2}$

11. $\sin(-\pi/6) = -\frac{1}{2}$, $\cos(-\pi/6) = \sqrt{3}/2$, $\tan(-\pi/6) = -1/\sqrt{3}$, $\cot(-\pi/6) = -\sqrt{3}$, $\sec(-\pi/6) = 2/\sqrt{3}$, $\csc(-\pi/6) = -2$

13. $\sin 480° = \sqrt{3}/2$, $\cos 480° = -\frac{1}{2}$, $\tan 480° = -\sqrt{3}$, $\cot 480° = -1/\sqrt{3}$, $\sec 480° = -2$, $\csc 480° = 2/\sqrt{3}$

15. $\sin 3\pi = 0$, $\cos 3\pi = -1$, $\tan 3\pi = 0$, $\cot 3\pi$ undefined, $\sec 3\pi = -1$, $\csc 3\pi$ undefined

17. **a.** $-1/\sqrt{2}$ **b.** $-\sqrt{3}$ **c.** 2 **d.** $-1/\sqrt{2}$ **e.** $\sqrt{3}$ **19.** **a.** $-\frac{1}{2}$ **b.** $\sqrt{3}/2$ **c.** -1 **d.** $-\sqrt{2}$ **e.** -1 **21.** **a.** $1/\sqrt{3}$ **b.** $\sqrt{2}$ **c.** $1/\sqrt{3}$ **d.** $\frac{1}{2}$ **e.** $2/\sqrt{3}$ **23.** **a.** $\cot 32°$ **b.** $-\sin 22°$ **c.** $\tan 80°$ **d.** $\cos 70°$ **e.** $-\csc 49°$ **25.** **a.** $-\tan(5\pi/12)$ **b.** $-\sec(\pi/8)$ **c.** $-\sin(\pi/5)$ **d.** $\csc(\pi/9)$ **e.** $-\cos(\pi/12)$

27. $\sin\theta = -\frac{12}{13}$, $\cos\theta = \frac{5}{13}$, $\tan\theta = -\frac{12}{5}$, $\cot\theta = -\frac{5}{12}$, $\sec\theta = \frac{13}{5}$, $\csc\theta = -\frac{13}{12}$

29. $\sin\theta = -3/\sqrt{13}$, $\cos\theta = 2/\sqrt{13}$, $\tan\theta = -\frac{3}{2}$, $\cot\theta = -\frac{2}{3}$, $\sec\theta = \sqrt{13}/2$, $\csc\theta = -\sqrt{13}/3$

31. $\sin\theta = -\frac{24}{25}$, $\cos\theta = -\frac{7}{25}$, $\tan\theta = \frac{24}{7}$, $\cot\theta = \frac{7}{24}$, $\sec\theta = -\frac{25}{7}$, $\csc\theta = -\frac{25}{24}$

33. -8

35. $\sin\theta = -\frac{12}{13}$, $\tan\theta = -\frac{12}{5}$, $\cot\theta = -\frac{5}{12}$, $\sec\theta = \frac{13}{5}$, $\csc\theta = -\frac{13}{12}$

37. $\sin\theta = \frac{1}{3}$, $\cos\theta = -2\sqrt{2}/3$, $\tan\theta = -1/2\sqrt{2}$, $\cot\theta = -2\sqrt{2}$, $\sec\theta = -3/2\sqrt{2}$

39. $\sin\theta = 3/\sqrt{10}$, $\cos\theta = -1/\sqrt{10}$, $\tan\theta = -3$, $\sec\theta = -\sqrt{10}$, $\csc\theta = \sqrt{10}/3$

41. $\sin\theta = t/\sqrt{1+t^2}$, $\cos\theta = 1/\sqrt{1+t^2}$, $\cot\theta = 1/t$, $\sec\theta = \sqrt{1+t^2}$, $\csc\theta = \sqrt{1+t^2}/t$

Solution 1

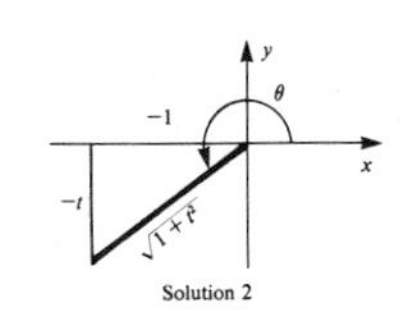

Solution 2

$\sin\theta = -t/\sqrt{1+t^2}$, $\cos\theta = -1/\sqrt{1+t^2}$, $\cot\theta = 1/t$, $\sec\theta = -\sqrt{1+t^2}$, $\csc\theta = -\sqrt{1+t^2}/t$

43. **a.** $\sin(\pi-\theta) = \sin\theta$, $\cos(\pi-\theta) = -\cos\theta$, $\tan(\pi-\theta) = -\tan\theta$
$\cot(\pi-\theta) = -\cot\theta$, $\sec(\pi-\theta) = -\sec\theta$, $\csc(\pi-\theta) = \csc\theta$

b. $\sin(\pi+\theta) = -\sin\theta$, $\cos(\pi+\theta) = -\cos\theta$, $\tan(\pi+\theta) = \tan\theta$,
$\cot(\pi+\theta) = \cot\theta$, $\sec(\pi+\theta) = -\sec\theta$, $\csc(\pi+\theta) = -\csc\theta$

c. $\sin(\pi/2+\theta) = \cos\theta$, $\cos(\pi/2+\theta) = -\sin\theta$, $\tan(\pi/2+\theta) = -\cot\theta$,
$\cot(\pi/2+\theta) = -\tan\theta$, $\sec(\pi/2+\theta) = -\csc\theta$, $\csc(\pi/2+\theta) = \sec\theta$

d. $\sin(3\pi/2 - \theta) = -\cos\theta$, $\cos(3\pi/2 - \theta) = -\sin\theta$, $\tan(3\pi/2 - \theta) = \cot\theta$, $\cot(3\pi/2 - \theta) = \tan\theta$, $\sec(3\pi/2 - \theta) = -\csc\theta$, $\csc(3\pi/2 - \theta) = -\sec\theta$
e. $\sin(2\pi - \theta) = -\sin\theta$, $\cos(2\pi - \theta) = \cos\theta$, $\tan(2\pi - \theta) = -\tan\theta$, $\cot(2\pi - \theta) = -\cot\theta$, $\sec(2\pi - \theta) = \sec\theta$, $\csc(2\pi - \theta) = -\csc\theta$

45. If n is even $(2n + 1)\pi/2$ is coterminal with $\pi/2$, and in this case $(-1)^n = 1$, so formula is correct. If n is odd, $(2n + 1)\pi/2$ is coterminal with $3\pi/2$, and in this case $(-1)^n = -1$, so formula is correct.

47. For $k = 0, \pm 4, \pm 8, \ldots$, $(1 + 2k)\pi/4$ is coterminal with $\pi/4$, and $(-1)^k = 1$, so formula is true.
For $k = \pm 2, \pm 6, \pm 10, \ldots$, $(1 + 2k)\pi/4$ is coterminal with $5\pi/4$, and $(-1)^k = 1$, so formula is true.
For $k = \pm 1, \pm 5, \pm 9, \ldots$, $(1 + 2k)\pi/4$ is coterminal with $3\pi/4$, and $(-1)^k = -1$, so formula is true.
For $k = \pm 3, \pm 7, \pm 11, \ldots$, $(1 + 2k)\pi/4$ is coterminal with $7\pi/4$, and $(-1)^k = -1$, so formula is true.

1.6 Review Exercise Set

1. **a.** $4\pi/3$ **b.** $7\pi/4$ **c.** $\pi/12$ **d.** 3π **e.** $8\pi/9$ **3.** **a.** 59.62° **b.** 205.38° **c.** −452.77° **d.** 14.57° **e.** 256.86° **5.** **a.** 0.601 **b.** 2.324 **c.** 0.267 **d.** −0.434 **e.** 2.356
7. $\theta = (240/\pi)^\circ \approx 76.39^\circ$ **9.** $\omega = \frac{80}{3}$ rad/sec; $800/\pi$ rpm ≈ 254.6 rpm **11.** $a = b = 2\sqrt{2}$
13. $a = 5/\sqrt{3}$, $c = 10/\sqrt{3}$ **15.** $a = \frac{5}{2}$, $b = 5\sqrt{3}/2$ **17.** $b = \sqrt{65}$ **19.** **a.** $1 - 1/\sqrt{3}$ **b.** $-1/\sqrt{6}$
21.

	0°	30°	45°	60°	90°	120°	135°	150°	180°	210°	225°	240°	270°	300°	315°	330°
sin	0	$1/2$	$1/\sqrt{2}$	$\sqrt{3}/2$	1	$\sqrt{3}/2$	$1/\sqrt{2}$	$1/2$	0	$-1/2$	$-1/\sqrt{2}$	$-\sqrt{3}/2$	-1	$-\sqrt{3}/2$	$-1/\sqrt{2}$	$-1/2$
cos	1	$\sqrt{3}/2$	$1/\sqrt{2}$	$1/2$	0	$-1/2$	$-1/\sqrt{2}$	$-\sqrt{3}/2$	-1	$-\sqrt{3}/2$	$-1/\sqrt{2}$	$-1/2$	0	$1/2$	$1/\sqrt{2}$	$\sqrt{3}/2$
tan	0	$1/\sqrt{3}$	1	$\sqrt{3}$	undef.	$-\sqrt{3}$	-1	$-1/\sqrt{3}$	0	$1/\sqrt{3}$	1	$\sqrt{3}$	undef.	$-\sqrt{3}$	-1	$-1/\sqrt{3}$
cot	undef.	$\sqrt{3}$	1	$1/\sqrt{3}$	0	$-1/\sqrt{3}$	-1	$-\sqrt{3}$	undef.	$\sqrt{3}$	1	$1/\sqrt{3}$	0	$-1/\sqrt{3}$	-1	$-\sqrt{3}$
sec	1	$2/\sqrt{3}$	$\sqrt{2}$	2	undef.	-2	$-\sqrt{2}$	$-2/\sqrt{3}$	-1	$-2/\sqrt{3}$	$-\sqrt{2}$	-2	undef.	2	$\sqrt{2}$	$2/\sqrt{3}$
csc	undef.	2	$\sqrt{2}$	$2/\sqrt{3}$	1	$2/\sqrt{3}$	$\sqrt{2}$	2	undef.	-2	$-\sqrt{2}$	$-2/\sqrt{3}$	-1	$-2/\sqrt{3}$	$-\sqrt{2}$	-2

23. **a.** $\sin\theta = -\frac{4}{5}$, $\cos\theta = -\frac{3}{5}$, $\tan\theta = \frac{4}{3}$, $\cot\theta = \frac{3}{4}$, $\sec\theta = -\frac{5}{3}$, $\csc\theta = -\frac{5}{4}$
b. $\sin\theta = -\frac{12}{13}$, $\cos\theta = \frac{5}{13}$, $\tan\theta = -\frac{12}{5}$, $\cot\theta = -\frac{5}{12}$, $\sec\theta = \frac{13}{5}$, $\csc\theta = -\frac{13}{12}$
c. $\sin\theta = 3/\sqrt{13}$, $\cos\theta = -2/\sqrt{13}$, $\tan\theta = -\frac{3}{2}$, $\cot\theta = -\frac{2}{3}$, $\sec\theta = -\sqrt{13}/2$, $\csc\theta = \sqrt{13}/3$
d. $\sin\theta = -\frac{1}{3}$, $\cos\theta = 2\sqrt{2}/3$, $\tan\theta = -1/2\sqrt{2}$, $\cot\theta = -2\sqrt{2}$, $\sec\theta = 3/2\sqrt{2}$, $\csc\theta = -3$

27. **a.** $\theta = 7\pi/6, 11\pi/6$ **b.** $\theta = \pi$ **c.** $\theta = 2\pi/3, 5\pi/3$ **d.** $\theta = \pi/3, 5\pi/3$ **e.** $\theta = \pi/2, 3\pi/2$

29. **a.** $9 + 3\sqrt{3}/4$ square feet **b.** $9 + 2\sqrt{3}$ feet

31. **a.** $\sin A = \frac{4}{5}$, $\tan A = \frac{4}{3}$, $\cot A = \frac{3}{4}$, $\sec A = \frac{5}{3}$, $\csc A = \frac{5}{4}$

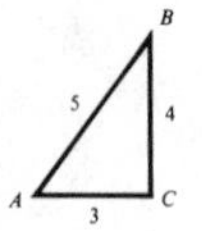

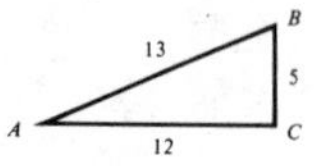

b. $\sin B = \frac{12}{13}$, $\cos B = \frac{5}{13}$, $\cot B = \frac{5}{12}$, $\sec B = \frac{13}{5}$, $\csc B = \frac{13}{12}$

33. **a.** $\sin\theta = -2/\sqrt{13}$, $\cos\theta = -3/\sqrt{13}$, $\cot\theta = \frac{3}{2}$, $\sec\theta = -\sqrt{13}/3$, $\csc\theta = -\sqrt{13}/2$

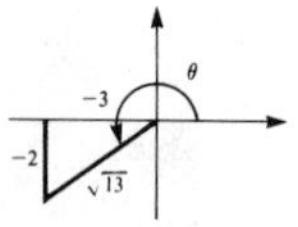

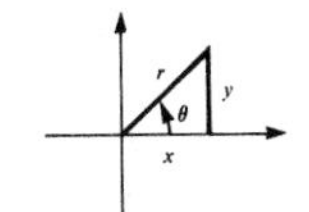

b. $\cos\theta = \sqrt{15}/4$, $\tan\theta = -1/\sqrt{15}$, $\cot\theta = -\sqrt{15}$, $\sec\theta = 4/\sqrt{15}$, $\csc\theta = -4$

35. $\omega = 264/13$ rad/sec; $7920/13\pi$ rpm ≈ 193.92 rpm **37.** $90\sqrt{2}$

39. $x^2 + y^2 = r^2$

a. Divide by r^2: $(x/r)^2 + (y/r)^2 = 1$, or $\sin^2\theta + \cos^2\theta = 1$

b. Divide by x^2 $(x \neq 0)$: $1 + (y/x)^2 = (r/x)^2$ or $1 + \tan^2\theta = \sec^2\theta$

41.

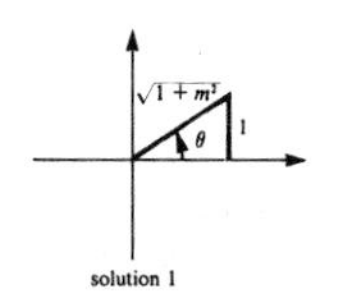

solution 1

$\sin\theta = 1/\sqrt{1+m^2}$, $\csc\theta = \sqrt{1+m^2}$,
$\cos\theta = m/\sqrt{1+m^2}$, $\sec\theta = \sqrt{1+m^2}/m$,
$\tan\theta = 1/m$

solution 2

$\sin\theta = -1/\sqrt{1+m^2}$, $\csc\theta = -\sqrt{1+m^2}$,
$\cos\theta = -m/\sqrt{1+m^2}$, $\sec\theta = -\sqrt{1+m^2}/m$,
$\tan\theta = 1/m$

Exercise Set 2.2

1. a. 0.6205 **b.** 0.6049 **c.** 0.2157 **d.** 1.0377 **3. a.** 2.8797 **b.** 1.1277 **c.** −0.9609 **d.** −0.5279 **5. a.** 1.1316 **b.** 4.6237 **c.** −0.7150 **d.** 1.8811 **7. a.** 29.59° **b.** 15.74° **9. a.** 22.87° **b.** 32.58° **11. a.** 63.43° **b.** 14.25° **13. a.** −51.07° **b.** 114.25° **15. a.** 0.272 **b.** 0.759 **17. a.** 0.415 **b.** 1.005 **19. a.** −0.461 **b.** 2.376 **21. a.** 0.5329 **b.** 0.9466 **c.** 0.4986 **d.** 1.4974 **23. a.** 3.3759 **b.** 1.1716 **c.** 0.8348 **d.** 1.5577 **25. a.** 0.2131 **b.** 0.8674 **c.** 1.127 **d.** 0.9916 **27. a.** 0.5396 **b.** 1.004 **c.** −0.4964 **d.** 1.072 **29. a.** 32° 42′ **b.** 63° 48′ **c.** 21° 36′ **d.** 42° 30′ **31. a.** 0.5943 **b.** 0.9046 **33. a.** 1.1598 **b.** 0.3099 **35. a.** 0.9894 **b.** 0.6008 **37. a.** 0.1336 **b.** 0.3276 **39. a.** 1.485 **b.** 1.935 **41. a.** 51° 55′ **b.** 19° 41′ **43. a.** 0.368 **b.** 0.673 **45. a.** 1.065 **b.** 0.961 **47. a.** −0.8889 **b.** 2.7155 **49. a.** 1.102 **b.** −0.7765 **51. a.** −0.2960 **b.** −1.198 **53. a.** 257° 08′ **b.** 4.488 **55. a.** −0.41421 **b.** −3.86370 **57. a.** 244.67° **b.** 144.38° **59.** 43° 35′

Exercise Set 2.3

1. a. 321.55 **b.** 2.11 **c.** 0.20 **d.** 1.04 **e.** 25.64 **3. a.** 2.40×10^4 **b.** 2.400×10^4 **c.** 5.2×10^{-4} **d.** 3.2015×10^3 **e.** 8.500×10^7 **5.** 219.5 **7.** 8.57 **9.** 0.32 **11.** 13.9 **13.** 126.6 **15.** 13.9 **17.** 1.03×10^4 **19. a.** 0.05, 0.0002 **b.** 0.0005, 0.02 **c.** 0.005, 0.0003 **d.** 5×10^{-10}, 9×10^{-5} **21.** 5.51×10^3 cu. in., min. vol. $\approx$ 5,464 cu. in., max. vol. $\approx$ 5,562 cu. in.

Exercise Set 2.4

1. $B = 32.6°$, $a = 23.8$, $c = 28.2$ **3.** $A = 22.7°$, $a = 5.65$, $c = 14.6$ **5.** $B = 40.3°$, $a = 18.8$, $b = 15.9$ **7.** $B = 57.5°$, $b = 38.8$, $c = 46.0$ **9.** $A = 25.53°$, $B = 64.47°$, $b = 28.24$ **11.** $A = 67° 20'$, $a = 6.13$, $c = 6.64$ **13.** $B = 54.43°$, $b = 4.841$, $c = 5.952$ **15.** $A = 48.3°$, $B = 41.7°$, $a = 28.5$ **17.** 381 ft. **19.** 127 ft. **21.** 45.1 ft., 49.9 ft. **23.** 19.2 mi., 51.3° **25.** 840 meters from B, 926 meters from A **27.** 6.142 miles and 8.331 miles **29.** 249° 10′, 219 mph **31.** 54.3 ft. **33.** 51.5 miles **35.** 4.18 km

Exercise Set 2.5

1. $|\mathbf{V}| = 7.62$, $\alpha = 23.2°$ **3.** $|\mathbf{V}| = 4.25$, $\alpha = 240.4°$ **5.** $V_x = 21.2$, $V_y = 13.2$ **7.** $V_x = 1.93$, $V_y = -0.621$
9. $V_x = 3$, $V_y = -3$ **11.** $V_x = -12$, $V_y = -1$
13.

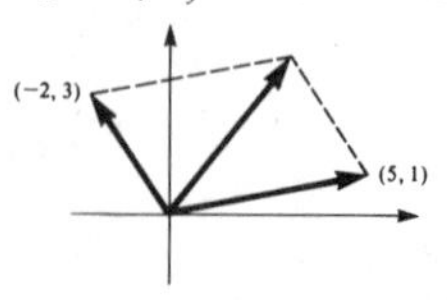

15.

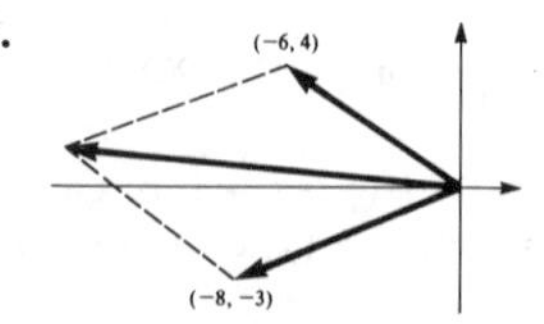

17.

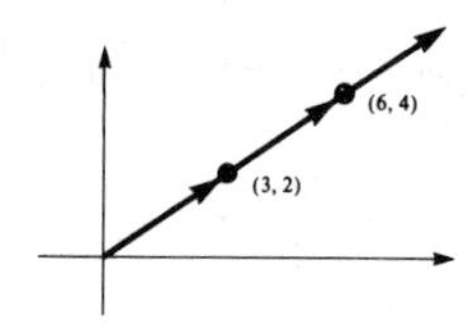

19. $\mathbf{U} \cdot \mathbf{V} = 10$, $\theta = 45°$ **21.** In each case $\mathbf{V} \cdot \mathbf{W} = 0$, so $\theta = 90°$ **23.** $b = -\frac{6}{5}$
25. course = 102.4°, speed = 205 mph **27.** $|\mathbf{V}| = 5.39$ mph, angle from bank = 68.2°, 40.9 sec.
29. 32.2° from vertical, 82.7 feet/sec.
31. **a.**

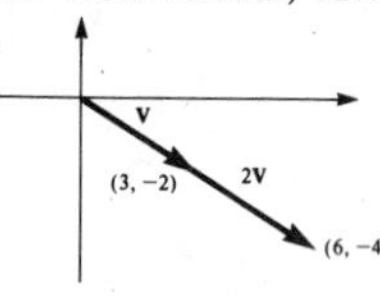

b.

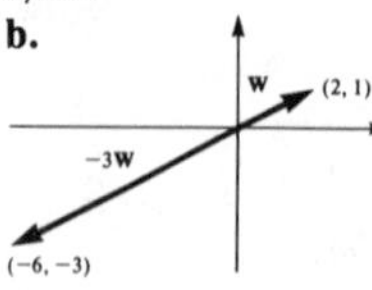

c.

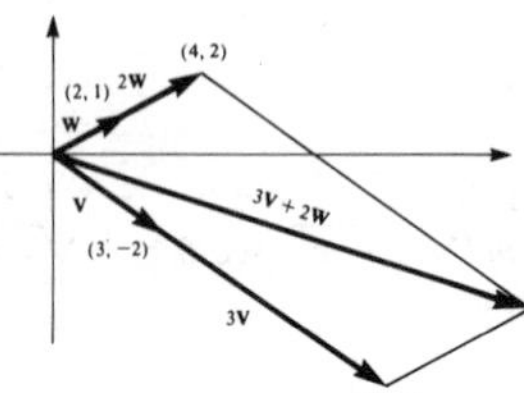

d.

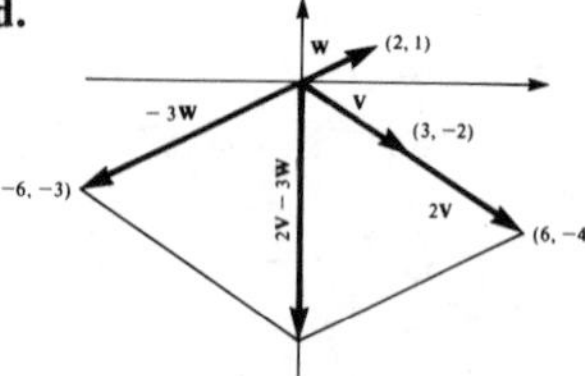

33.

35. $|(cV_x, cV_y)| = \sqrt{(cV_x)^2 + (cV_y)^2} = \sqrt{c^2(V_x^{\ 2} + V_y^{\ 2})} = |c|\sqrt{V_x^{\ 2} + V_y^{\ 2}} = |c|\ |\mathbf{V}|$.
Let θ = angle $\mathbf{V}$ makes with x axis. Then $\tan\theta = V_y/V_x = cV_y/cV_x$. So both magnitude and direction of (cV_x, cV_y) agree with $c\mathbf{V}$. Thus, $c\mathbf{V} = (cV_x, cV_y)$.

37. Let $\mathbf{V} = (V_x, V_y)$, $\mathbf{U} = (U_x, U_y)$, and $\mathbf{W} = (W_x, W_y)$.
a. $\mathbf{V} \cdot \mathbf{W} = V_xW_x + V_yW_y = W_xV_x + W_yV_y = \mathbf{W} \cdot \mathbf{V}$
b. $\mathbf{U} \cdot (\mathbf{V} + \mathbf{W}) = (U_x, U_y) \cdot (V_x + W_x, V_y + W_y)$
$= U_x(V_x + W_x) + U_y(V_y + W_y)$
$= (U_xV_x + U_yV_y) + (U_xW_x + U_yW_y)$
$= \mathbf{U} \cdot \mathbf{V} + \mathbf{V} \cdot \mathbf{W}$
c. $c(U \cdot V) = c(U_xV_x + U_yV_y) = (cU_x)V_x + (cU_y)V_y$
$= (cU_x, cU_y) \cdot (V_x, V_y) = (c\mathbf{U}) \cdot \mathbf{V}$
Also, $c(U_xV_x + U_yV_y) = U_x(cV_x) + U_y(cV_y) = (U_x, U_y) \cdot (cV_x, cV_y)$
$= \mathbf{U} \cdot (c\mathbf{V})$

39. Approx. 634 pounds **41.** 19,359 pounds in AC, 16,644 pounds in BC.

2.6 Review Exercise Set

1. **a.** 0.7360 **b.** −2.8444 **c.** 0.9717 **d.** −0.8109 **e.** 0.9150 **3.** **a.** 71° 40′ **b.** 60° 44′
c. 25° 13′ **d.** 14° 27′ **e.** 73° 37′ **5.** **a.** 0.47270 **b.** −0.012407 **c.** −0.36461 **d.** −1.0792
e. −14.783 **7.** **a.** 2.035 **b.** 0.0805 **c.** 19.54 **d.** 1.834 **e.** 343.0 **9.** **a.** 34.8
b. 174.8 **11.** $A = 57.3°$, $a = 16.1$, $b = 10.3$ **13.** $B = 43° 49'$, $c = 17.2$, $b = 11.9$
15. $B = 37° 48'$, $a = 41.33$, $b = 32.06$ **17.** $c = 17$, $A = 28°$, $B = 62°$ **19.** $c = 85.38$, $A = 28.17°$, $B = 61.83°$
21. $B = 22.5°$, $b = 5.22$, $c = 13.6$ **23.** $A = 17.7°$, $b = 7.93$, $c = 8.32$ **25.** 193 feet **27.** 87.9 feet
29. **a.** $t = \frac{8}{3}$ **b.** $t = \frac{15}{4}$ **31.** true course = 210.6°, true speed = 163 miles per hour
33. $V_y = 71.9$ pounds, $\theta = 64.1°$ **35.** **a.** 0.0654 **b.** 0.875 **37.** 45.6 m, 24.6 m
39. heading: A to B = 68.9°, B to C = 180°, C to A = 321.3°; total time = 6.59 hours **41.** $|\mathbf{W}| = 8.06$, $\theta = 60.26°$

Exercise Set 3.1

1. $\sin\theta = -2\sqrt{2}/3$, $\cos\theta = -\frac{1}{3}$, $\tan\theta = 2\sqrt{2}$, $\csc\theta = -3/2\sqrt{2}$, $\sec\theta = -3$, $\cot\theta = 1/2\sqrt{2}$

3. $\sin\theta = -\frac{12}{13}$, $\cos\theta = \frac{5}{13}$, $\tan\theta = -\frac{12}{5}$, $\csc\theta = -\frac{13}{12}$, $\sec\theta = \frac{13}{5}$, $\cot\theta = -\frac{5}{12}$

5. $\sin\theta = -\sqrt{15}/4$, $\cos\theta = \frac{1}{4}$, $\tan\theta = -\sqrt{15}$, $\csc\theta = -4/\sqrt{15}$, $\sec\theta = 4$, $\cot\theta = -1\sqrt{15}$

7. $\sin\theta = \frac{3}{5}$, $\cos\theta = -\frac{4}{5}$, $\tan\theta = -\frac{3}{4}$, $\csc\theta = \frac{5}{3}$, $\sec\theta = -\frac{5}{4}$, $\cot\theta = -\frac{4}{3}$

9. $\sin\theta = -2\sqrt{2}/3$, $\cos\theta = \frac{1}{3}$, $\tan\theta = -2\sqrt{2}$, $\csc\theta = -3/2\sqrt{2}$, $\sec\theta = 3$, $\cot\theta = -1/2\sqrt{2}$

11. **a.** $(0, 1)$ **b.** $(-1, 0)$ **c.** $(0, -1)$ **d.** $(-\sqrt{3}/2, \frac{1}{2})$ **e.** $(-\frac{1}{2}, -\sqrt{3}/2)$

13. **a.**

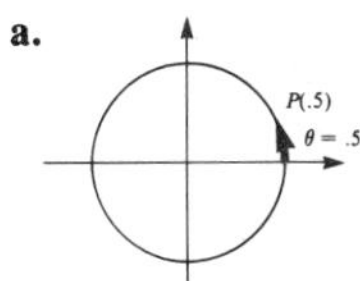

b.

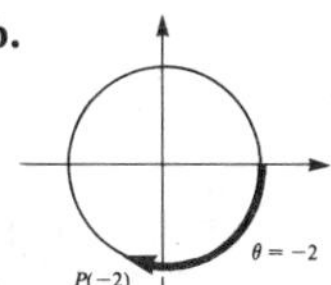

c.

θ = 4
P(4)

d.

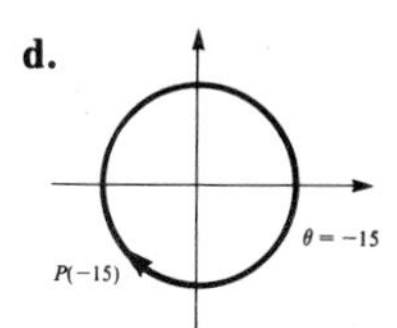

15. **a.** $\pi/6$ **b.** $\pi/4$ **c.** $\pi/3$ **d.** $\pi/2$ **e.** 0

17. **a.** $\pi/2$ **b.** 0 **c.** $\pi/4, 5\pi/4$ **d.** $0, \pi$ **e.** $\pi/2, 3\pi/2$ **f.** $0, \pi$ **g.** $3\pi/2$ **h.** π **i.** $3\pi/4, 7\pi/4$

19.

P(π − θ) P(θ) P(π + θ)

a. ordinate of $P(\pi - \theta)$ = ordinate of $P(\theta)$
b. abscissa of $P(\pi - \theta)$ = −abscissa of $P(\theta)$
c. ordinate of $P(\pi + \theta)$ = −ordinate of $P(\theta)$
d. abscissa of $P(\pi + \theta)$ = −abscissa of $P(\theta)$
e. ratio of ordinate to abscissa of $P(\pi + \theta)$ = ratio of ordinate to abscissa of $P(\theta)$

21. **a.** For all integers n, $2n + 1$ is an odd number, so θ is an odd multiple of $\pi/2$, hence is coterminal with $\pi/2$ or $3\pi/2$. So the abscissa of $P(\theta) = 0$, and hence $\tan\theta$ is not defined.
b. For all integers n, $n\pi$ is coterminal with 0 or π, so the ordinate of θ is 0. Hence $\cot\theta$ is undefined.
c. Since the abscissa of θ is 0, $\sec\theta$ is undefined.
d. Since the ordinate of θ is 0, $\csc\theta$ is undefined.

23. Let $P(\theta) = (x, y)$. Then $\tan\theta = y/x$. Since $x^2 + y^2 = 1$, $\tan\theta = \pm\sqrt{1 - x^2}/x$. Let k designate any real number. We wish to find x such that $\pm\sqrt{1 - x^2}/x = k$ or $1 - x^2 = k^2x^2$. So $x^2 = 1/(1 + k^2)$ and we may choose $x = 1/\sqrt{1 + k^2}$. If $k > 0$, choose $y = \sqrt{1 - x^2}$ and if $k < 0$, choose $y = -\sqrt{1 - x^2}$. Thus, a point (x, y) is determined on the unit circle for which $y/x = k$. If θ is such that $P(\theta) = (x, y)$, then $\tan\theta = k$. Thus, the tangent function assumes all real values. For the cotangent reverse the roles of x and y.

Exercise Set 3.2

1. $\cos\theta = -\frac{4}{5}$, $\tan\theta = \frac{3}{4}$, $\cot\theta = \frac{4}{3}$, $\sec\theta = -\frac{5}{4}$, $\csc\theta = -\frac{5}{3}$

3. $\sin\theta = \frac{4}{5}$, $\cos\theta = -\frac{3}{5}$, $\cot\theta = -\frac{3}{4}$, $\sec\theta = -\frac{5}{3}$, $\csc\theta = \frac{5}{4}$

5. $\sin\theta = -1/\sqrt{5}$, $\tan\theta = -\frac{1}{2}$, $\cot\theta = -2$, $\sec\theta = \sqrt{5}/2$, $\csc\theta = -\sqrt{5}$

7. $\cos\theta = -2\sqrt{2}/3$, $\tan\theta = -1/2\sqrt{2}$, $\cot\theta = -2\sqrt{2}$, $\sec\theta = -3/2\sqrt{2}$, $\csc\theta = 3$

9. $$\frac{\tan\theta}{\sin\theta} = \frac{\sin\theta}{\cos\theta}\cdot\frac{1}{\sin\theta} = \frac{1}{\cos\theta} = \sec\theta$$

11. $$\frac{\sin\theta}{\cot\theta} = \frac{\sin\theta}{\dfrac{\cos\theta}{\sin\theta}} = \frac{\sin^2\theta}{\cos\theta} = \frac{1 - \cos^2\theta}{\cos\theta} = \frac{1}{\cos\theta} - \cos\theta = \sec\theta - \cos\theta$$

13. $$\cot\theta\sec\theta = \frac{\cos\theta}{\sin\theta}\cdot\frac{1}{\cos\theta} = \frac{1}{\sin\theta} = \csc\theta$$

15. $\dfrac{\sec\theta}{\csc\theta} = \dfrac{\dfrac{1}{\cos\theta}}{\dfrac{1}{\sin\theta}} = \dfrac{\sin\theta}{\cos\theta} = \tan\theta$

17. $\dfrac{1}{\sec^2\theta} + \dfrac{1}{\csc^2\theta} = \left(\dfrac{1}{\sec\theta}\right)^2 + \left(\dfrac{1}{\csc\theta}\right)^2 = \cos^2\theta + \sin^2\theta = 1$

19. $\dfrac{\sqrt{1+t^2}}{t} = \dfrac{\sqrt{1+\tan^2\theta}}{\tan\theta} = \dfrac{\sqrt{\sec^2\theta}}{\tan\theta} = \dfrac{\sec\theta}{\tan\theta} = \dfrac{\dfrac{1}{\cos\theta}}{\dfrac{\sin\theta}{\cos\theta}} = \dfrac{1}{\sin\theta} = \csc\theta$ (Note that $\sec\theta > 0$ for $-\pi/2 < \theta < \pi/2$.)

21. $\dfrac{\sqrt{4t^2-9}}{t} = \dfrac{\sqrt{4\cdot\frac{9}{4}\sec^2\theta - 9}}{\frac{3}{2}\sec\theta} = \dfrac{3\sqrt{\sec^2\theta-1}}{\frac{3}{2}\sec\theta} = \dfrac{2\sqrt{\tan^2\theta}}{\sec\theta} = \dfrac{2\tan\theta}{\sec\theta} = \dfrac{2\dfrac{\sin\theta}{\cos\theta}}{\dfrac{1}{\cos\theta}} = 2\sin\theta$

(Note that for θ as specified, $\tan\theta > 0$.)

23. Since $\cot\theta = t$, $\tan\theta = 1/t$. Thus, $\sec\theta = \pm\sqrt{1+1/t^2} = \pm\sqrt{1+t^2}/|t|$ and $\csc\theta = \pm\sqrt{1+t^2}$. If $t>0$, θ is in quadrant I or III, so $\sec\theta$ and $\csc\theta$ are like in sign. Thus, $\sec\theta\csc\theta = (1+t^2)/|t| = (1+t^2)/t$. If $t<0$, θ is in quadrant II or IV, so $\sec\theta$ and $\csc\theta$ are opposite in sign. Thus, $\sec\theta\csc\theta = -(1+t^2)/|t| = -(1+t^2)/(-t) = (1+t^2)/t$.

Exercise Set 3.4

1. $(\sqrt{2}+\sqrt{6})/4$, $(\sqrt{2}-\sqrt{6})/4$ **3. a.** $-(\sqrt{6}+\sqrt{2})/4$ **b.** $-(\sqrt{6}+\sqrt{2})/4$ **5. a.** $(\sqrt{2}-\sqrt{6})/4$ **b.** $-(\sqrt{6}+\sqrt{2})/4$ **7. a.** $(\sqrt{6}+\sqrt{2})/4$ **b.** $(\sqrt{6}+\sqrt{2})/4$ **9. a.** $-\frac{56}{65}$ **b.** $-\frac{63}{65}$
11. a. $(2\sqrt{10}-2)/9$ **b.** $(-\sqrt{5}+4\sqrt{2})/9$ **13. a.** $31/17\sqrt{5}$ **b.** $38/17\sqrt{5}$ **15. a.** $-\frac{65}{33}$ **b.** $-\frac{65}{16}$
17. a. $\sin(\pi/2+\theta) = \sin(\pi/2)\cos\theta + \cos(\pi/2)\sin\theta = \cos\theta$
b. $\cos(\pi/2+\theta) = \cos(\pi/2)\cos\theta - \sin(\pi/2)\sin\theta = -\sin\theta$
19. a. $\sin(\pi-\theta) = \sin\pi\cos\theta - \cos\pi\sin\theta = \sin\theta$
b. $\cos(\pi-\theta) = \cos\pi\cos\theta + \sin\pi\sin\theta = -\cos\theta$
31. $0, \pi$
33. a. $\sin\alpha\cos\beta\cos\gamma + \cos\alpha\sin\beta\cos\gamma + \cos\alpha\cos\beta\sin\gamma - \sin\alpha\sin\beta\sin\gamma$
b. $\cos\alpha\cos\beta\cos\gamma - \sin\alpha\sin\beta\cos\gamma - \sin\alpha\cos\beta\sin\gamma - \cos\alpha\sin\beta\sin\gamma$

Exercise Set 3.5

1. $\sin 2\theta = -\frac{24}{25}$, $\cos 2\theta = -\frac{7}{25}$ **3.** $\sin 2\theta = \frac{120}{169}$, $\cos 2\theta = \frac{119}{169}$ **5.** $\sin 2\theta = -\frac{4}{5}$, $\cos 2\theta = \frac{3}{5}$
7. $\sin 2\theta = 2x\sqrt{1-x^2}$, $\cos 2\theta = 1-2x^2$ **9.** $\sin\theta = \sqrt{3}/3$, $\cos\theta = -\sqrt{6}/3$ **11.** $\sin\theta = -\frac{3}{5}$, $\cos\theta = \frac{4}{5}$
13. a. $\sqrt{2-\sqrt{2}}/2$ **b.** $\sqrt{2+\sqrt{3}}/2$ **c.** $\sqrt{2+\sqrt{3}}/2$ **d.** $\sqrt{2-\sqrt{2}}/2$
15. $\sin(\alpha/2) = 3/\sqrt{13}$, $\cos(\alpha/2) = -2/\sqrt{13}$ **17.** $\sin(\alpha/2) = -\sqrt{3}/3$, $\cos(\alpha/2) = \sqrt{6}/3$ **19.** $P(\theta) = (\frac{3}{5}, \frac{4}{5})$
21. a. $-\sin\theta$ **b.** $-\sin\theta$ **c.** $\cos\theta$ **d.** $-\cos\theta$ **e.** $\cos\theta$
39. $\sin 2\theta = 2x/(1+x^2)$, $\cos 2\theta = (1-x^2)/(1+x^2)$
41. $\sin 3\theta = 3\sin\theta - 4\sin^3\theta$
$\cos 3\theta = 4\cos^3\theta - 3\cos\theta$

Exercise Set 3.6

1. a. $2+\sqrt{3}$ **b.** $2-\sqrt{3}$ **3.** $\frac{24}{7}$ **5.** $(9+5\sqrt{2})/2$
7. $\tan 2\alpha = \frac{120}{119}$, $\tan 2\beta = -\frac{240}{161}$, $\tan(\alpha/2) = -\frac{2}{3}$, $\tan(\beta/2) = 4$ **9.** $\tan 2\theta = -\frac{24}{7}$, $\tan(\theta/2) = 3$ **11. a.** $\sqrt{6}/2$ **b.** $-\sqrt{6}/2$ **13. a.** $\sqrt{2}/2$ **b.** $-\sqrt{2}/2$ **15. a.** $2\sin 4x\cos x$ **b.** $\frac{1}{2}(\sin 8x + \sin 2x)$

Exercise Set 3.8

1. $\{\pi/6, 11\pi/6\}$ **3.** $\{\pi/3, 5\pi/3\}$ **5.** $\{\pi/3, \pi/2, 3\pi/2, 5\pi/3\}$ **7.** $\{\pi/3, \pi, 5\pi/3\}$ **9.** $\{\pi/2, 3\pi/2\}$
11. $\{\pi/6, \pi/2, 5\pi/6, 3\pi/2\}$ **13.** $\{\pi/3, 2\pi/3, 4\pi/3, 5\pi/3\}$ **15.** no solution **17.** $\{\pi/2, 3\pi/2\}$
19. $\{0, 2\pi/9, \pi/3, 4\pi/9, 2\pi/3, 8\pi/9, \pi, 10\pi/9, 4\pi/3, 14\pi/9, 5\pi/3, 16\pi/9\}$ **21.** $\{0, \pi/2\}$
23. $\{\pi/24, 11\pi/24, 13\pi/24, 23\pi/24, 25\pi/24, 35\pi/24, 37\pi/24, 47\pi/24\}$ **25.** no solution
27. $\{2\pi/9, \pi/4, 5\pi/9, 3\pi/4, 8\pi/9, 11\pi/9, 5\pi/4, 14\pi/9, 7\pi/4, 17\pi/9\}$ **29.** no solution
31. $\{0.848, 2.29, 3.87, 5.55\}$ **33.** $\{0.464, 1.11, 3.61, 4.25\}$ **35.** $\{0.285, 2.86\}$
37. $\{\pi/3, 2\pi/3, 3\pi/4, 4\pi/3, 5\pi/3, 7\pi/4\}$ **39.** $\{\pi/6, 5\pi/6, 3\pi/2\}$

3.9 Review Exercise Set

1. **a.** $\frac{3}{2}$ **b.** $-2/\sqrt{5}$ **c.** $-\frac{1}{9}$ **d.** $4\sqrt{5}$ **3.** **a.** $-\cos\theta$ **b.** $\tan\theta$ **c.** $\sec\theta$ **d.** $\sin\theta$
e. $\sec\theta$ **5.** **a.** $-\frac{24}{25}$ **b.** $-\frac{7}{25}$ **c.** $2/\sqrt{5}$ **d.** 2 **7.** **a.** $-\frac{13}{85}$ **b.** $\frac{36}{85}$ **c.** $\frac{13}{84}$
9. **a.** $5/\sqrt{34}$ **b.** $-3/\sqrt{34}$ **c.** $-\frac{5}{3}$ **11.** **a.** $\sqrt{6}/2$ **b.** $\sqrt{2}/4$ **c.** $\frac{1}{4}$
27. $\{0, \pi/6, 5\pi/6, \pi, 7\pi/6, 11\pi/6\}$ **29.** $\{\pi/6, 5\pi/6, 3\pi/2\}$ **31.** $\{\pi/2, 7\pi/6, 11\pi/6\}$ **33.** **a.** $(2 - x^2)/x^2$
b. $-2\sqrt{x^2 - 1}/x^2$ **c.** $\sqrt{(x + 1)/2x}$ **d.** $-\sqrt{x^2 - 1}/(x + 1)$ **41.** $\{\pi/3, \pi/2, 2\pi/3, 4\pi/3, 3\pi/2, 5\pi/3\}$
43. $\{\pi/6, \pi/2, 5\pi/6, 3\pi/2\}$ **45.** $\{7\pi/4\}$ **47.** $\{0.464, 2.03, 3.61, 5.18\}$ **49.** $\{1.05, 2.42, 3.86, 5.24\}$

Exercise Set 4.1

1. amplitude = 1, period = π

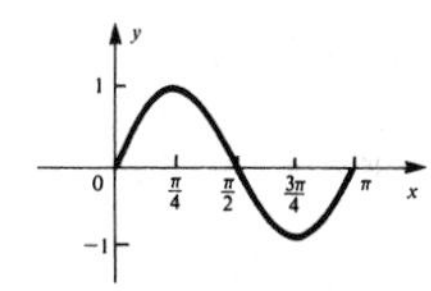

3. amplitude = 2, period = 2π

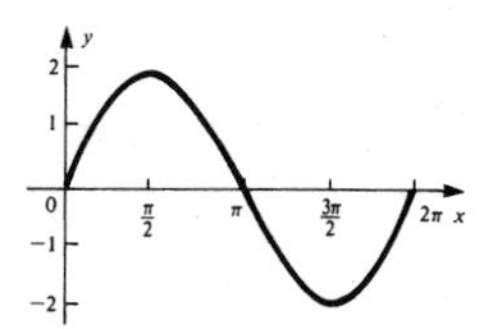

5. amplitude = 3, period = 4

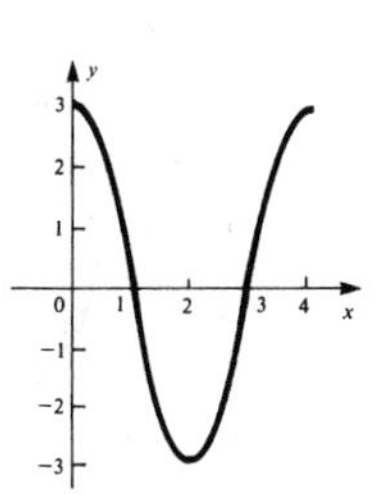

7. amplitude = $\frac{1}{2}$, period = 2

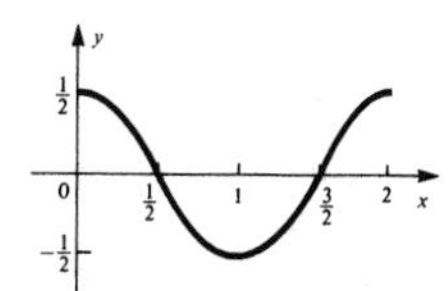

9. amplitude = 2, period = $2\pi/3$

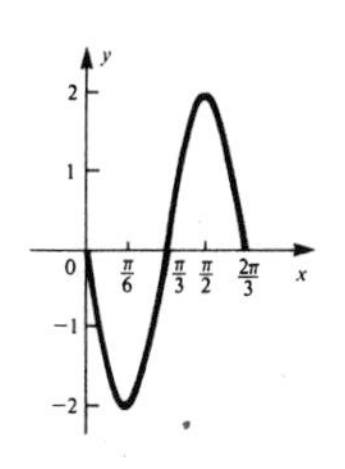

11. amplitude = 1, period = 4π

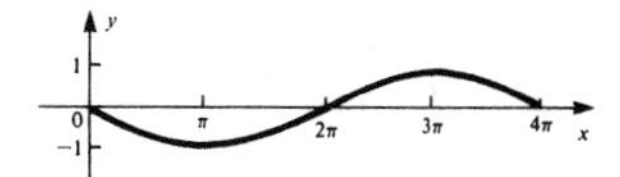

13. amplitude = 1, period = π

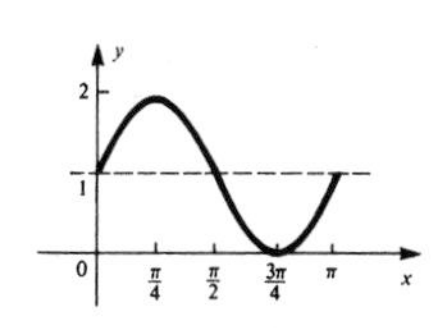

15. amplitude = 2, period = $2\pi/3$

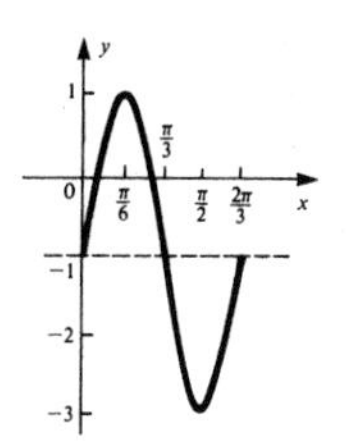

17. amplitude = 1,
period = 2π,
phase shift = $\pi/3$

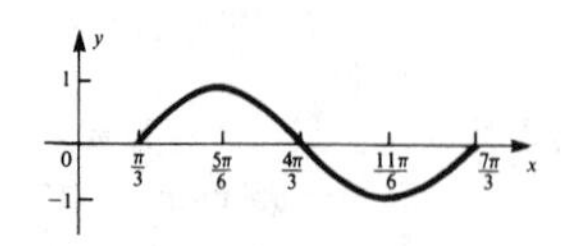

19. amplitude = 1,
period = π,
phase shift = $-\pi/4$

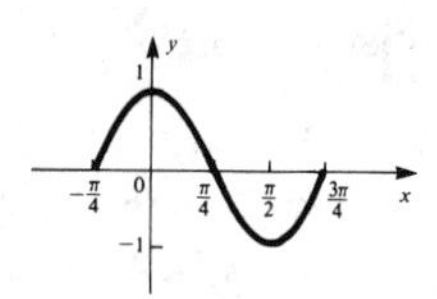

21. amplitude = 2,
period = $2\pi/3$,
phase shift = $-\pi/6$

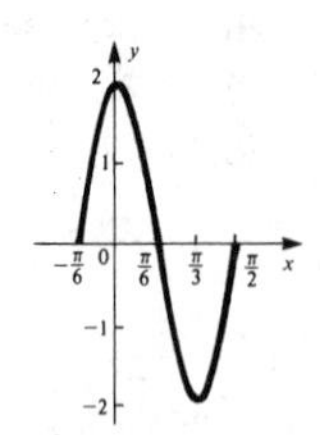

23. amplitude = 1,
period = 2,
phase shift = 3

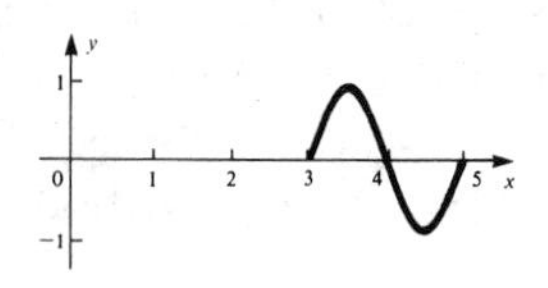

25. amplitude = 2,
period = $2\pi/3$,
phase shift = $-\frac{2}{3}$

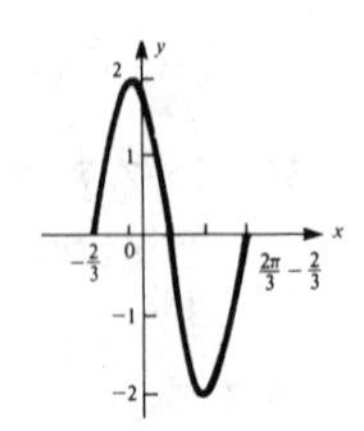

27. amplitude = 2,
period = 2,
phase shift = $\frac{1}{3}$

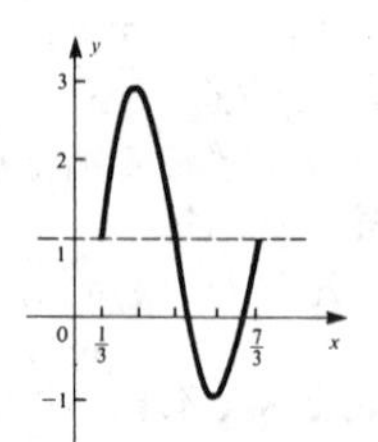

29. amplitude = $\frac{3}{2}$,
period = π,
phase shift = $\frac{2}{3}$

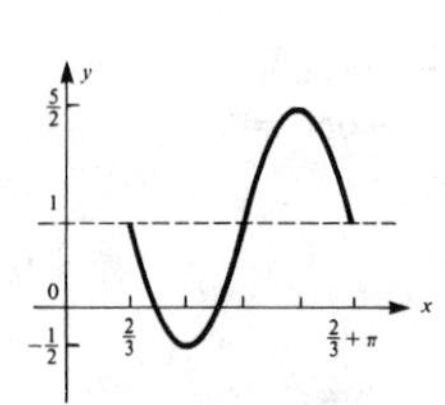

31. amplitude = $\frac{3}{4}$,
period = 2,
phase shift = $-\frac{1}{3}$

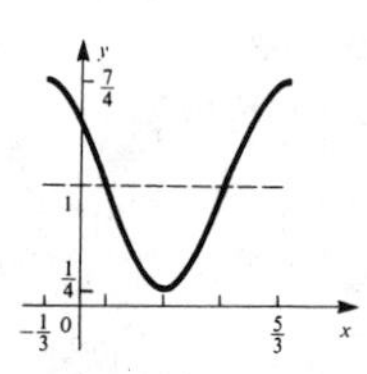

Exercise Set 4.2

1.

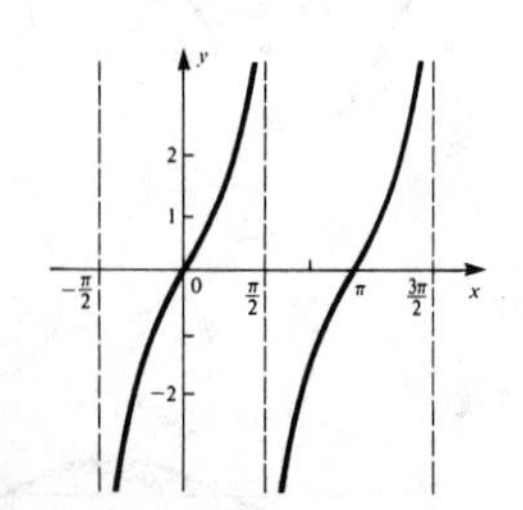

3.

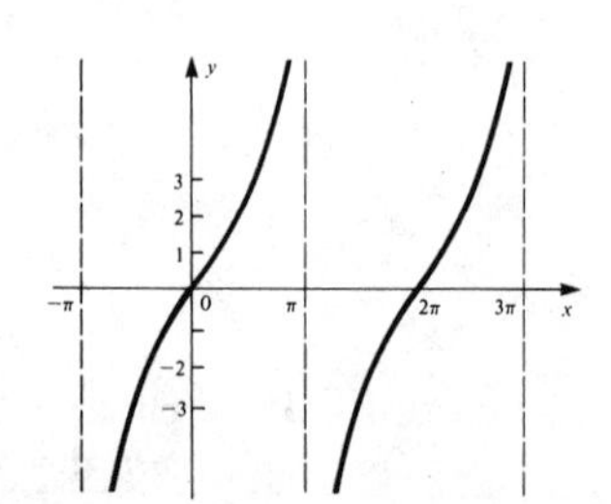

5.

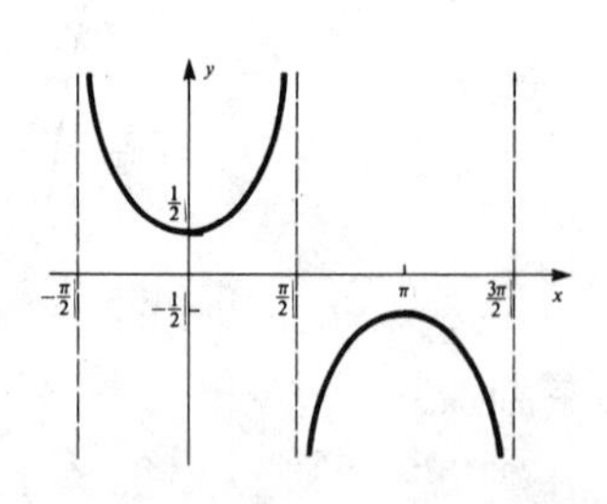

7.

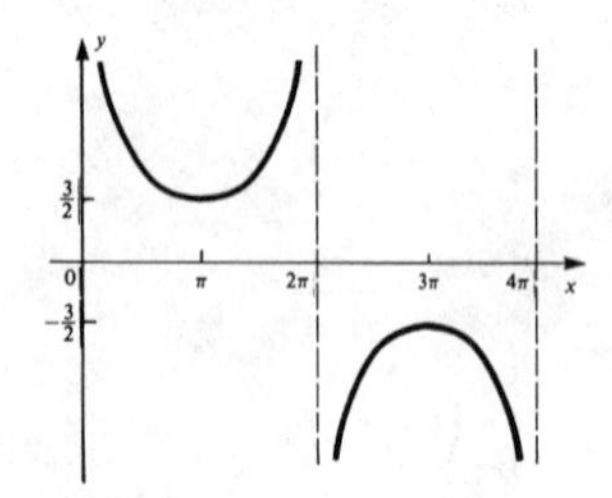

9.

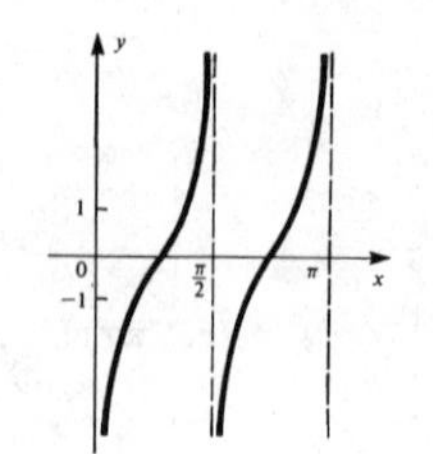

11.

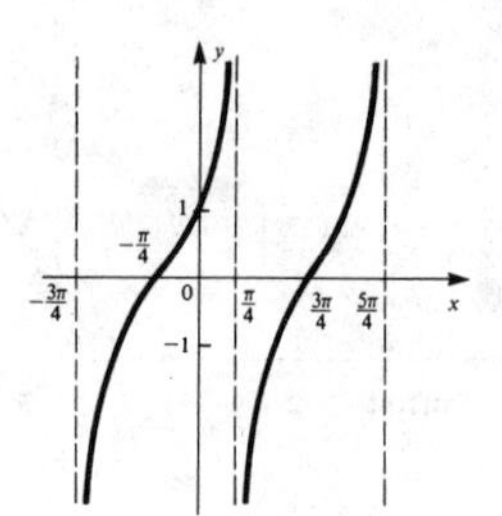

13.

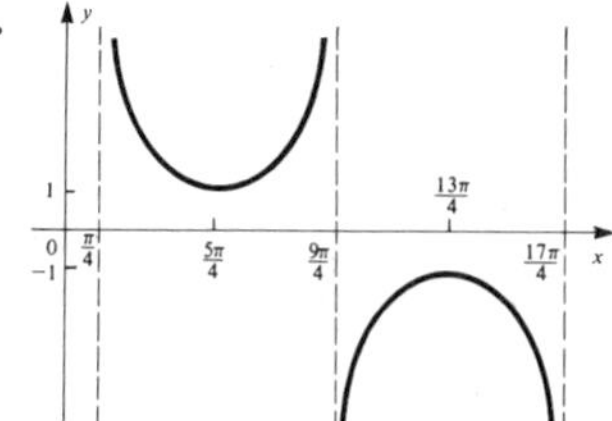

15.

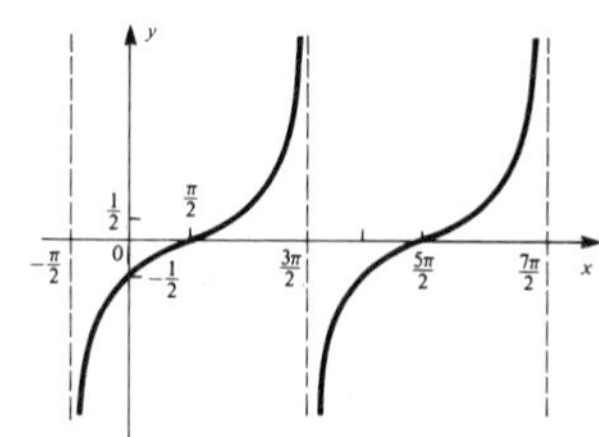

17.

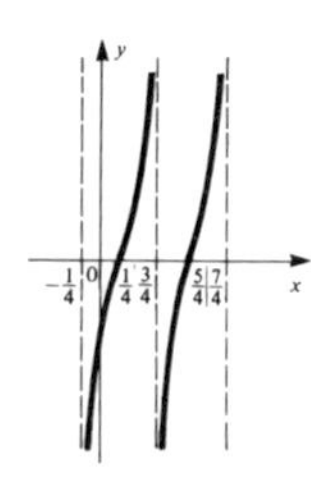

19.

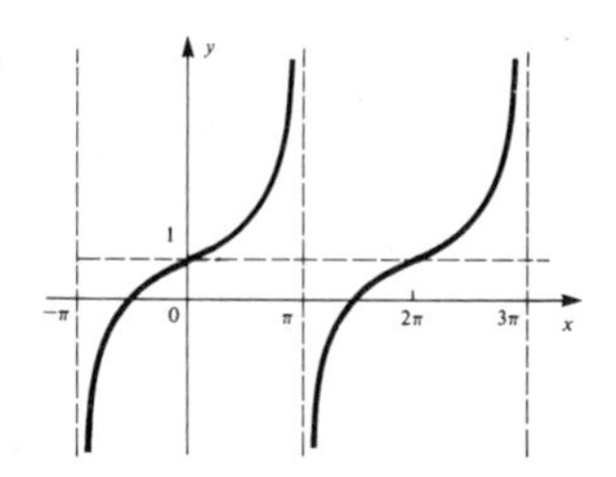

21.

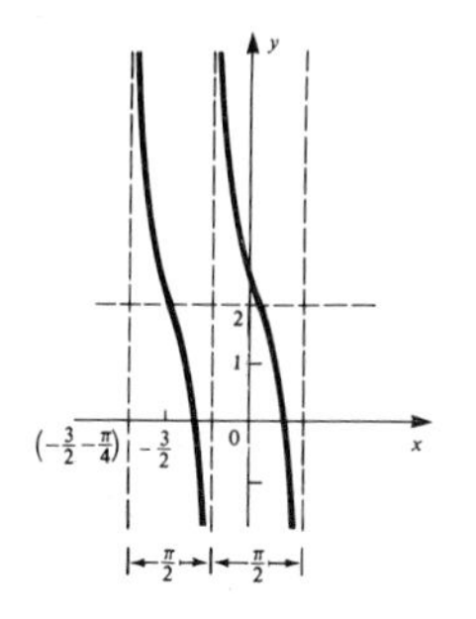

23.

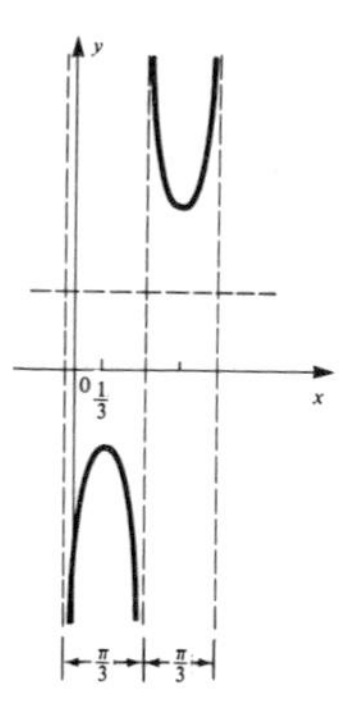

25.

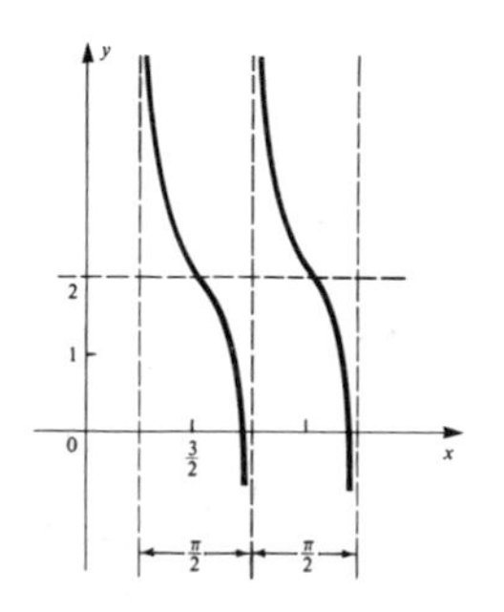

Exercise Set 4.3

1.

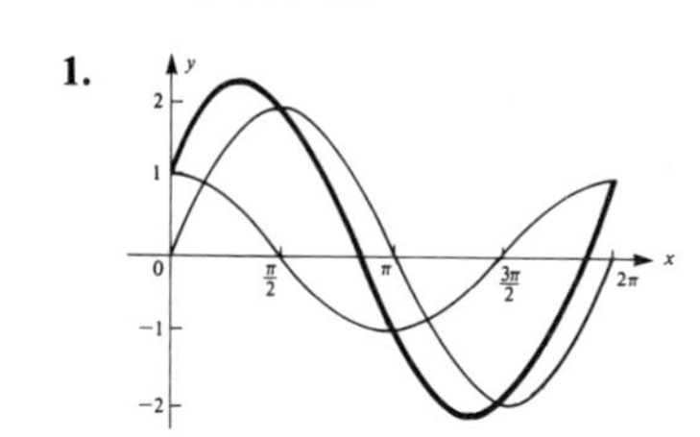

3.

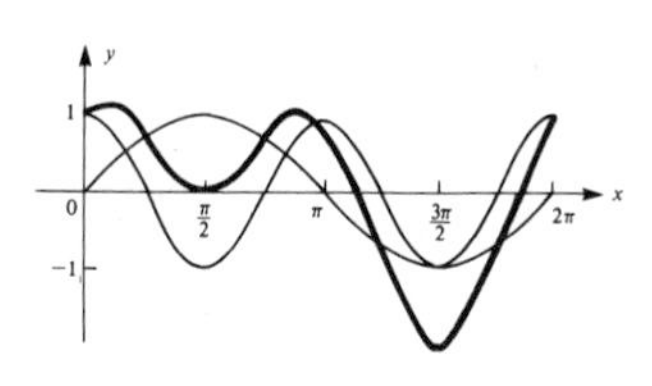

5.

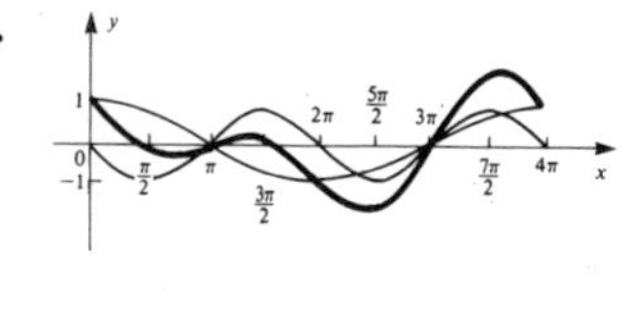

7.

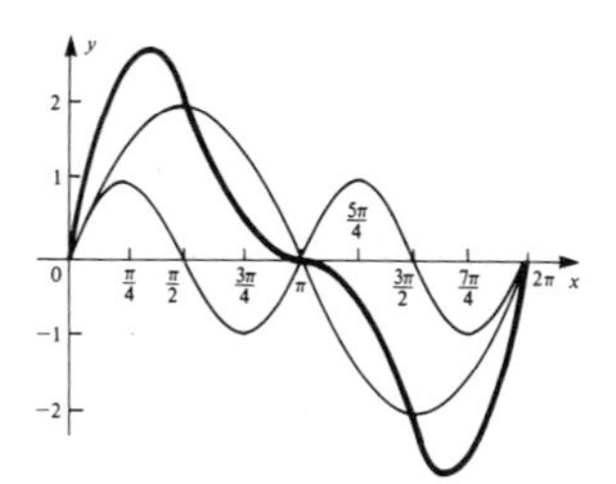

9.

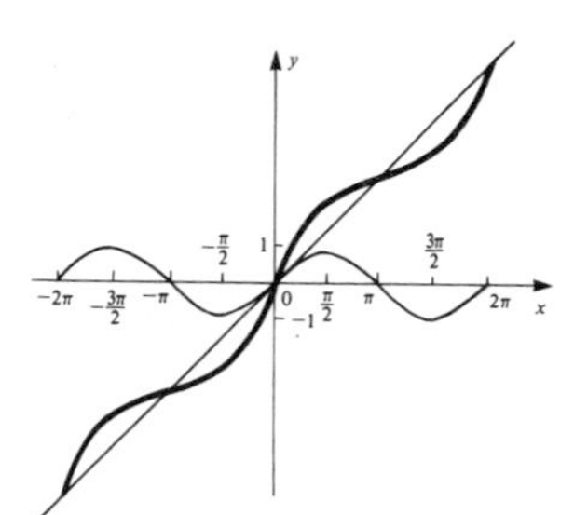

11. $y = \sqrt{2}\sin(x - \pi/4)$
amplitude $= \sqrt{2}$, period $= 2\pi$,
phase shift $= \pi/4$

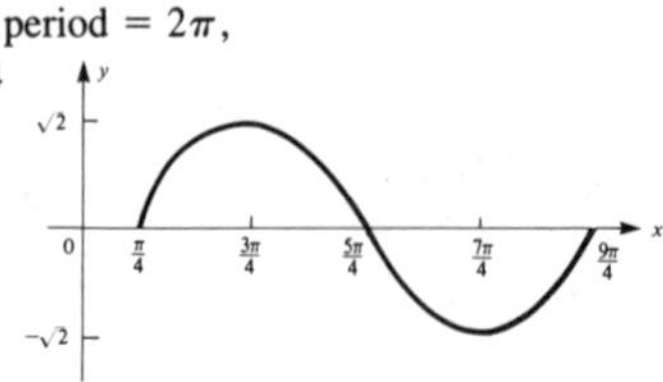

13. $y = 2\sin 2(x + \pi/6)$
amplitude $= 2$, period $= \pi$,
phase shift $= -\pi/6$

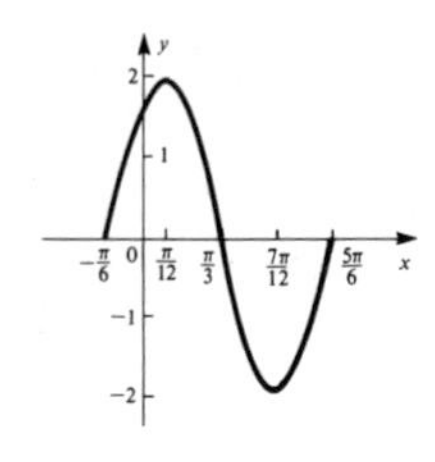

15. $y = 5\sin(x - 0.93)$
amplitude $= 5$, period $= 2\pi$,
phase shift $= 0.93$

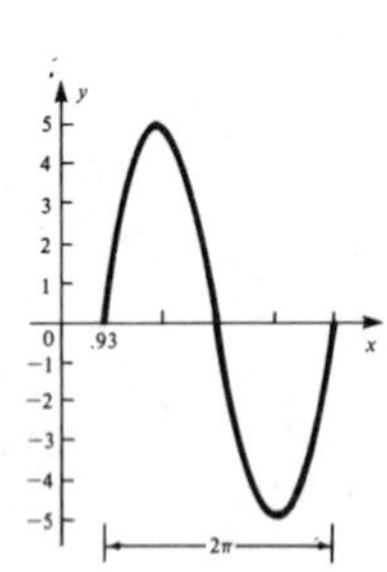

17. $y = -\sqrt{5}\sin(x - 0.46)$
amplitude $= \sqrt{5}$, period $= 2\pi$,
phase shift $= 0.46$

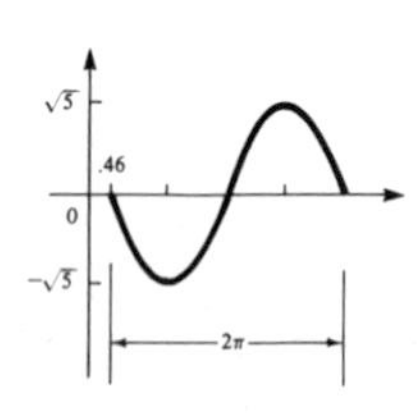

19.

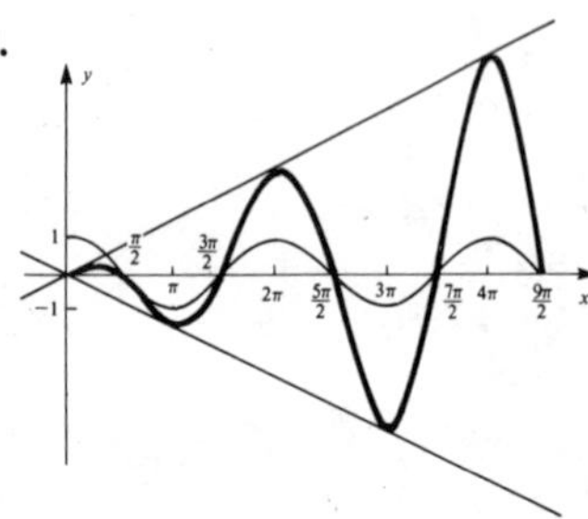

21.

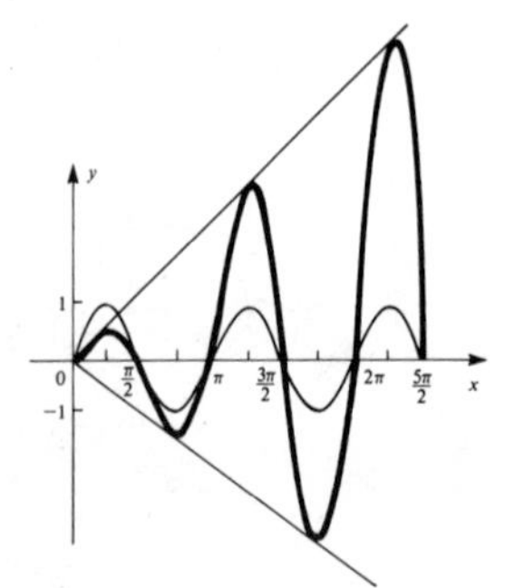

23.

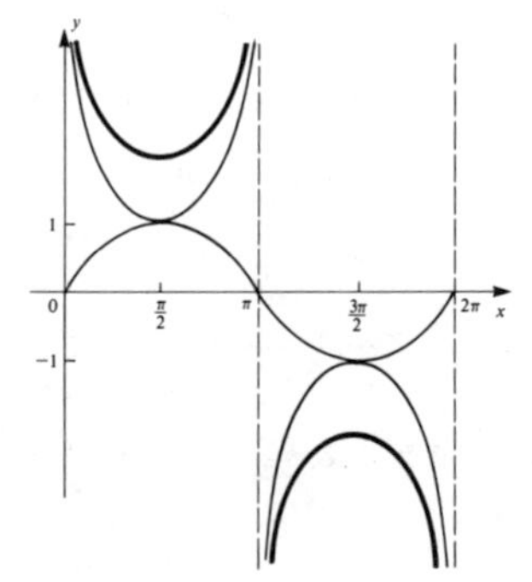

25.

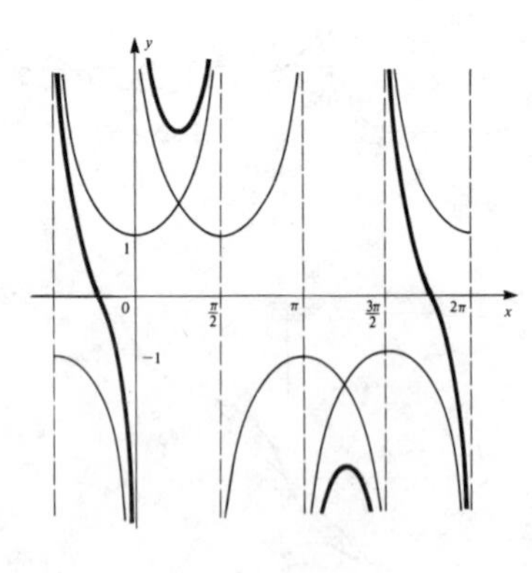

27.

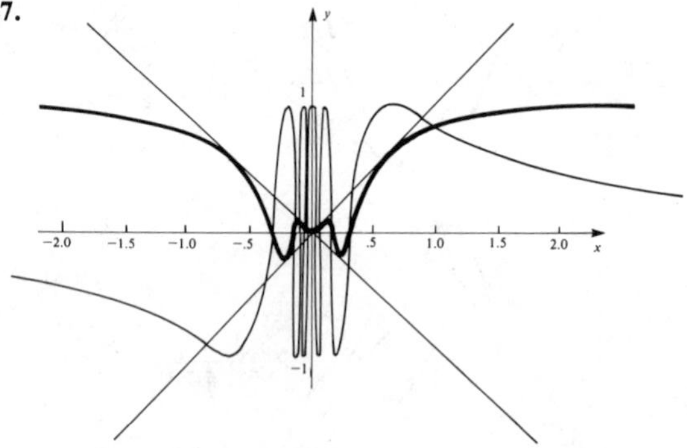

Exercise Set 4.4

1. a. $-\pi/3$ **b.** $5\pi/6$ **3. a.** $3\pi/4$ **b.** $\pi/2$ **5. a.** 0 **b.** $-\pi/6$ **7. a.** $\pi/2$ **b.** $7\pi/6$
9. a. $-2\sqrt{2}/3$ **b.** $\frac{5}{4}$ **11. a.** $\frac{4}{3}$ **b.** $\frac{3}{5}$ **13. a.** $2\sqrt{2}/3$ **b.** $-\frac{7}{24}$ **15. a.** $\pi/5$ **b.** $\pi/6$
17. a. $\pi/4$ **b.** $\pi/3$ **19.** $-\frac{63}{65}$ **21.** $-\frac{63}{16}$ **23.** $\frac{7}{9}$ **25.** $-\frac{1}{9}$ **27.** $-12\sqrt{10}/49$ **29.** $\frac{3}{5}$ **31.** $-\frac{2}{3}$

33. $\sin(\alpha/2) = 1/\sqrt{5}$, $\cos(\alpha/2) = 2/\sqrt{5}$, $\tan(\alpha/2) = \frac{1}{2}$ **35.** $\frac{87}{425}$ **37. a.** 0.7275 **b.** 1.150 **39. a.** −1.464 **b.** −0.5007 **41.** $\pi/4$ **47.** −0.9721 **49.** −1.940

4.5 Review Exercise Set

1. period = $2\pi/3$, amplitude = 2

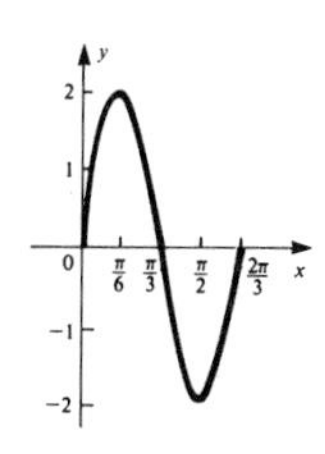

3. period = 1

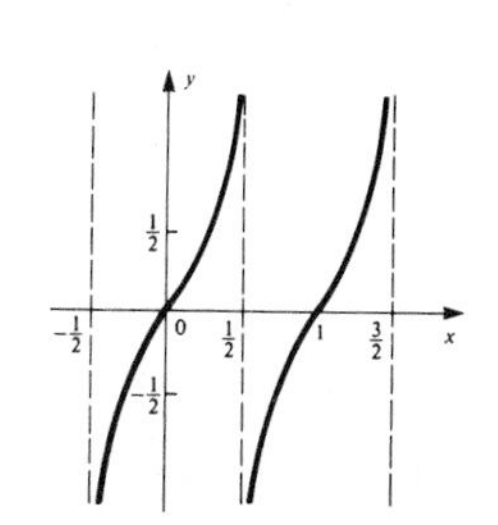

5. period = 6

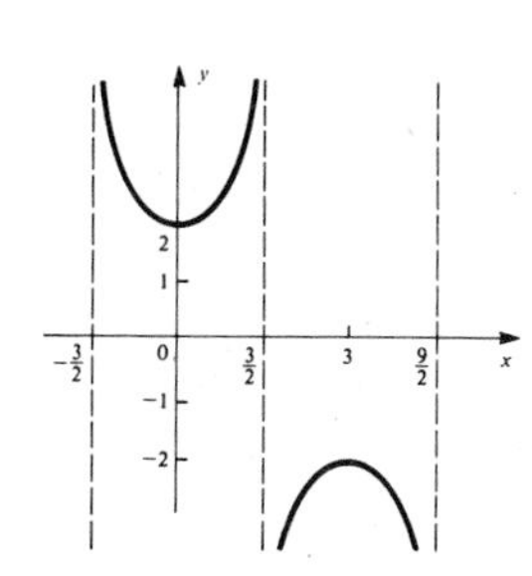

7. amplitude = 2, period = 2π

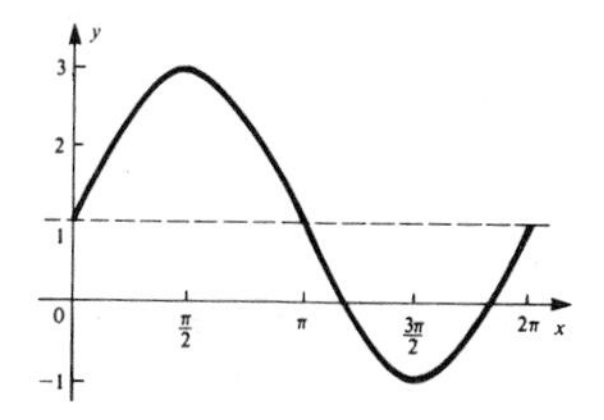

9. amplitude = 1, period = 2, phase shift = $\frac{1}{3}$

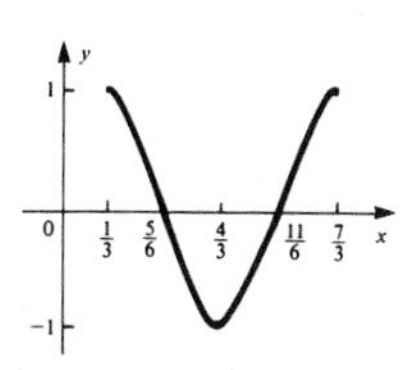

11. period = π, phase shift = $-\pi/4$

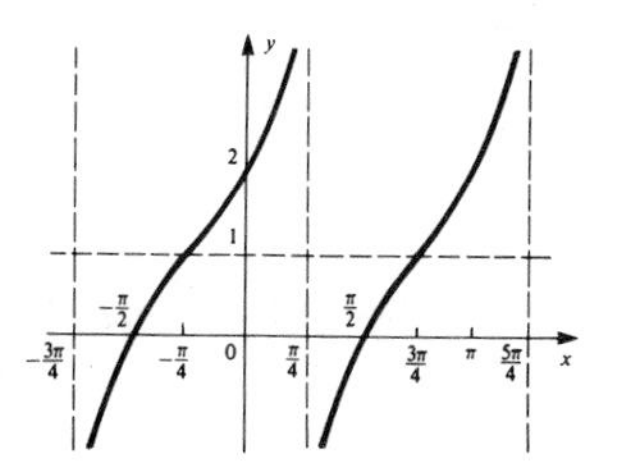

13. period = 1, phase shift = $\frac{1}{6}$

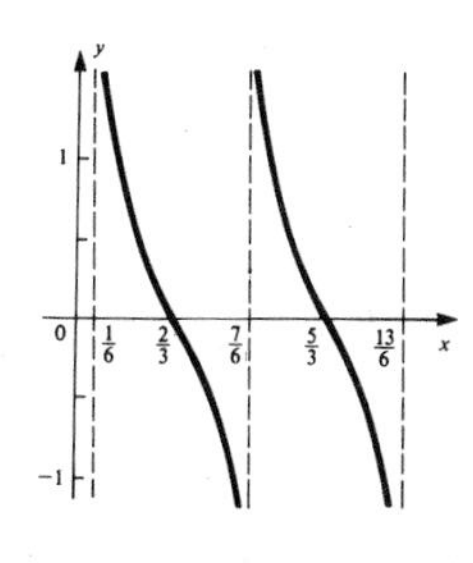

15. amplitude = 1, period = π, phase shift = $\pi/6$

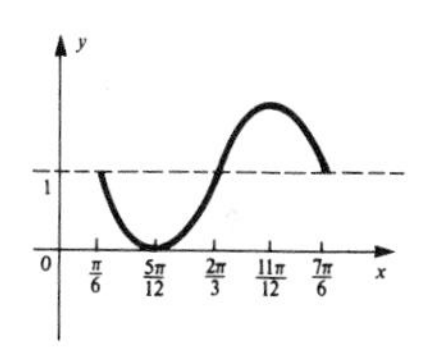

17.

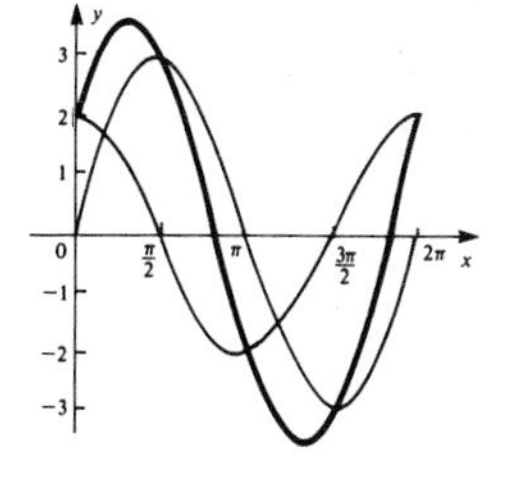

19.

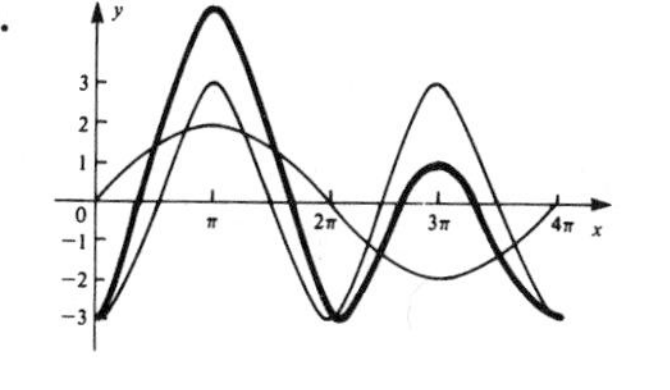

21.

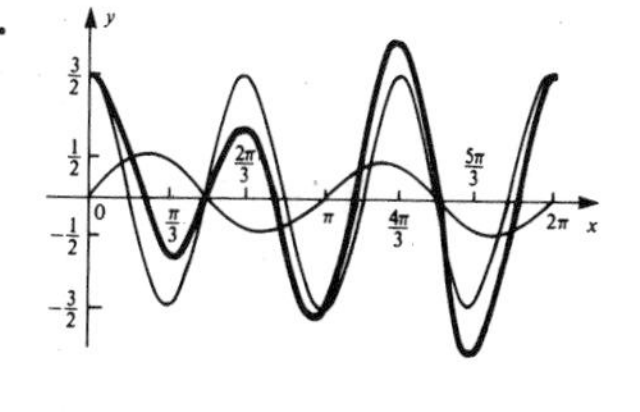

23. $y = 2\sin(x + \pi/3)$
amplitude $= 2$,
period $= 2\pi$,
phase shift $= -\pi/3$

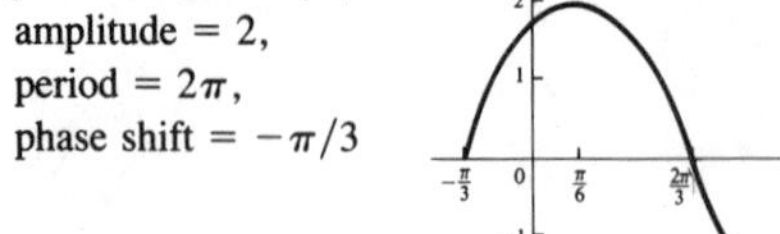

25. $y = 5\sin(x + 0.644)$
amplitude $= 5$,
period $= 2\pi$,
phase shift $= -0.644$

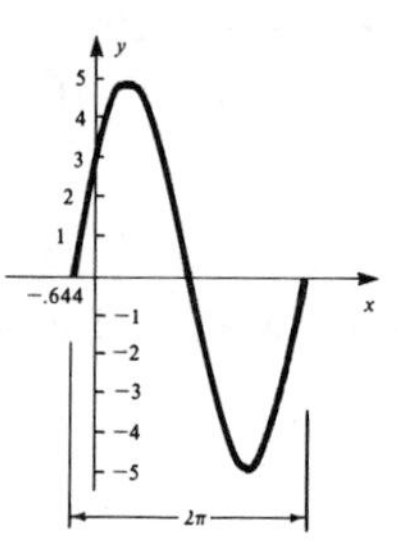

27.

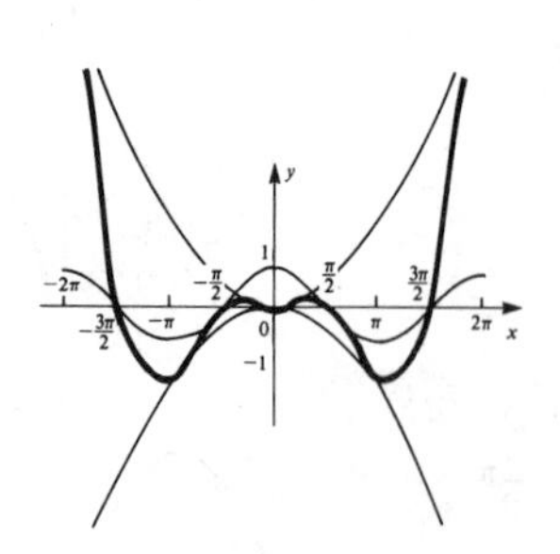

29.

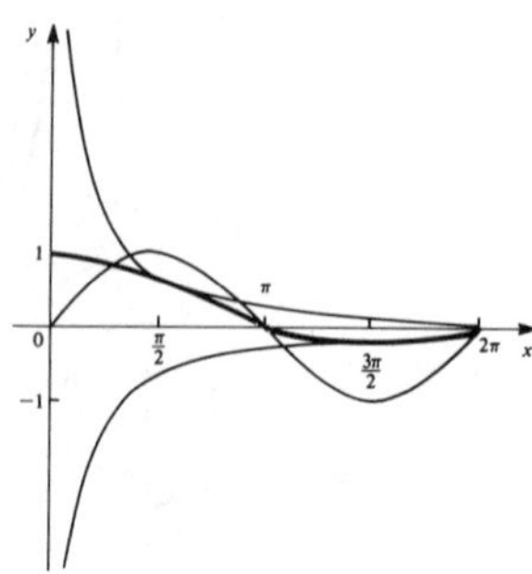

31. **a.** $\pi/3$ **b.** $3\pi/4$ **c.** $7\pi/6$ **d.** $\pi/2$ **e.** π **33.** **a.** $-\frac{24}{7}$ **b.** $2/\sqrt{7}$ **35.** **a.** $-\pi/4$ **b.** $-\pi/6$ **37.** **a.** $\frac{63}{65}$ **b.** $-\frac{56}{65}$ **39.** **a.** $-\frac{56}{65}$ **b.** $\frac{9}{7}$ **41.** **a.** 0.8913 **b.** -0.3201

43. amplitude $= \frac{3}{2}$,
period $= 2\pi/3$,
phase shift $= \frac{2}{3}$

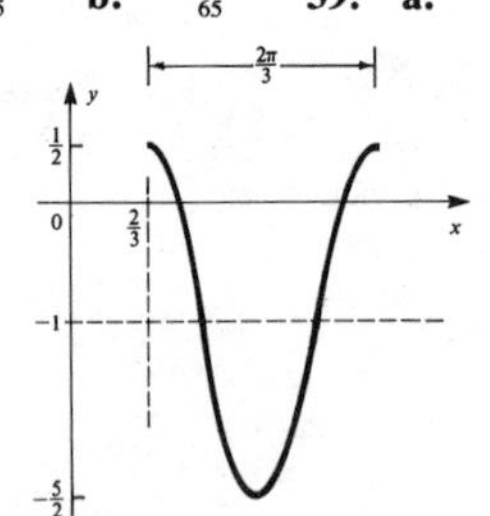

45.

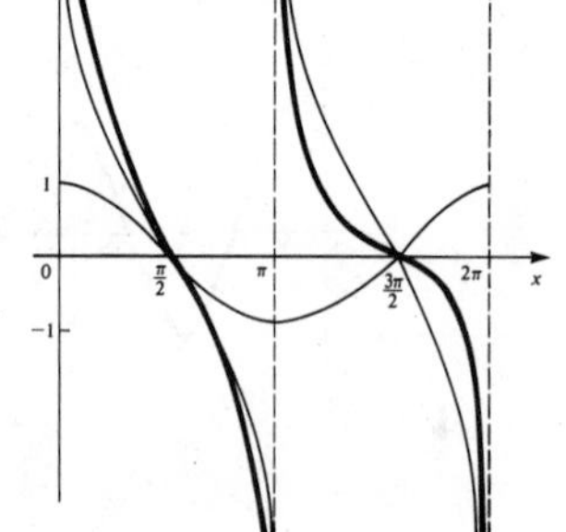

47. $-\pi/4$

Exercise Set 5.1

1. **a.** $x = 16$ **b.** $x = 2$ **3.** **a.** $x = 0$ **b.** $x = e$ **5.** **a.** $x = 10$ **b.** $x = 9$
7. **a.** $\log_{10} 10{,}000 = 4$ **b.** $\log_{10} 0.001 = -3$ **9.** **a.** $2^4 = 16$ **b.** $10^2 = 100$ **11.** **a.** $a^y = x$
b. $x^t = z$ **13.** **a.** -2 **b.** -2 **15.** **a.** 4 **b.** 4 **17.** **a.** $\log(x + 1) + \log(x + 2)$
b. $\log(x + 1) - \log(x - 2)$ **19.** **a.** $3\log x - \log(3x + 1)$ **b.** $\log x + \log(x + 2) - 2\log(x + 3)$

21. **a.** $\log(2x/y)$ **b.** $\log x^2(x + 1)^3$ **23.** $\log \dfrac{\sqrt{5x + 3}}{(x + 2)^2\,(3x + 1)}$ **25.** 2 **27.** 4 **29.** 8 **31.** 1

33. **a.** $x = \dfrac{\ln 3}{\ln 4}$ **b.** $x = \dfrac{\ln 5}{\ln 5 - \ln 3}$ **35.** **a.** $x = \dfrac{\ln 8}{1 + \ln 2}$ **b.** $x = \dfrac{\ln 2 - \ln 3}{\ln 16 + \ln 3}$

Exercise Set 5.2

1. 1.5145 3. $0.8837 - 1$ **5.** 5.4082 **7.** 7.5315 **9.** 4,530 **11.** 7.81 **13.** 0.00219
15. 0.00446 **17.** 0.7538 **19.** $0.5782 - 3$ **21.** 5.6778 **23.** 6.8297 **25.** 686.4 **27.** 0.04246

29. 0.02764 **31.** 6.527 **33.** 235 **35.** 3.21 **37.** 119 **39.** 24,700 **41.** 3.77 **43.** 915
45. 0.7970 **47.** 0.01508 **49.** 377.3 **51.** **a.** 116.7 cm^3 **b.** 114.6 cm^2 **53.** 129.7 cm^3 **55.** 3.690
57. 4.207 **59.** -1.048 **61.** 0.6096

Exercise Set 5.3

1. $9.7160 - 10$ **3.** $9.5798 - 10$ **5.** 0.6424 **7.** $24° 30'$ **9.** $57° 40'$ **11.** $85° 00'$ **13.** $9.4440 - 10$
15. $9.8649 - 10$ **17.** $9.5622 - 10$ **19.** $38° 03'$ **21.** $27° 14'$ **23.** $71° 55'$
25. $B = 47° 39'$, $b = 28.2$, $c = 38.2$ **27.** $B = 68° 24'$, $a = 14.12$, $c = 38.36$
29. $A = 39° 35'$, $B = 50° 25'$, $c = 392.5$ **31.** $B = 35.38°$, $a = 108.0$, $b = 76.67$
33. $B = 26.25°$, $b = 20.36$, $c = 46.02$ **35.** 21.53 **37.** 85.92 meters **39.** 21.8 feet

5.4 Review Exercise Set

1. **a.** $5^2 = 25$ **b.** $4^4 = 256$ **c.** $2^{-2} = 0.25$ **d.** $r^z = s$ **e.** $k^v = y$ **3.** **a.** $x = 16$ **b.** $x = \frac{1}{4}$
c. $x = 4$ **5.** **a.** $x = 8$ **b.** $x = 2$ **c.** $x = -7$ **7.** **a.** $\log(x - 1) - [2 \log(x + 2) + \log(x - 3)]$
b. $\frac{1}{2}[\log(2x - 1) - \log 3 - \log x]$ **9.** **a.** $\log \dfrac{2x^3}{\sqrt[3]{(x-1)(x+2)}}$ **b.** $\log \dfrac{(x+2)^3 \sqrt{2x-1}}{\sqrt{(2-x)^3}}$
11. **a.** $0.1222 - 1$ **b.** 0.8770 **13.** **a.** $9.9838 - 10$ **b.** 0.5325 **15.** **a.** 0.02067 **b.** 0.005809
17. **a.** $49° 35'$ **b.** $29° 54'$ **19.** 2.31 **21.** 138.1 **23.** $A = 76° 10'$, $b = 1.86$, $c = 7.77$
25. $B = 70° 23'$, $b = 128.3$, $c = 136.2$ **27.** $A = 26.25°$, $a = 1.894$, $b = 3.841$ **29.** $x = 4$ **31.** $x = \frac{2}{3}$
33. **a.** $x = \dfrac{\ln 7}{\ln 3}$ **b.** $x = \dfrac{\ln 2}{\ln 9 - \ln 2}$ **35.** **a.** 0.6962 **b.** -0.1254 **37.** 505.6 meters **39.** 0.9450

Exercise Set 6.2

1. $C = 120°$, $b = 2.7$, $c = 6.7$
3. Solution 1: $A = 82° 30'$, $B = 61° 00'$, $a = 142$
Solution 2: $A = 24° 30'$, $B = 119° 00'$, $a = 59.3$
5. $B = 12.9°$, $C = 138.6°$, $c = 30.2$ **7.** $A = 107.0°$, $a = 16.1$, $b = 12.4$ **9.** $C = 102° 02'$, $a = 9.900$, $c = 15.93$
11. $A = 113.07°$, $C = 23.68°$, $a = 23.52$ **13.** $A = 33° 16'$, $C = 38° 12'$, $a = 65.14$
15. Solution 1: $A = 39° 29'$, $B = 118° 01'$, $b = 8.930$
Solution 2: $A = 140° 31'$, $B = 16° 59'$, $b = 2.955$
17. $\theta = 68° 00'$, $|\mathbf{F}_1| = 156.3$ pounds **19.** $x \approx 1{,}030$ ft. **21.** S 66° 20′ W, 37 minutes
23. 3.51 miles from A, 2.77 miles from B, 1.88 miles from nearest point on shore

Exercise Set 6.3

1. $B = 79.1°$, $C = 40.9°$, $a = 13.2$ **3.** $A = 23° 05'$, $B = 59° 35'$, $c = 12.65$
5. $A = 52.8°$, $B = 32.1°$, $C = 95.1°$ **7.** $b = 313$, $A = 46.7°$, $C = 29.8°$ **9.** $A = 49.8°$, $B = 93.1°$, $C = 37.1°$
11. $a = 37.70$, $B = 54° 16'$, $C = 41° 21'$ **13.** 13.56 cm **15.** 53.9 pounds, 10.4° from 30-pound force
17. 23.4 miles per hour, S 55.7° E **19.** 74.98° **21.** 58.27 feet **23.** 477.7 miles
25. $|\mathbf{R}| = 84.5$ pounds, $\theta = 61.7°$ **27.** 36.2 miles

Exercise Set 6.4

1. 216 **3.** 64.9 **5.** 4.45 **7.** 25.98 **9.** 48.92 **11.** 996 **13.** 0.7784
15. Let A be obtuse as shown. Then, $h = b \sin(180° - A) = b \sin A$.
Area $= \frac{1}{2} hc = \frac{1}{2} bc \sin A$.

C
180° − A
A
B

17. 48.7 **19.** 97.6

6.5 Review Exercise Set

1. $A = 24.19°$, $B = 45.81°$, $c = 9.17$, area $= 13.16$ **3.** $A = 37.36°$, $B = 43.05°$, $C = 99.59°$, area $= 35.50$
5. $C = 118.9°$, $b = 10.7$, $c = 20.8$, area $= 62.9$ **7.** $A = 100.2°$, $B = 52.2°$, $C = 27.6°$, area $= 92.5$
9. $A = 131°\ 03'$, $C = 12°\ 37'$, $a = 43.96$, area $= 165.8$ **11.** solution 1: $B = 104.2°$, $C = 50.6°$, $b = 34.4$, area $= 200.5$ solution 2: $B = 25.4°$, $C = 129.4°$, $b = 15.2$, area $= 88.7$
13. $A = 25°\ 51'$, $B = 21°\ 56'$, $c = 55.23$, area $= 335.4$
15. solution 1: $A = 59.8°$, $B = 69.9°$, $a = 18.9$, area $= 149$ solution 2: $A = 19.6°$, $B = 110.1°$, $a = 7.32$, area $= 57.7$
17. third side $= 103$ feet, area $= 4{,}915$ **19.** $|\mathbf{V}_2| = 15.0$, $|\mathbf{V}_1 + \mathbf{V}_2| = 25.0$ **21.** $|\mathbf{R}| = 80.6$ pounds, $\theta = 53°\ 38'$
23. 342.7 feet **25.** first leg: speed $= 411$ miles per hour, direction $= 131.6°$; second leg: speed $= 439$ miles per hour, direction $= 40.6°$; distance $= 1{,}006$ miles **27.** heading going $= 227°\ 03'$, heading returning $= 22°\ 57'$, total time $= 4$ hours 43 minutes **29.** 6.65 miles and 9.38 miles **31.** $h = 11.76$, $\theta = 35°\ 21'$

Exercise Set 7.2

1. a. $9 - 2i$ **b.** $-2 - 5i$ **c.** $-1 + 9i$ **d.** $6 - 3i$ **3. a.** $18 + i$ **b.** 5 **c.** $-83 - i$ **d.** $-9 + 13i$ **5. a.** $(1 - i)/2$ **b.** $(3 + 2i)/13$ **c.** $-i$ **d.** $(7 - 8i)/113$ **7. a.** i **b.** $-i$ **c.** $(3 + 29i)/34$ **d.** $(9 + 23i)/61$ **9. a.** $x = 2$, $y = -3$ **b.** $x = 3$, $y = -4$ **11.** $x = 3$, $y = -1$
13. a. $5i\sqrt{3}$ **b.** $-i\sqrt{3}/6$ **c.** $-3\sqrt{2}$ **d.** $-2i$ **e.** $4i$ **15. a.** -1 **b.** i **c.** $-i$ **d.** $-i$ **e.** -1 **17. a.** $x = \pm 3i$ **b.** $x = \pm 5i/2$ **19. a.** $x = 1 \pm i\sqrt{5}$ **b.** $t = (-3 \pm i\sqrt{7})/2$
21. a. $s = (5 \pm i\sqrt{23})/4$ **b.** $s = (-1 \pm i\sqrt{2})/3$ **23. a.** $x = (4 \pm i\sqrt{19})/7$ **b.** $x = (4 \pm 2i)/5$
25. a. real and unequal **b.** real and equal **27. a.** $-46 + 9i$ **b.** $-4 - 4i$ **29.** 0 **31. a.** $2i$, $2i/3$ **b.** i, $1 - i$ **33.** $k > \frac{1}{5}$ or $k < -\frac{1}{3}$

Exercise Set 7.3

1. a. **b.** 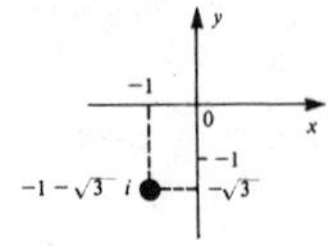 **c.** 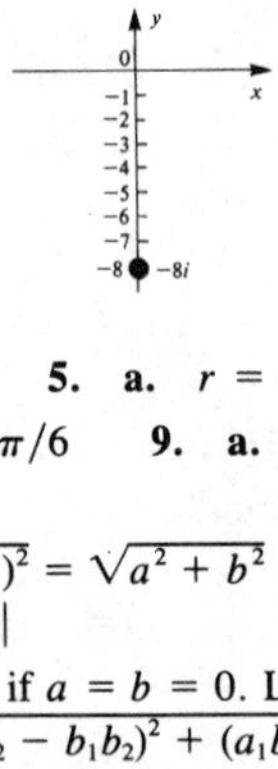 **d.** 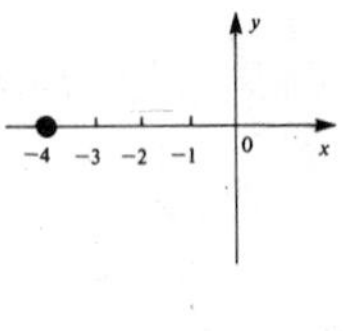 **e.**

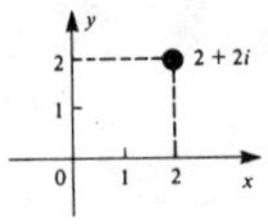

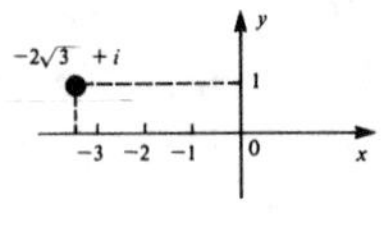

3. a. $r = 4$, $\theta = 11\pi/6$ **b.** $r = 2$, $\theta = \pi/4$ **5. a.** $r = 4\sqrt{2}$, $\theta = 3\pi/4$ **b.** $r = 16$, $\theta = \pi$
7. a. $r = 4\sqrt{2}$, $\theta = \pi/3$ **b.** $r = 2\sqrt{3}$, $\theta = 5\pi/6$ **9. a.** $-\sqrt{3} + i$ **b.** $-12i$
11. a. $-2 - 2i\sqrt{3}$ **b.** $-4\sqrt{3} + 4i$
19. a. $|a + bi| = \sqrt{a^2 + b^2}$; $|a - bi| = \sqrt{a^2 + (-b)^2} = \sqrt{a^2 + b^2}$
b. $\sqrt{z\bar{z}} = \sqrt{(a + bi)(a - bi)} = \sqrt{a^2 + b^2} = |z|$
21. $|z| = \sqrt{a^2 + b^2} \geq 0$ and equality holds if and only if $a = b = 0$. Let $z_1 = a_1 + b_1 i$, $z_2 = a_2 + b_2 i$. Then $z_1 z_2 = (a_1a_2 - b_1b_2) + (a_1b_2 + a_2b_1)i$, and $|z_1z_2| = \sqrt{(a_1a_2 - b_1b_2)^2 + (a_1b_2 + a_2b_1)^2}$
$= \sqrt{a_1^2a_2^2 + b_1^2b_2^2 + a_1^2b_2^2 + a_1^2b_2^2 + a_2^2b_1^2}$
$|z_1|\,|z_2| = \sqrt{a_1^2 + b_1^2} \cdot \sqrt{a_2^2 + b_2^2} = \sqrt{a_1^2a_2^2 + b_1^2b_2^2 + a_1^2b_2^2 + a_2^2b_1^2}$
23. $\dfrac{1}{z} = \dfrac{1}{r(\cos\theta + i\sin\theta)} \cdot \dfrac{\cos\theta - i\sin\theta}{\cos\theta - i\sin\theta} = \dfrac{\cos\theta - i\sin\theta}{r(\cos^2\theta + \sin^2\theta)} = \dfrac{\cos\theta - i\sin\theta}{r}$
25. a. $|z| = 5.725$, $\arg z = 5.210$, $1/z = 0.08336 + 0.1535\,i$
b. $|z| = 90.99$, $\arg z = 3.790$, $1/z = -0.008758 + 0.006638\,i$
c. $|z| = 0.5291$, $\arg z = 2.690$, $1/z = -1.700 - 0.8252\,i$

Exercise Set 7.4

1. $-6i$ **3.** $-4\sqrt{2} - 4i\sqrt{2}$ **5.** $-15 + 15i\sqrt{3}$ **7.** $-18\sqrt{2} + 18i\sqrt{2}$ **9.** $16\sqrt{2}$ **11.** $-\frac{2}{3}$
13. $\sqrt{2} - i\sqrt{2}$ **15.** $(-3\sqrt{2}/4) + (3i\sqrt{2}/4)$ **17.** $(3\sqrt{2}/2) - (3i\sqrt{2}/2)$ **19.** $i\sqrt{2}/2$ **21.** $-8 + 8i\sqrt{3}$

23. $-16 + 16i\sqrt{3}$ **25.** $-32i$ **27.** $16\sqrt{2} + 16i\sqrt{2}$ **29.** $32 - 32i\sqrt{3}$

35. **a.** $z_1z_2z_3 = r_1r_2r_3\,[\cos(\theta_1 + \theta_2 + \theta_3) + i\,\sin(\theta_1 + \theta_2 + \theta_3)]$
$z_1z_2z_3 \ldots z_n = r_1r_2r_3 \ldots r_n[\cos(\theta_1 + \theta_2 + \theta_3 + \ldots + \theta_n) + i\,\sin(\theta_1 + \theta_2 + \theta_3 + \ldots + \theta_n)]$
b. $-12 + 12i\sqrt{3}$

Exercise Set 7.5

1. $2^{1/6}[\cos(\pi/4) + i\,\sin(\pi/4)] = 2^{1/6}(1 + i)$, $2^{1/6}[\cos(11\pi/12) + i\,\sin(11\pi/12]$, $2^{1/6}[\cos(19\pi/12) + i\,\sin(19\pi/12)]$

3. $2, 1 + i\sqrt{3}, -1 + i\sqrt{3}, -2, -1 - i\sqrt{3}, 1 - i\sqrt{3}$ **5.** $\sqrt{2}\,(1 + i), \sqrt{2}(-1 + i), \sqrt{2}\,(-1 - i), \sqrt{2}\,(1 - i)$

7. $\frac{1}{2} + i\sqrt{3}/2, (-\sqrt{3}/2) + (i/2), -\frac{1}{2} - i\sqrt{3}/2, (\sqrt{3}/2) - (i/2)$ **9.** $\{1, -\frac{1}{2} + i\sqrt{3}/2, -\frac{1}{2} - i\sqrt{3}/2\}$

11. $\{2[\cos(2k\pi/5) + i\,\sin(2k\pi/5)]: k = 0, 1, 2, 3, 4\}$

13. $\zeta_0 = 1, \zeta_1 = \frac{1}{2} + i\sqrt{3}/2, \zeta_2 = -\frac{1}{2} + i\sqrt{3}/2, \zeta_3 = -1, \zeta_4 = -\frac{1}{2} - i\sqrt{3}/2, \zeta_5 = \frac{1}{2} - i\sqrt{3}/2$

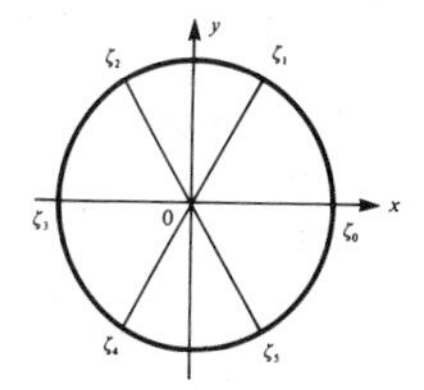

15. $(\sqrt{6}/2) + (i\sqrt{2}/2)$, $0.366 + 1.366i$, $(-\sqrt{2}/2) + (i\sqrt{6}/2)$, $-1.366 + 0.366\,i$, $(-\sqrt{6}/2) - (i\sqrt{2}/2)$, $0.366 - 1.366\,i$, $(\sqrt{2}/2) - (i\sqrt{6}/2)$, $1.366 - 0.366\,i$

17. $1.3833 + 1.0424\,i$, $-0.2111 + 1.7191\,i$,
$-1.5944 + 0.6768\,i$, $-1.3833 - 1.0424\,i$,
$0.2111 - 1.7191\,i$, $1.5944 - 0.6768\,i$

19. $\{2 + 2i\sqrt{3}, -2 + 2i\sqrt{3}, -4i\}$

Exercise Set 7.6

1.

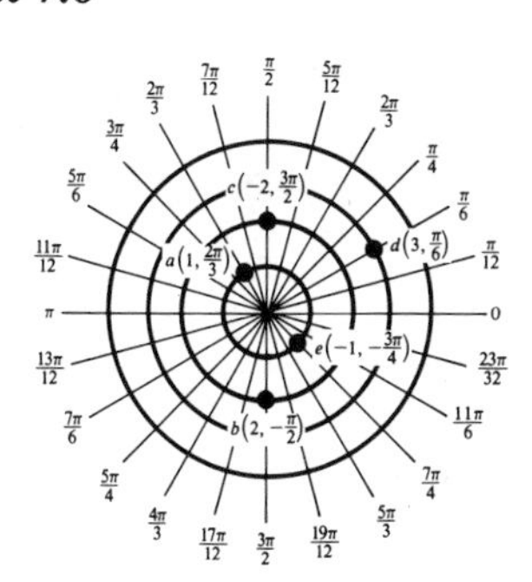

3. $(2, 8\pi/3), (-2, 5\pi/3),$
$(2, -4\pi/3), (-2, -\pi/3),$
$(2, 14\pi/3)$

5. **a.** $(2, 2)$ **b.** $(3\sqrt{3}, -3)$ **c.** $(-5\sqrt{2}, 5\sqrt{2})$ **d.** $(-4, 0)$ **e.** $(2, 2)$ **7.** **a.** $(4\sqrt{2}, 7\pi/4)$
b. $(4, 5\pi/6)$ **c.** $(2, 3\pi/2)$ **d.** $(1, 5\pi/3)$ **e.** $(2\sqrt{6}, 7\pi/6)$ **9.** $\phi = 3\pi/4$ **11.** $r\cos\theta = 3$

13. $r = 4\sin\theta$ **15.** $r^2 = 4/\sin 2\theta$ **17.** $r = -2\cos\theta$ **19.** $x^2 + y^2 = 16$ **21.** $x = 2$

23. $2x - 3y = 5$ **25.** $x = 3$ **27.** $x^2 + y^2 + 2x = 0$ **29.** $y^2 = 1 + 2x$

31.

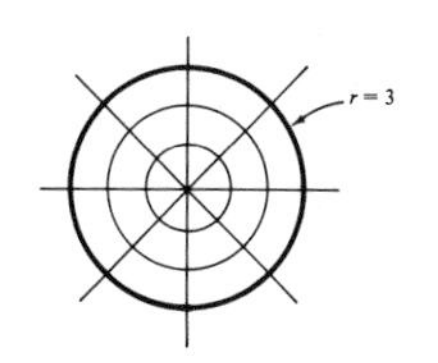

33.

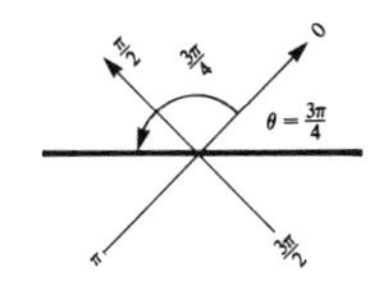

35.

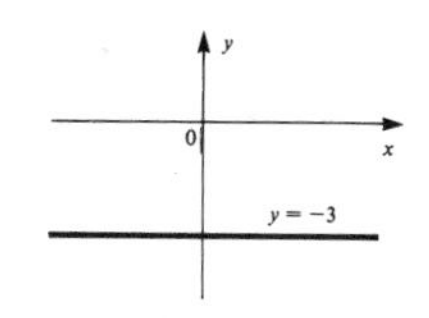

37. cardioid

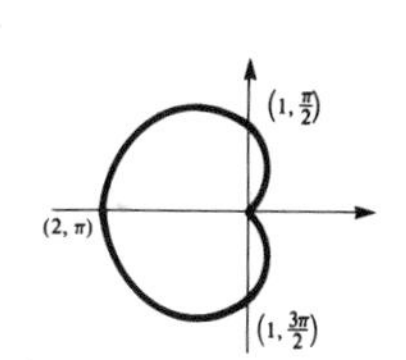

39. limaçon

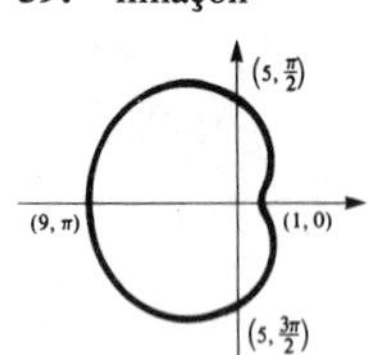

41. lemniscate

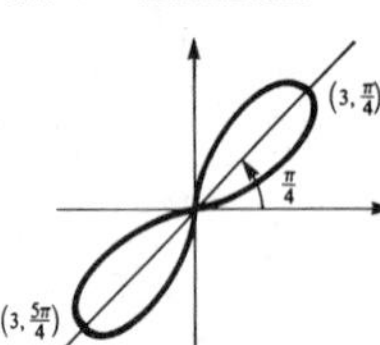

43. 3-leaf rose

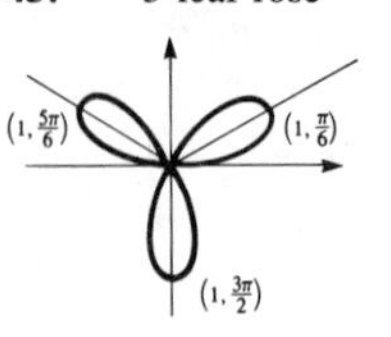

45. 4-leaf rose

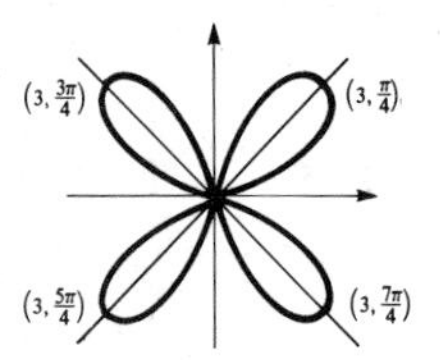

47. limaçon

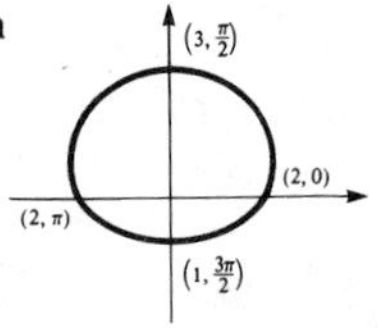

49. circle

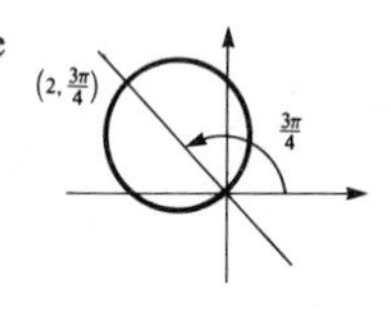

51. cardioid

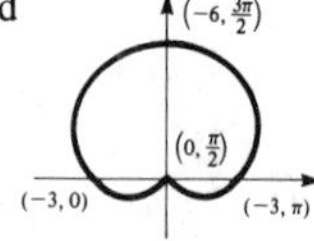

53. **a.** $r = 16 \csc^2 2\theta$, together with $r = 0$
b. $r = 4 \cos \theta \sec 2\theta$

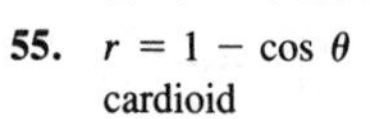

55. $r = 1 - \cos \theta$
cardioid

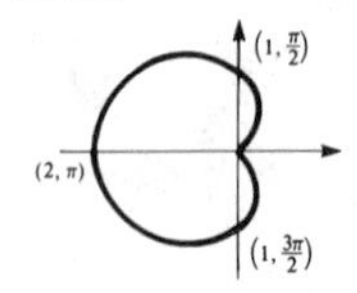

57. $r = \sqrt{2} \cos(\theta - \pi/4)$
circle

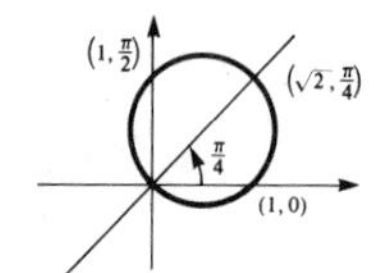

59.

θ	0	$\pi/6$	$\pi/4$	$\pi/3$	$\pi/2$	$2\pi/3$	$3\pi/4$	$5\pi/6$	π	$7\pi/6$	$5\pi/4$	$4\pi/3$	$3\pi/2$	$5\pi/3$	$7\pi/4$	$11\pi/6$
r	2	1.93	1.85	1.73	1.41	1	0.77	0.52	0	−0.52	−0.77	−1	−1.41	−1.73	−1.85	−1.93

Use symmetry

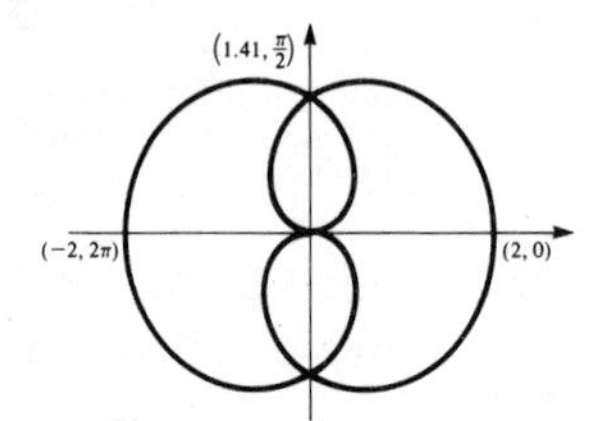

61. 5-leaf rose

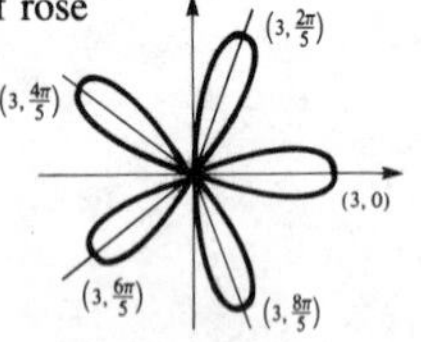

63.

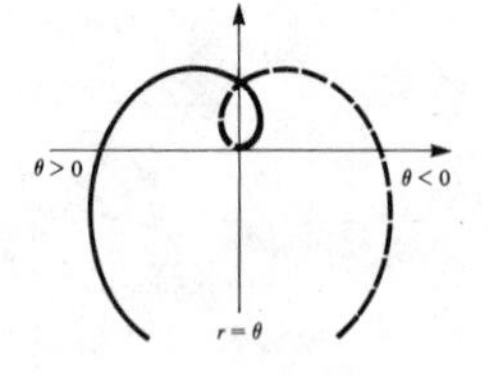

65.

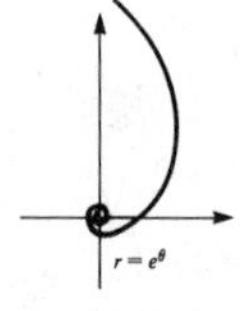

67. $(2, \pi/2)$, $(2, 3\pi/2)$

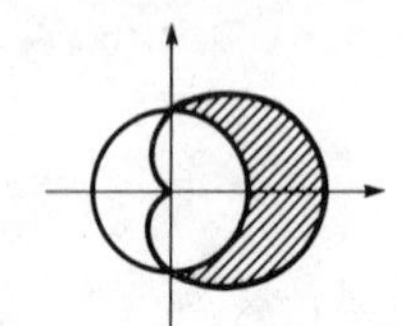

69. $(\frac{3}{2}, \pi/3)$, $(\frac{3}{2}, 5\pi/3)$

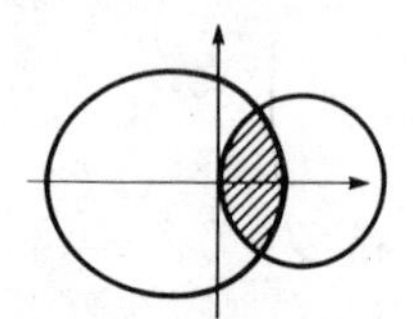

71. $(\frac{5}{2}, \pi/3)$, $(\frac{5}{2}, 5\pi/3)$

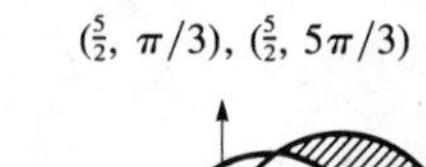

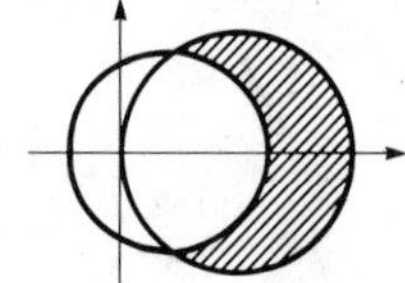

7.7 Review Exercise Set

1. **a.** $8 + i$ **b.** $5 - 7i$ **3.** **a.** $(2 + 3i)/13$ **b.** $-i$ **5.** **a.** $x = 3, y = -\frac{4}{3}$ **b.** $x = -\frac{5}{3}, y = \frac{4}{3}$

7. **a.** $-i$ **b.** $-i$ **c.** i **9.** $x = (-1 \pm i\sqrt{7})/4$ **11.** $y = (3 \pm i\sqrt{11})/2$ **13.** $x = -2 \pm i$

15. **a.** $r = 4, \theta = 11\pi/6$, $4[\cos(11\pi/6) + i \sin(11\pi/6)]$

b. $r = 4\sqrt{2}, \theta = 7\pi/4$, $4\sqrt{2}[\cos(7\pi/4) + i \sin(7\pi/4)]$

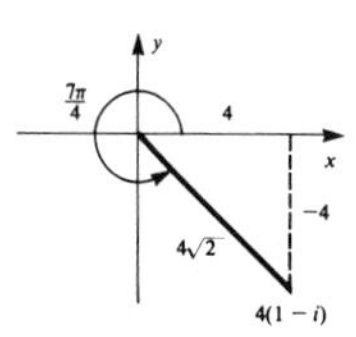

c. $r = 2\sqrt{3}, \theta = 4\pi/3$, $2\sqrt{3}[\cos(4\pi/3) + i \sin(4\pi/3)]$

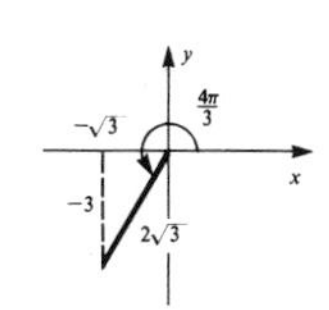

d. $r = 8, \theta = 3\pi/2$, $8[\cos(3\pi/2) + i \sin(3\pi/2)]$

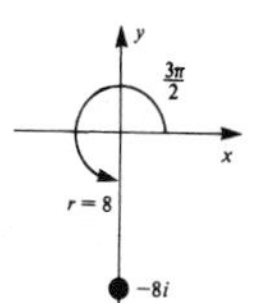

e. $r = 4, \theta = \pi/6$, $4[\cos(\pi/6) + i \sin(\pi/6)]$

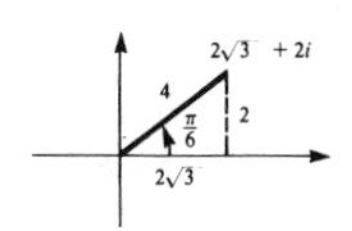

17. **a.** $-6\sqrt{2} - 6i\sqrt{2}$ **b.** $3i$ **19.** **a.** $\frac{1}{16}(-\sqrt{3} + i)$ **b.** $\sqrt{3}(1 + i)$ **21.** **a.** $128(-1 + i)$

b. $-\frac{1}{2} + i\sqrt{3}/2$ **23.** $3, 3i, -3, -3i$ **25.** $(\sqrt{3}/2) + (i/2), (-\sqrt{3}/2) + (i/2), -i$

27. **a.** $(-2, 2)$ **b.** $(2\sqrt{3}, -2)$ **c.** $(3, -3\sqrt{3})$

d. $(0, -5)$ **e.** $(2, 2)$

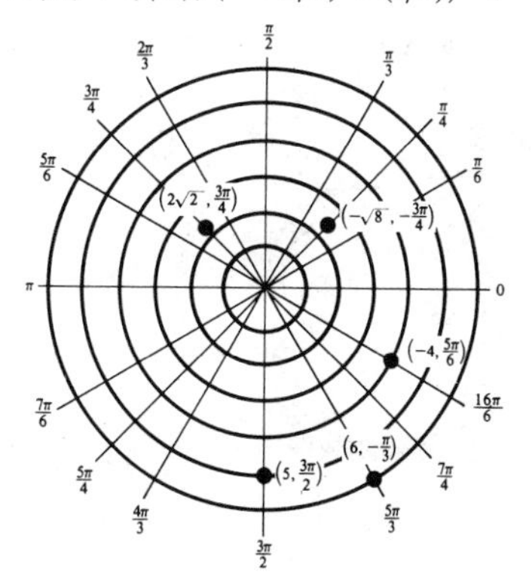

29. **a.** $r = 4 \sin \theta$ **b.** $\theta = 2\pi/3$

31. **a.** $r = 4/(\sin \theta - 1)$ **b.** $r^2 = \tan 2\theta$ **33.** **a.** $x^2 + y^2 + 2y = 0$ **b.** $x + 2\sqrt{2}y = 0$

35. circle **37.** circle **39.** limaçon **41.** cardioid **43.** cardioid

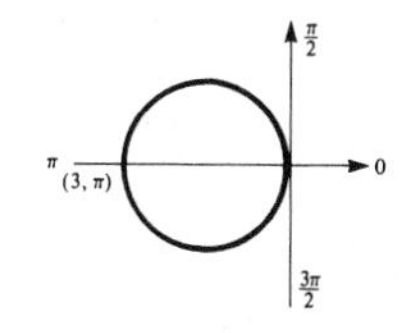

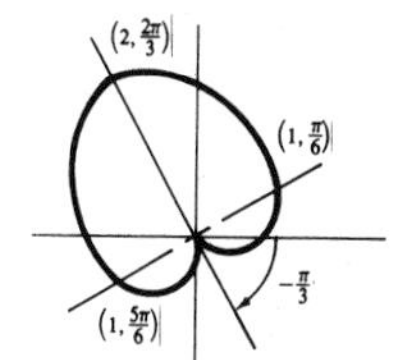

45. $\pm 256\sqrt{2}(1 + i\sqrt{3})$ **49.** **a.** $-0.3691 + 0.9294\,i$

b. $1.356 - 0.2544\,i$, $0.6610 + 1.211\,i$, $-0.9475 + 1.003\,i$, $-1.247 - 0.5912\,i$, $0.1771 - 1.368\,i$

51. $r = 2\cos(\theta - \pi/6)$
circle

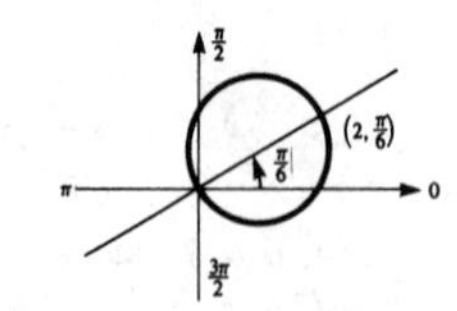

53. $r = -2(1 + \cos\theta)$
cardioid

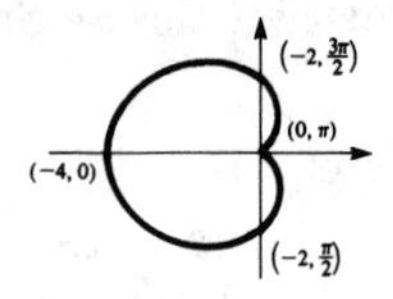

55.

θ	0	$\pi/6$	$\pi/4$	$\pi/3$	$2\pi/3$	$3\pi/4$	$5\pi/6$	π
r	0	0.58	1.4	3	-3	-1.4	-0.58	0

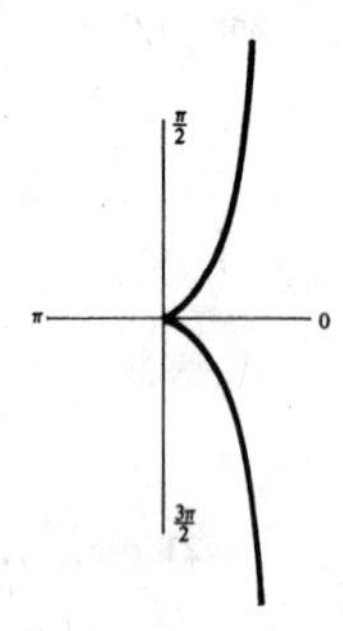

Symmetric to polar axis. As $\theta \to (\pi/2)^-$, $r \to +\infty$. As $\theta \to (\pi/2)^+$, $r \to -\infty$.

57. $(3 + 2\sqrt{3}, 5\pi/6)$, $(3 + 2\sqrt{3}, 7\pi/6)$, pole

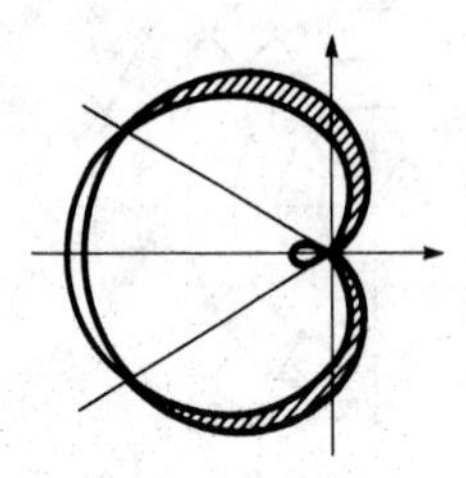

Tables

Table I Values of Trigonometric Functions (Angle in Degrees)

θ deg	θ deg-min	sin θ	cos θ	tan θ	csc θ	sec θ	cot θ		
0.0°	0°00′	0.0000	1.0000	0.0000	no value	1.0000	no value	90°0′	90.0°
0.1	0 06	0.0017	1.0000	0.0017	572.96	1.0000	572.96	89 54	89.9
0.2	0 12	0.0035	1.0000	0.0035	286.48	1.0000	286.48	89 48	89.8
0.3	0 18	0.0052	1.0000	0.0052	190.99	1.0000	190.98	89 42	89.7
0.4	0 24	0.0070	1.0000	0.0070	143.24	1.0000	143.24	89 36	89.6
0.5	0 30	0.0087	1.0000	0.0087	114.59	1.0000	114.59	89 30	89.5
0.6	0 36	0.0105	0.9999	0.0105	95.495	1.0001	95.490	89 24	89.4
0.7	0 42	0.0122	0.9999	0.0122	81.853	1.0001	81.847	89 18	89.3
0.8	0 48	0.0140	0.9999	0.0140	71.622	1.0001	71.615	89 12	89.2
0.9	0 54	0.0157	0.9999	0.0157	63.665	1.0001	63.657	89 06	89.1
1.0°	1°00′	0.0175	0.9998	0.0175	57.299	1.0002	57.290	89°00′	89.0°
1.1	1 06	0.0192	0.9998	0.0192	52.090	1.0002	52.081	88 54	88.9
1.2	1 12	0.0209	0.9998	0.0209	47.750	1.0002	47.740	88 48	88.8
1.3	1 18	0.0227	0.9997	0.0227	44.077	1.0003	44.066	88 42	88.7
1.4	1 24	0.0244	0.9997	0.0244	40.930	1.0003	40.917	88 36	88.6
1.5	1 30	0.0262	0.9997	0.0262	38.202	1.0003	38.188	88 30	88.5
1.6	1 36	0.0279	0.9996	0.0279	35.815	1.0004	35.801	88 24	88.4
1.7	1 42	0.0297	0.9996	0.0297	33.708	1.0004	33.694	88 18	88.3
1.8	1 48	0.0314	0.9995	0.0314	31.836	1.0005	31.821	88 12	88.2
1.9	1 54	0.0332	0.9995	0.0332	30.161	1.0005	30.145	88 06	88.1
2.0°	2°00′	0.0349	0.9994	0.0349	28.654	1.0006	28.636	88°00′	88.0°
2.1	2 06	0.0366	0.9993	0.0367	27.290	1.0007	27.271	87 54	87.9
2.2	2 12	0.0384	0.9993	0.0384	26.050	1.0007	26.031	87 48	87.8
2.3	2 18	0.0401	0.9992	0.0402	24.918	1.0008	24.898	87 42	87.7
2.4	2 24	0.0419	0.9991	0.0419	23.880	1.0009	23.859	87 36	87.6
2.5	2 30	0.0436	0.9990	0.0437	22.926	1.0010	22.904	87 30	87.5
2.6	2 36	0.0454	0.9990	0.0454	22.044	1.0010	22.022	87 24	87.4
2.7	2 42	0.0471	0.9989	0.0472	21.229	1.0011	21.205	87 18	87.3
2.8	2 48	0.0488	0.9988	0.0489	20.471	1.0012	20.446	87 12	87.2
2.9	2 54	0.0506	0.9987	0.0507	19.766	1.0013	19.740	87 06	87.1
3.0°	3°00′	0.0523	0.9986	0.0524	19.107	1.0014	19.081	87°00′	87.0°
3.1	3 06	0.0541	0.9985	0.0542	18.492	1.0015	18.464	86 54	86.9
3.2	3 12	0.0558	0.9984	0.0559	17.914	1.0016	17.886	86 48	86.8
3.3	3 18	0.0576	0.9983	0.0577	17.372	1.0017	17.343	86 42	86.7
3.4	3 24	0.0593	0.9982	0.0594	16.862	1.0018	16.832	86 36	86.6
3.5	3 30	0.0610	0.9981	0.0612	16.380	1.0019	16.350	86 30	86.5
3.6	3 36	0.0628	0.9980	0.0629	15.926	1.0020	15.895	86 24	86.4
3.7	3 42	0.0645	0.9979	0.0647	15.496	1.0021	15.464	86 18	86.3
3.8	3 48	0.0663	0.9978	0.0664	15.089	1.0022	15.056	86 12	86.2
3.9	3 54	0.0680	0.9977	0.0682	14.703	1.0023	14.669	86 06	86.1
4.0°	4°00′	0.0698	0.9976	0.0699	14.336	1.0024	14.301	86°00′	86.0°
4.1	4 06	0.0715	0.9974	0.0717	13.987	1.0026	13.951	85 54	85.9
4.2	4 12	0.0732	0.9973	0.0734	13.654	1.0027	13.617	85 48	85.8
4.3	4 18	0.0750	0.9972	0.0752	13.337	1.0028	13.300	85 42	85.7
4.4	4 24	0.0767	0.9971	0.0769	13.035	1.0030	12.996	85 36	85.6
4.5	4 30	0.0785	0.9969	0.0787	12.746	1.0031	12.706	85 30	85.5
4.6	4 36	0.0802	0.9968	0.0805	12.469	1.0032	12.429	85 24	85.4
4.7	4 42	0.0819	0.9966	0.0822	12.204	1.0034	12.163	85 18	85.3
4.8	4 48	0.0837	0.9965	0.0840	11.951	1.0035	11.909	85 12	85.2
4.9	4 54	0.0854	0.9963	0.0857	11.707	1.0037	11.665	85°06′	85.1°
		cos θ	sin θ	cot θ	sec θ	csc θ	tan θ	θ deg-min	θ deg

Table I Values of Trigonometric Functions (Angle in Degrees) (continued)

θ deg	θ deg-min	sin θ	cos θ	tan θ	csc θ	sec θ	cot θ		
5.0°	5°00′	0.0872	0.9962	0.0875	11.474	1.0038	11.430	85°00′	85.0°
5.1	5 06	0.0889	0.9960	0.0892	11.249	1.0040	11.205	84 54	84.9
5.2	5 12	0.0906	0.9959	0.0910	11.034	1.0041	10.988	84 48	84.8
5.3	5 18	0.0924	0.9957	0.0928	10.826	1.0043	10.780	84 42	84.7
5.4	5 24	0.0941	0.9956	0.0945	10.626	1.0045	10.579	84 36	84.6
5.5	5 30	0.0958	0.9954	0.0963	10.433	1.0046	10.385	84 30	84.5
5.6	5 36	0.0976	0.9952	0.0981	10.248	1.0048	10.199	84 24	84.4
5.7	5 42	0.0993	0.9951	0.0998	10.069	1.0050	10.019	84 18	84.3
5.8	5 48	0.1011	0.9949	0.1016	9.8955	1.0051	9.8448	84 12	84.2
5.9	5 54	0.1028	0.9947	0.1033	9.7283	1.0053	9.6768	84 06	84.1
6.0°	6°00′	0.1045	0.9945	0.1051	9.5668	1.0055	9.5144	84°00′	84.0°
6.1	6 06	0.1063	0.9943	0.1069	9.4105	1.0057	9.3573	83 54	83.9
6.2	6 12	0.1080	0.9942	0.1086	9.2593	1.0059	9.2052	83 48	83.8
6.3	6 18	0.1097	0.9940	0.1104	9.1129	1.0061	9.0579	83 42	83.7
6.4	6 24	0.1115	0.9938	0.1122	8.9711	1.0063	8.9152	83 36	83.6
6.5	6 30	0.1132	0.9936	0.1139	8.8337	1.0065	8.7769	83 30	83.5
6.6	6 36	0.1149	0.9934	0.1157	8.7004	1.0067	8.6428	83 24	83.4
6.7	6 42	0.1167	0.9932	0.1175	8.5711	1.0069	8.5126	83 18	83.3
6.8	6 48	0.1184	0.9930	0.1192	8.4457	1.0071	8.3863	83 12	83.2
6.9	6 54	0.1201	0.9928	0.1210	8.3238	1.0073	8.2636	83 06	83.1
7.0°	7°00′	0.1219	0.9925	0.1228	8.2055	1.0075	8.1444	83°00′	83.0°
7.1	7 06	0.1236	0.9923	0.1246	8.0905	1.0077	8.0285	82 54	82.9
7.2	7 12	0.1253	0.9921	0.1263	7.9787	1.0079	7.9158	82 48	82.8
7.3	7 18	0.1271	0.9919	0.1281	7.8700	1.0082	7.8062	82 42	82.7
7.4	7 24	0.1288	0.9917	0.1299	7.7642	1.0084	7.6996	82 36	82.6
7.5	7 30	0.1305	0.9914	0.1317	7.6613	1.0086	7.5958	82 30	82.5
7.6	7 36	0.1323	0.9912	0.1334	7.5611	1.0089	7.4947	82 24	82.4
7.7	7 42	0.1340	0.9910	0.1352	7.4635	1.0091	7.3962	82 18	82.3
7.8	7 48	0.1357	0.9907	0.1370	7.3684	1.0093	7.3002	82 12	82.2
7.9	7 54	0.1374	0.9905	0.1388	7.2757	1.0096	7.2066	82 06	82.1
8.0°	8°00′	0.1392	0.9903	0.1405	7.1853	1.0098	7.1154	82°00′	82.0°
8.1	8 06	0.1409	0.9900	0.1423	7.0972	1.0101	7.0264	81 54	81.9
8.2	8 12	0.1426	0.9898	0.1441	7.0112	1.0103	6.9395	81 48	81.8
8.3	8 18	0.1444	0.9895	0.1459	6.9273	1.0106	6.8548	81 42	81.7
8.4	8 24	0.1461	0.9893	0.1477	6.8454	1.0108	6.7720	81 36	81.6
8.5	8 30	0.1478	0.9890	0.1495	6.7655	1.0111	6.6912	81 30	81.5
8.6	8 36	0.1495	0.9888	0.1512	6.6874	1.0114	6.6122	81 24	81.4
8.7	8 42	0.1513	0.9885	0.1530	6.6111	1.0116	6.5350	81 18	81.3
8.8	8 48	0.1530	0.9882	0.1548	6.5366	1.0119	6.4596	81 12	81.2
8.9	8 54	0.1547	0.9880	0.1566	6.4637	1.0122	6.3859	81 06	81.1
9.0°	9°00′	0.1564	0.9877	0.1584	6.3925	1.0125	6.3138	81°00′	81.0°
9.1	9 06	0.1582	0.9874	0.1602	6.3228	1.0127	6.2432	80 54	80.9
9.2	9 12	0.1599	0.9871	0.1620	6.2547	1.0130	6.1742	80 48	80.8
9.3	9 18	0.1616	0.9869	0.1638	6.1880	1.0133	6.1066	80 42	80.7
9.4	9 24	0.1633	0.9866	0.1655	6.1227	1.0136	6.0405	80 36	80.6
9.5	9 30	0.1650	0.9863	0.1673	6.0589	1.0139	5.9758	80 30	80.5
9.6	9 36	0.1668	0.9860	0.1691	5.9963	1.0142	5.9124	80 24	80.4
9.7	9 42	0.1685	0.9857	0.1709	5.9351	1.0145	5.8502	80 18	80.3
9.8	9 48	0.1702	0.9854	0.1727	5.8751	1.0148	5.7894	80 12	80.2
9.9	9 54	0.1719	0.9851	0.1745	5.8164	1.0151	5.7297	80°06′	80.1°
		cos θ	sin θ	cot θ	sec θ	csc θ	tan θ	θ deg-min	θ deg

Table I Values of Trigonometric Functions (Angle in Degrees) (continued)

θ deg	θ deg-min	sin θ	cos θ	tan θ	csc θ	sec θ	cot θ		
10.0°	10°00′	0.1736	0.9848	0.1763	5.7588	1.0154	5.6713	80°00′	80.0°
10.1	10 06	0.1754	0.9845	0.1781	5.7023	1.0157	5.6140	79 54	79.9
10.2	10 12	0.1771	0.9842	0.1799	5.6470	1.0161	5.5578	79 48	79.8
10.3	10 18	0.1788	0.9839	0.1817	5.5928	1.0164	5.5027	79 42	79.7
10.4	10 24	0.1805	0.9836	0.1835	5.5396	1.0167	5.4486	79 36	79.6
10.5	10 30	0.1822	0.9833	0.1853	5.4874	1.0170	5.3955	79 30	79.5
10.6	10 36	0.1840	0.9829	0.1871	5.4362	1.0174	5.3435	79 24	79.4
10.7	10 42	0.1857	0.9826	0.1890	5.3860	1.0177	5.2924	79 18	79.3
10.8	10 48	0.1874	0.9823	0.1908	5.3367	1.0180	5.2422	79 12	79.2
10.9	10 54	0.1891	0.9820	0.1926	5.2883	1.0184	5.1929	79 06	79.1
11.0°	11°00′	0.1908	0.9816	0.1944	5.2408	1.0187	5.1446	79°00′	79.0°
11.1	11 06	0.1925	0.9813	0.1962	5.1942	1.0191	5.0970	78 54	78.9
11.2	11 12	0.1942	9.9810	0.1980	5.1484	1.0194	5.0504	78 48	78.8
11.3	11 18	0.1959	0.9806	0.1998	5.1034	1.0198	5.0045	78 42	78.7
11.4	11 24	0.1977	0.9803	0.2016	5.0593	1.0201	4.9595	78 36	78.6
11.5	11 30	0.1994	0.9799	0.2035	5.0159	1.0205	4.9152	78 30	78.5
11.6	11 36	0.2011	0.9796	0.2053	4.9732	1.0209	4.8716	78 24	78.4
11.7	11 42	0.2028	0.9792	0.2071	4.9313	1.0212	4.8288	78 18	78.3
11.8	11 48	0.2045	0.9789	0.2089	4.8901	1.0216	4.7867	78 12	78.2
11.9	11 54	0.2062	0.9785	0.2107	4.8496	1.0220	4.7453	78 06	78.1
12.0°	12°00′	0.2079	0.9781	0.2126	4.8097	1.0223	4.7046	78°00′	78.0°
12.1	12 06	0.2096	0.9778	0.2144	4.7706	1.0227	4.6646	77 54	77.9
12.2	12 12	0.2113	0.9774	0.2162	4.7321	1.0231	4.6252	77 48	77.8
12.3	12 18	0.2130	0.9770	0.2180	4.6942	1.0235	4.5864	77 42	77.7
12.4	12 24	0.2147	0.9767	0.2199	4.6569	1.0239	4.5483	77 36	77.6
12.5	12 30	0.2164	0.9763	0.2217	4.6202	1.0243	4.5107	77 30	77.5
12.6	12 36	0.2181	0.9759	0.2235	4.5841	1.0247	4.4737	77 24	77.4
12.7	12 42	0.2198	0.9755	0.2254	4.5486	1.0251	4.4374	77 18	77.3
12.8	12 48	0.2215	0.9751	0.2272	4.5137	1.0255	4.4015	77 12	77.2
12.9	12 54	0.2232	0.9748	0.2290	4.4793	1.0259	4.3662	77 06	77.1
13.0°	13°00′	0.2250	0.9744	0.2309	4.4454	1.0263	4.3315	77°00′	77.0°
13.1	13 06	0.2267	0.9740	0.2327	4.4121	1.0267	4.2972	76 54	76.9
13.2	13 12	0.2284	0.9736	0.2345	4.3792	1.0271	4.2635	76 48	76.8
13.3	13 18	0.2300	0.9732	0.2364	4.3469	1.0276	4.2303	76 42	76.7
13.4	13 24	0.2317	0.9728	0.2382	4.3150	1.0280	4.1976	76 36	76.6
13.5	13 30	0.2334	0.9724	0.2401	4.2837	1.0284	4.1653	76 30	76.5
13.6	13 36	0.2351	0.9720	0.2419	4.2528	1.0288	4.1335	76 24	76.4
13.7	13 42	0.2368	0.9715	0.2438	4.2223	1.0293	4.1022	76 18	76.3
13.8	13 48	0.2385	0.9711	0.2456	4.1923	1.0297	4.0713	76 12	76.2
13.9	13 54	0.2402	0.9707	0.2475	4.1627	1.0302	4.0408	76 06	76.1
14.0°	14°00′	0.2419	0.9703	0.2493	4.1336	1.0306	4.0108	76°00′	76.0°
14.1	14 06	0.2436	0.9699	0.2512	4.1048	1.0311	3.9812	75 54	75.9
14.2	14 12	0.2453	0.9694	0.2530	4.0765	1.0315	3.9520	75 48	75.8
14.3	14 18	0.2470	0.9690	0.2549	4.0486	1.0320	3.9232	75 42	75.7
14.4	14 24	0.2487	0.9686	0.2568	4.0211	1.0324	3.8947	75 36	75.6
14.5	14 30	0.2504	0.9681	0.2586	3.9939	1.0329	3.8667	75 30	75.5
14.6	14 36	0.2521	0.9677	0.2605	3.9672	1.0334	3.8391	75 24	75.4
14.7	14 42	0.2538	0.9673	0.2623	3.9408	1.0338	3.8118	75 18	75.3
14.8	14 48	0.2554	0.9668	0.2642	3.9147	1.0343	3.7849	75 12	75.2
14.9	14 54	0.2571	0.9664	0.2661	3.8890	1.0348	3.7583	75°06′	75.1°
		cos θ	sin θ	cot θ	sec θ	csc θ	tan θ	θ deg-min	θ deg

Table I Values of Trigonometric Functions (Angle in Degrees) (continued)

θ deg	θ deg-min	$\sin\theta$	$\cos\theta$	$\tan\theta$	$\csc\theta$	$\sec\theta$	$\cot\theta$		
15.0°	15°00′	0.2588	0.9659	0.2679	3.8637	1.0353	3.7321	75°00′	75.0°
15.1	15 06	0.2605	0.9655	0.2698	3.8387	1.0358	3.7062	74 54	74.9
15.2	15 12	0.2622	0.9650	0.2717	3.8140	1.0363	3.6806	74 48	74.8
15.3	15 18	0.2639	0.9646	0.2736	3.7897	1.0367	3.6554	74 42	74.7
15.4	15 24	0.2656	0.9641	0.2754	3.7657	1.0372	3.6305	74 36	74.6
15.5	15 30	0.2672	0.9636	0.2773	3.7420	1.0377	3.6059	74 30	74.5
15.6	15 36	0.2689	0.9632	0.2792	3.7186	1.0382	3.5816	74 24	74.4
15.7	15 42	0.2706	0.9627	0.2811	3.6955	1.0388	3.5576	74 18	74.3
15.8	15 48	0.2723	0.9622	0.2830	3.6727	1.0393	3.5339	74 12	74.2
15.9	15 54	0.2740	0.9617	0.2849	3.6502	1.0398	3.5105	74 06	74.1
16.0°	16°00′	0.2756	0.9613	0.2867	3.6280	1.0403	3.4874	74°00′	74.0°
16.1	16 06	0.2773	0.9608	0.2886	3.6060	1.0408	3.4646	73 54	73.9
16.2	16 12	0.2790	0.9603	0.2905	3.5843	1.0413	3.4420	73 48	73.8
16.3	16 18	0.2807	0.9598	0.2924	3.5629	1.0419	3.4197	73 42	73.7
16.4	16 24	0.2823	0.9593	0.2943	3.5418	1.0424	3.3977	73 36	73.6
16.5	16 30	0.2840	0.9588	0.2962	3.5209	1.0429	3.3759	73 30	73.5
16.6	16 36	0.2857	0.9583	0.2981	3.5003	1.0435	3.3544	73 24	74.4
16.7	16 42	0.2874	0.9578	0.3000	3.4800	1.0440	3.3332	73 18	73.3
16.8	16 48	0.2890	0.9573	0.3019	3.4598	1.0446	3.3122	73 12	73.2
16.9	16 54	0.2907	0.9568	0.3038	3.4399	1.0451	3.2914	73 06	73.1
17.0°	17 00′	0.2924	0.9563	0.3057	3.4203	1.0457	3.2709	73°00′	73.0°
17.1	17 06	0.2940	0.9558	0.3076	3.4009	1.0463	3.2506	72 54	72.9
17.2	17 12	0.2957	0.9553	0.3096	3.3817	1.0468	3.2305	72 48	72.8
17.3	17 18	0.2974	0.9548	0.3115	3.3628	1.0474	3.2106	72 42	72.7
17.4	17 24	0.2990	0.9542	0.3134	3.3440	1.0480	3.1910	72 36	72.6
17.5	17 30	0.3007	0.9537	0.3153	3.3255	1.0485	3.1716	72 30	72.5
17.6	17 36	0.3024	0.9532	0.3172	3.3072	1.0491	3.1524	72 24	72.4
17.7	17 42	0.3040	0.9527	0.3191	3.2891	1.0497	3.1334	72 18	72.3
17.8	17 48	0.3057	0.9521	0.3211	3.2712	1.0503	3.1146	72 12	72.2
17.9	17 54	0.3074	0.9516	0.3230	3.2536	1.0509	3.0961	72 06	72.1
18.0°	18°00′	0.3090	0.9511	0.3249	3.2361	1.0515	3.0777	72°00′	72.0°
18.1	18 06	0.3107	0.9505	0.3268	3.2188	1.0521	3.0595	71 54	71.9
18.2	18 12	0.3123	0.9500	0.3288	3.2017	1.0527	3.0415	71 48	71.8
18.3	18 18	0.3140	0.9494	0.3307	3.1848	1.0533	3.0237	71 42	71.7
18.4	18 24	0.3156	0.9489	0.3327	3.1681	1.0539	3.0061	71 36	71.6
18.5	18 30	0.3173	0.9483	0.3346	3.1515	1.0545	2.9887	71 30	71.5
18.6	18 36	0.3190	0.9478	0.3365	3.1352	1.0551	2.9714	71 24	71.4
18.7	18 42	0.3206	0.9472	0.3385	3.1190	1.0557	2.9544	71 18	71.3
18.8	18 48	0.3223	0.9466	0.3404	3.1030	1.0564	2.9375	71 12	71.2
18.9	18 54	0.3239	0.9461	0.3424	3.0872	1.0570	2.9208	71 06	71.1
19.0°	19°00′	0.3256	0.9455	0.3443	3.0716	1.0576	2.9042	71°00′	71.0°
19.1	19 06	0.3272	0.9449	0.3463	3.0561	1.0583	2.8878	70 54	70.9
19.2	19 12	0.3289	0.9444	0.3482	3.0407	1.0589	2.8716	70 48	70.8
19.3	19 18	0.3305	0.9438	0.3502	3.0256	1.0595	2.8556	70 42	70.7
19.4	19 24	0.3322	0.9432	0.3522	3.0106	1.0602	2.8397	70 36	70.6
19.5	19 30	0.3338	0.9426	0.3541	2.9957	1.0608	2.8239	70 30	70.5
19.6	19 36	0.3355	0.9421	0.3561	2.9811	1.0615	2.8083	70 24	70.4
19.7	19 42	0.3371	0.9415	0.3581	2.9665	1.0622	2.7929	70 18	70.3
19.8	19 48	0.3387	0.9409	0.3600	2.9521	1.0628	2.7776	70 12	70.2
19.9	19 54	0.3404	0.9403	0.3620	2.9379	1.0635	2.7625	70°06′	70.1°
		$\cos\theta$	$\sin\theta$	$\cot\theta$	$\sec\theta$	$\csc\theta$	$\tan\theta$	θ deg-min	θ deg

Table I Values of Trigonometric Functions (Angle in Degrees) (continued)

θ deg	θ deg-min	sin θ	cos θ	tan θ	csc θ	sec θ	cot θ		
20.0°	20°00′	0.3420	0.9397	0.3640	2.9238	1.0642	2.7475	70°00′	70.0°
20.1	20 06	0.3437	0.9391	0.3659	2.9099	1.0649	2.7326	69 54	69.9
20.2	20 12	0.3453	0.9385	0.3679	2.8960	1.0655	2.7179	69 48	69.8
20.3	20 18	0.3469	0.9379	0.3699	2.8824	1.0662	2.7034	69 42	69.7
20.4	20 24	0.3486	0.9373	0.3719	2.8688	1.0669	2.6889	69 36	69.6
20.5	20 30	0.3502	0.9367	0.3739	2.8555	1.0676	2.6746	69 30	69.5
20.6	20 36	0.3518	0.9361	0.3759	2.8422	1.0683	2.6605	69 24	69.4
20.7	20 42	0.3535	0.9354	0.3779	2.8291	1.0690	2.6464	69 18	69.3
20.8	20 48	0.3551	0.9348	0.3799	2.8161	1.0697	2.6325	69 12	69.2
20.9	20 54	0.3567	0.9342	0.3819	2.8032	1.0704	2.6187	69 06	69.1
21.0°	21°00′	0.3584	0.9336	0.3839	2.7904	1.0711	2.6051	69°00′	69.0°
21.1	21 06	0.3600	0.9330	0.3859	2.7778	1.0719	2.5916	68 54	68.9
21.2	21 12	0.3616	0.9323	0.3879	2.7653	1.0726	2.5782	68 48	68.8
21.3	21 18	0.3633	0.9317	0.3899	2.7529	1.0733	2.5649	68 42	68.7
21.4	21 24	0.3649	0.9311	0.3919	2.7407	1.0740	2.5517	68 36	68.6
21.5	21 30	0.3665	0.9304	0.3939	2.7285	1.0748	2.5386	68 30	68.5
21.6	21 36	0.3681	0.9298	0.3959	2.7165	1.0755	2.5257	68 24	68.4
21.7	21 42	0.3697	0.9291	0.3979	2.7046	1.0763	2.5129	68 18	68.3
21.8	21 48	0.3714	0.9285	0.4000	2.6927	1.0770	2.5002	68 12	68.2
21.9	21 54	0.3730	0.9278	0.4020	2.6811	1.0778	2.4876	68 06	68.1
22.0°	22°00′	0.3746	0.9272	0.4040	2.6695	1.0785	2.4751	68°00′	68.0°
22.1	22 06	0.3762	0.9265	0.4061	2.6580	1.0793	2.4627	67 54	67.9
22.2	22 12	0.3778	0.9259	0.4081	2.6466	1.0801	2.4504	67 48	67.8
22.3	22 18	0.3795	0.9252	0.4101	2.6354	1.0808	2.4383	67 42	67.7
22.4	22 24	0.3811	0.9245	0.4122	2.6242	1.0816	2.4262	67 36	67.6
22.5	22 30	0.3827	0.9239	0.4142	2.6131	1.0824	2.4142	67 30	67.5
22.6	22 36	0.3843	0.9232	0.4163	2.6022	1.0832	2.4023	67 24	67.4
22.7	22 42	0.3859	0.9225	0.4183	2.5913	1.0840	2.3906	67 18	67.3
22.8	22 48	0.3875	0.9219	0.4204	2.5805	1.0848	2.3789	67 12	67.2
22.9	22 54	0.3891	0.9212	0.4224	2.5699	1.0856	2.3673	67 06	67.1
23.0°	23°00′	0.3907	0.9205	0.4245	2.5593	1.0864	2.3559	67°00′	67.0°
23.1	23 06	0.3923	0.9198	0.4265	2.5488	1.0872	2.3445	66 54	66.9
23.2	23 12	0.3939	0.9191	0.4286	2.5384	1.0880	2.3332	66 48	66.8
23.3	23 18	0.3955	0.9184	0.4307	2.5282	1.0888	2.3220	66 42	66.7
23.4	23 24	0.3971	0.9178	0.4327	2.5180	1.0896	2.3109	66 36	66.6
23.5	23 30	0.3987	0.9171	0.4348	2.5078	1.0904	2.2998	66 30	66.5
23.6	23 36	0.4003	0.9164	0.4369	2.4978	1.0913	2.2889	66 24	66.4
23.7	23 42	0.4019	0.9157	0.4390	2.4879	1.0921	2.2781	66 18	66.3
23.8	23 48	0.4035	0.9150	0.4411	2.4780	1.0929	2.2673	66 12	66.2
23.9	23 54	0.4051	0.9143	0.4431	2.4683	1.0938	2.2566	66 06	66.1
24.0°	24°00′	0.4067	0.9135	0.4452	2.4586	1.0946	2.2460	66°00′	66.0°
24.1	24 06	0.4083	0.9128	0.4473	2.4490	1.0955	2.2355	65 54	65.9
24.2	24 12	0.4099	0.9121	0.4494	2.4395	1.0963	2.2251	65 48	65.8
24.3	24 18	0.4115	0.9114	0.4515	2.4301	1.0972	2.2148	65 42	65.7
24.4	24 24	0.4131	0.9107	0.4536	2.4207	1.0981	2.2045	65 36	65.6
24.5	24 30	0.4147	0.9100	0.4557	2.4114	1.0989	2.1943	65 30	65.5
24.6	24 36	0.4163	0.9092	0.4578	2.4022	1.0998	2.1842	65 24	65.4
24.7	24 42	0.4179	0.9085	0.4599	2.3931	1.1007	2.1742	65 18	65.3
24.8	24 48	0.4195	0.9078	0.4621	2.3841	1.1016	2.1642	65 12	65.2
24.9	24 54	0.4210	0.9070	0.4642	2.3751	1.1025	2.1543	65°06′	65.1°
		cos θ	sin θ	cot θ	sec θ	csc θ	tan θ	θ deg-min	θ deg

Table I Values of Trigonometric Functions (Angle in Degrees) (continued)

θ deg	deg-min	sin θ	cos θ	tan θ	csc θ	sec θ	cot θ		
25.0°	25°00′	0.4226	0.9063	0.4663	2.3662	1.1034	2.1445	65°00′	65.0°
25.1	25 06	0.4242	0.9056	0.4684	2.3574	1.1043	2.1348	64 54	64.9
25.2	25 12	0.4258	0.9048	0.4706	2.3486	1.1052	2.1251	64 48	64.8
25.3	25 18	0.4274	0.9041	0.4727	2.3400	1.1061	2.1155	64 42	64.7
25.4	25 24	0.4289	0.9033	0.4748	2.3314	1.1070	2.1060	64 36	64.6
25.5	25 30	0.4305	0.9026	0.4770	2.3228	1.1079	2.0965	64 30	64.5
25.6	25 36	0.4321	0.9018	0.4791	2.3144	1.1089	2.0872	64 24	64.4
25.7	25 42	0.4337	0.9011	0.4813	2.3060	1.1098	2.0778	64 18	64.3
25.8	25 48	0.4352	0.9003	0.4834	2.2976	1.1107	2.0686	64 12	64.2
25.9	25 54	0.4368	0.8996	0.4856	2.2894	1.1117	2.0594	64 06	64.1
26.0°	26°00′	0.4384	0.8988	0.4877	2.2812	1.1126	2.0503	64°00′	64.0°
26.1	26 06	0.4399	0.8980	0.4899	2.2730	1.1136	2.0413	63 54	63.9
26.2	26 12	0.4415	0.8973	0.4921	2.2650	1.1145	2.0323	63 48	63.8
26.3	26 18	0.4431	0.8965	0.4942	2.2570	1.1155	2.0233	63 42	63.7
26.4	26 24	0.4446	0.8957	0.4964	2.2490	1.1164	2.0145	63 36	63.6
26.5	26 30	0.4462	0.8949	0.4986	2.2412	1.1174	2.0057	63 30	63.5
26.6	26 36	0.4478	0.8942	0.5008	2.2333	1.1184	1.9970	63 24	63.4
26.7	26 42	0.4493	0.8934	0.5029	2.2256	1.1194	1.9883	63 18	63.3
26.8	26 48	0.4509	0.8926	0.5051	2.2179	1.1203	1.9797	63 12	63.2
26.9	26 54	0.4524	0.8918	0.5073	2.2103	1.1213	1.9711	63 06	63.1
27.0°	27°00′	0.4540	0.8910	0.5095	2.2027	1.1223	1.9626	63°00′	63.0°
27.1	27 06	0.4555	0.8902	0.5117	2.1952	1.1233	1.9542	62 54	62.9
27.2	27 12	0.4571	0.8894	0.5139	2.1877	1.1243	1.9458	62 48	62.8
27.3	27 18	0.4586	0.8886	0.5161	2.1803	1.1253	1.9375	62 42	62.7
27.4	27 24	0.4602	0.8878	0.5184	2.1730	1.1264	1.9292	62 36	62.6
27.5	27 30	0.4617	0.8870	0.5206	2.1657	1.1274	1.9210	62 30	62.5
27.6	27 36	0.4633	0.8862	0.5228	2.1584	1.1284	1.9128	62 24	62.4
27.7	27 42	0.4648	0.8854	0.5250	2.1513	1.1294	1.9047	62 18	62.3
27.8	27 48	0.4664	0.8846	0.5272	2.1441	1.1305	1.8967	62 12	62.2
27.9	27 54	0.4679	0.8838	0.5295	2.1371	1.1315	1.8887	62 06	62.1
28.0°	28°00′	0.4695	0.8829	0.5317	2.1301	1.1326	1.8807	62°00′	62.0°
28.1	28 06	0.4710	0.8821	0.5339	2.1231	1.1336	1.8728	61 54	61.9
28.2	28 12	0.4726	0.8813	0.5362	2.1162	1.1347	1.8650	61 48	61.8
28.3	28 18	0.4741	0.8805	0.5384	2.1093	1.1357	1.8572	61 42	61.7
28.4	28 24	0.4756	0.8796	0.5407	2.1025	1.1368	1.8495	61 36	61.6
28.5	28 30	0.4772	0.8788	0.5430	2.0957	1.1379	1.8418	61 30	61.5
28.6	28 36	0.4787	0.8780	0.5452	2.0890	1.1390	1.8341	61 24	61.4
28.7	28 42	0.4802	0.8771	0.5475	2.0824	1.1401	1.8265	61 18	61.3
28.8	28 48	0.4818	0.8763	0.5498	2.0758	1.1412	1.8190	61 12	61.2
28.9	28 54	0.4833	0.8755	0.5520	2.0692	1.1423	1.8115	61 06	61.1
29.0°	29°00′	0.4848	0.8746	0.5543	2.0627	1.1434	1.8040	61°00′	61.0°
29.1	29 06	0.4863	0.8738	0.5566	2.0562	1.1445	1.7966	60 54	60.9
29.2	29 12	0.4879	0.8729	0.5589	2.0598	1.1456	1.7893	60 48	60.8
29.3	29 18	0.4894	0.8721	0.5612	2.0434	1.1467	1.7820	60 42	60.7
29.4	29 24	0.4909	0.8712	0.5635	2.0371	1.1478	1.7747	60 36	60.6
29.5	29 30	0.4924	0.8704	0.5658	2.0308	1.1490	1.7675	60 30	60.5
29.6	29 36	0.4939	0.8695	0.5681	2.0245	1.1501	1.7603	60 24	60.4
29.7	29 42	0.4955	0.8686	0.5704	2.0183	1.1512	1.7532	60 18	60.3
29.8	29 48	0.4970	0.8678	0.5727	2.0122	1.1524	1.7461	60 12	60.2
29.9	29 54	0.4985	0.8669	0.5750	2.0061	1.1535	1.7391	60°06′	60.1°
		cos θ	sin θ	cot θ	sec θ	csc θ	tan θ	θ deg-min	θ deg

Table I Values of Trigonometric Functions (Angle in Degrees) (continued)

θ deg	θ deg-min	sin θ	cos θ	tan θ	csc θ	sec θ	cot θ		
30.0°	30°00′	0.5000	0.8660	0.5774	2.0000	1.1547	1.7321	60°00′	60.0°
30.1	30 06	0.5015	0.8652	0.5797	1.9940	1.1559	1.7251	59 54	59.9
30.2	30 12	0.5030	0.8643	0.5820	1.9880	1.1570	1.7182	59 48	59.8
30.3	30 18	0.5045	0.8634	0.5844	1.9821	1.1582	1.7113	59 42	59.7
30.4	30 24	0.5060	0.8625	0.5867	1.9762	1.1594	1.7045	59 36	59.6
30.5	30 30	0.5075	0.8616	0.5890	1.9703	1.1606	1.6977	59 30	59.5
30.6	30 36	0.5090	0.8607	0.5914	1.9645	1.1618	1.6909	59 24	59.4
30.7	30 42	0.5105	0.8599	0.5938	1.9587	1.1630	1.6842	59 18	59.3
30.8	30 48	0.5120	0.8590	0.5961	1.9530	1.1642	1.6775	59 12	59.2
30.9	30 54	0.5135	0.8581	0.5985	1.9473	1.1654	1.6709	59 06	59.1
31.0°	31°00′	0.5150	0.8572	0.6009	1.9416	1.1666	1.6643	59°00′	59.0°
31.1	31 06	0.5165	0.8563	0.6032	1.9360	1.1679	1.6577	58 54	58.9
31.2	31 12	0.5180	0.8554	0.6056	1.9304	1.1691	1.6512	58 48	58.8
31.3	31 18	0.5195	0.8545	0.6080	1.9249	1.1703	1.6447	58 42	58.7
31.4	31 24	0.5210	0.8536	0.6104	1.9194	1.1716	1.6383	58 36	58.6
31.5	31 30	0.5225	0.8526	0.6128	1.9139	1.1728	1.6319	58 30	58.5
31.6	31 36	0.5240	0.8517	0.6152	1.9084	1.1741	1.6255	58 24	58.4
31.7	31 42	0.5255	0.8508	0.6176	1.9031	1.1753	1.6191	58 18	58.3
31.8	31 48	0.5270	0.8499	0.6200	1.8977	1.1766	1.6128	58 12	58.2
31.9	31 54	0.5284	0.8490	0.6224	1.8924	1.1779	1.6066	58 06	58.1
32.0°	32°00′	0.5299	0.8480	0.6249	1.8871	1.1792	1.6003	58°00′	58.0°
32.1	32 06	0.5314	0.8471	0.6273	1.8818	1.1805	1.5941	57 54	57.9
32.2	32 12	0.5329	0.8462	0.6297	1.8766	1.1818	1.5880	57 48	57.8
32.3	32 18	0.5344	0.8453	0.6322	1.8714	1.1831	1.5818	57 42	57.7
32.4	32 24	0.5358	0.8443	0.6346	1.8663	1.1844	1.5757	57 36	57.6
32.5	32 30	0.5373	0.8434	0.6371	1.8612	1.1857	1.5697	57 30	57.5
32.6	32 36	0.5388	0.8425	0.6395	1.8561	1.1870	1.5637	57 24	57.4
32.7	32 42	0.5402	0.8415	0.6420	1.8510	1.1883	1.5577	57 18	57.3
32.8	32 48	0.5417	0.8406	0.6445	1.8460	1.1897	1.5517	57 12	57.2
32.9	32 54	0.5432	0.8396	0.6469	1.8410	1.1910	1.5458	57 06	57.1
33.0°	33°00′	0.5446	0.8387	0.6494	1.8361	1.1924	1.5399	57°00′	57.0°
33.1	33 06	0.5461	0.8377	0.6519	1.8312	1.1937	1.5340	56 54	56.9
33.2	33 12	0.5476	0.8368	0.6544	1.8263	1.1951	1.5282	56 48	56.8
33.3	33 18	0.5490	0.8358	0.6569	1.8214	1.1964	1.5224	56 42	56.7
33.4	33 24	0.5505	0.8348	0.6594	1.8166	1.1978	1.5166	56 36	56.6
33.5	33 30	0.5519	0.8339	0.6619	1.8118	1.1992	1.5108	56 30	56.5
33.6	33 36	0.5534	0.8329	0.6644	1.8070	1.2006	1.5051	56 24	56.4
33.7	33 42	0.5548	0.8320	0.6669	1.8023	1.2020	1.4994	56 18	56.3
33.8	33 48	0.5563	0.8310	0.6694	1.7976	1.2034	1.4938	56 12	56.2
33.9	33 54	0.5577	0.8300	0.6720	1.7929	1.2048	1.4882	56 06	56.1
34.0°	34°00′	0.5592	0.8290	0.6745	1.7883	1.2062	1.4826	56°00′	56.0°
34.1	34 06	0.5606	0.8281	0.6771	1.7837	1.2076	1.4770	55 54	55.9
34.2	34 12	0.5621	0.8271	0.6796	1.7791	1.2091	1.4715	55 48	55.8
34.3	34 18	0.5635	0.8261	0.6822	1.7745	1.2105	1.4659	55 42	55.7
34.4	34 24	0.5650	0.8251	0.6847	1.7700	1.2120	1.4605	55 36	55.6
34.5	34 30	0.5664	0.8241	0.6873	1.7655	1.2134	1.4550	55 30	55.5
34.6	34 36	0.5678	0.8231	0.6899	1.7610	1.2149	1.4496	55 24	55.4
34.7	34 42	0.5693	0.8221	0.6924	1.7566	1.2163	1.4442	55 18	55.3
34.8	34 48	0.5707	0.8211	0.6950	1.7522	1.2178	1.4388	55 12	55.2
34.9	34 54	0.5721	0.8202	0.6976	1.7478	1.2193	1.4335	55°06′	55.1°
		cos θ	sin θ	cot θ	sec θ	csc θ	tan θ	θ deg-min	θ deg

Table I Values of Trigonometric Functions (Angle in Degrees) (continued)

θ deg	θ deg-min	sin θ	cos θ	tan θ	csc θ	sec θ	cot θ		
35.0°	35°00′	0.5736	0.8192	0.7002	1.7434	1.2208	1.4281	55°00′	55.0°
35.1	35 06	0.5750	0.8181	0.7028	1.7391	1.2223	1.4229	54 54	54.9
35.2	35 12	0.5764	0.8171	0.7054	1.7348	1.2238	1.4176	54 48	54.8
35.3	35 18	0.5779	0.8161	0.7080	1.7305	1.2253	1.4124	54 42	54.7
35.4	35 24	0.5793	0.8151	0.7107	1.7263	1.2268	1.4071	54 36	54.6
35.5	35 30	0.5807	0.8141	0.7133	1.7221	1.2283	1.4019	54 30	54.5
35.6	35 36	0.5821	0.8131	0.7159	1.7179	1.2299	1.3968	54 24	54.4
35.7	35 42	0.5835	0.8121	0.7186	1.7137	1.2314	1.3916	54 18	54.3
35.8	35 48	0.5850	0.8111	0.7212	1.7095	1.2329	1.3865	54 12	54.2
35.9	35 54	0.5864	0.8100	0.7239	1.7054	1.2345	1.3814	54 06	54.1
36.0°	36°00′	0.5878	0.8090	0.7265	1.7013	1.2361	1.3764	54°00′	54.0°
36.1	36 06	0.5892	0.8080	0.7292	1.6972	1.2376	1.3713	53 54	53.9
36.2	36 12	0.5906	0.8070	0.7319	1.6932	1.2392	1.3663	53 48	53.8
36.3	36 18	0.5920	0.8059	0.7346	1.6892	1.2408	1.3613	53 42	53.7
36.4	36 24	0.5934	0.8049	0.7373	1.6852	1.2424	1.3564	53 36	53.6
36.5	36 30	0.5948	0.8039	0.7400	1.6812	1.2440	1.3514	53 30	53.5
36.6	36 36	0.5962	0.8028	0.7427	1.6772	1.2456	1.3465	53 24	53.4
36.7	36 42	0.5976	0.8018	0.7454	1.6733	1.2472	1.3416	53 18	53.3
36.8	36 48	0.5990	0.8007	0.7481	1.6694	1.2489	1.3367	53 12	53.2
36.9	36 54	0.6004	0.7997	0.7508	1.6655	1.2505	1.3319	53 06	53.1
37.0°	37°00′	0.6018	0.7986	0.7536	1.6616	1.2521	1.3270	53°00′	53.0°
37.1	37 06	0.6032	0.7976	0.7563	1.6578	1.2538	1.3222	52 54	52.9
37.2	37 12	0.6046	0.7965	0.7590	1.6540	1.2554	1.3175	52 48	52.8
37.3	37 18	0.6060	0.7955	0.7618	1.6502	1.2571	1.3127	52 42	52.7
37.4	37 24	0.6074	0.7944	0.7646	1.6464	1.2588	1.3079	52 36	52.6
37.5	37 30	0.6088	0.7934	0.7673	1.6427	1.2605	1.3032	52 30	52.5
37.6	37 36	0.6101	0.7923	0.7701	1.6390	1.2622	1.2985	52 24	52.4
37.7	37 42	0.6115	0.7912	0.7729	1.6353	1.2639	1.2938	52 18	52.3
37.8	37 48	0.6129	0.7902	0.7757	1.6316	1.2656	1.2892	52 12	52.2
37.9	37 54	0.6143	0.7891	0.7785	1.6279	1.2673	1.2846	52 06	52.1
38.0°	38°00′	0.6157	0.7880	0.7813	1.6243	1.2690	1.2799	52°00′	52.0°
38.1	38 06	0.6170	0.7869	0.7841	1.6207	1.2708	1.2753	51 54	51.9
38.2	38 12	0.6184	0.7859	0.7869	1.6171	1.2725	1.2708	51 48	51.8
38.3	38 18	0.6198	0.7848	0.7898	1.6135	1.2742	1.2662	51 42	51.7
38.4	38 24	0.6211	0.7837	0.7926	1.6099	1.2760	1.2617	51 36	51.6
38.5	38 30	0.6225	0.7826	0.7954	1.6064	1.2778	1.2572	51 30	51.5
38.6	38 36	0.6239	0.7815	0.7983	1.6029	1.2796	1.2527	51 24	51.4
38.7	38 42	0.6252	0.7804	0.8012	1.5994	1.2813	1.2482	51 18	51.3
38.8	38 48	0.6266	0.7793	0.8040	1.5959	1.2831	1.2437	51 12	51.2
38.9	38 54	0.6280	0.7782	0.8069	1.5925	1.2849	1.2393	51 06	51.1
39.0°	39°00′	0.6293	0.7771	0.8098	1.5890	1.2868	1.2349	51°00′	51.0°
39.1	39 06	0.6307	0.7760	0.8127	1.5856	1.2886	1.2305	50 54	50.9
39.2	39 12	0.6320	0.7749	0.8156	1.5822	1.2904	1.2261	50 48	50.8
39.3	39 18	0.6334	0.7738	0.8185	1.5788	1.2923	1.2218	50 42	50.7
39.4	39 24	0.6347	0.7727	0.8214	1.5755	1.2941	1.2174	50 36	50.6
39.5	39 30	0.6361	0.7716	0.8243	1.5721	1.2960	1.2131	50 30	50.5
39.6	39 36	0.6374	0.7705	0.8273	1.5688	1.2978	1.2088	50 24	50.4
39.7	39 42	0.6388	0.7694	0.8302	1.5655	1.2997	1.2045	50 18	50.3
39.8	39 48	0.6401	0.7683	0.8332	1.5622	1.3016	1.2002	50 12	50.2
39.9	39 54	0.6414	0.7672	0.8361	1.5590	1.3035	1.1960	50°06′	50.1°
		cos θ	sin θ	cot θ	sec θ	csc θ	tan θ	θ deg-min	θ deg

Table I Values of Trigonometric Functions (Angle in Degrees) (continued)

θ deg	θ deg-min	sin θ	cos θ	tan θ	csc θ	sec θ	cot θ		
40.0°	40°00′	0.6428	0.7660	0.8391	1.5557	1.3054	1.1918	50°00′	50.0°
40.1	40 06	0.6441	0.7649	0.8421	1.5525	1.3073	1.1875	49 54	49.9
40.2	40 12	0.6455	0.7638	0.8451	1.5493	1.3092	1.1833	49 48	49.8
40.3	40 18	0.6468	0.7627	0.8481	1.5461	1.3112	1.1792	49 42	49.7
40.4	40 24	0.6481	0.7615	0.8511	1.5429	1.3131	1.1750	49 36	49.6
40.5	40 30	0.6494	0.7604	0.8541	1.5398	1.3151	1.1708	49 30	49.5
40.6	40 36	0.6508	0.7593	0.8571	1.5366	1.3171	1.1667	49 24	49.4
40.7	40 42	0.6521	0.7581	0.8601	1.5335	1.3190	1.1626	49 18	49.3
40.8	40 48	0.6534	0.7570	0.8632	1.5304	1.3210	1.1585	49 12	49.2
40.9	40 54	0.6547	0.7559	0.8662	1.5273	1.3230	1.1544	49 06	49.1
41.0°	41°00′	0.6561	0.7547	0.8693	1.5243	1.3250	1.1504	49°00′	49.0°
41.1	41 06	0.6574	0.7536	0.8724	1.5212	1.3270	1.1463	48 54	48.9
41.2	41 12	0.6587	0.7524	0.8754	1.5182	1.3291	1.1423	48 48	48.8
41.3	41.18	0.6600	0.7513	0.8785	1.5151	1.3311	1.1383	48 42	48.7
41.4	41 24	0.6613	0.7501	0.8816	1.5121	1.3331	1.1343	48 36	48.6
41.5	41 30	0.6626	0.7490	0.8847	1.5092	1.3352	1.1303	48 30	48.5
41.6	41 36	0.6639	0.7478	0.8878	1.5062	1.3373	1.1263	48 24	48.4
41.7	41 42	0.6652	0.7466	0.8910	1.5032	1.3393	1.1224	48 18	48.3
41.8	41 48	0.6665	0.7455	0.8941	1.5003	1.3414	1.1184	48 12	48.2
41.9	41 54	0.6678	0.7443	0.8972	1.4974	1.3435	1.1145	48 06	48.1
42.0°	42°00′	0.6691	0.7431	0.9004	1.4945	1.3456	1.1106	48°00′	48.0°
42.1	42 06	0.6704	0.7420	0.9036	1.4916	1.3478	1.1067	47 54	47.9
42.2	42 12	0.6717	0.7408	0.9067	1.4887	1.3499	1.1028	47 48	47.8
42.3	42 18	0.6730	0.7396	0.9099	1.4859	1.3520	1.0990	47 42	47.7
42.4	42 24	0.6743	0.7385	0.9131	1.4830	1.3542	1.0951	47 36	47.6
42.5	42 30	0.6756	0.7373	0.9163	1.4802	1.3563	1.0913	47 30	47.5
42.6	42 36	0.6769	0.7361	0.9195	1.4774	1.3585	1.0875	47 24	47.4
42.7	42 42	0.6782	0.7349	0.9228	1.4746	1.3607	1.0837	47 18	47.3
42.8	42 48	0.6794	0.7337	0.9260	1.4718	1.3629	1.0799	47 12	47.2
42.9	42 54	0.6807	0.7325	0.9293	1.4690	1.3651	1.0761	47 06	47.1
43.0°	43°00′	0.6820	0.7314	0.9325	1.4663	1.3673	1.0724	47°00′	47.0°
43.1	43 06	0.6833	0.7302	0.9358	1.4635	1.3696	1.0686	46 54	46.9
43.2	43 12	0.6845	0.7290	0.9391	1.4608	1.3718	1.0649	46 48	46.8
43.3	43 18	0.6858	0.7278	0.9424	1.4581	1.3741	1.0612	46 42	46.7
43.4	43 24	0.6871	0.7266	0.9457	1.4554	1.3763	1.0575	46 36	46.6
43.5	43 30	0.6884	0.7254	0.9490	1.4527	1.3786	1.0538	46 30	46.5
43.6	43 36	0.6896	0.7242	0.9523	1.4501	1.3809	1.0501	46 24	46.4
43.7	43 42	0.6909	0.7230	0.9556	1.4474	1.3832	1.0464	46 18	46.3
43.8	43 48	0.6921	0.7218	0.9590	1.4448	1.3855	1.0428	46 12	46.2
43.9	43 54	0.6934	0.7206	0.9623	1.4422	1.3878	1.0392	46 06	46.1
44.0°	44°00′	0.6947	0.7193	0.9657	1.4396	1.3902	1.0355	46°00′	46.0°
44.1	44 06	0.6959	0.7181	0.9691	1.4370	1.3925	1.0319	45 54	45.9
44.2	44 12	0.6972	0.7169	0.9725	1.4344	1.3949	1.0283	45 48	45.8
44.3	44 18	0.6984	0.7157	0.9759	1.4318	1.3972	1.0247	45 42	45.7
44.4	44 24	0.6997	0.7145	0.9793	1.4293	1.3996	1.0212	45 36	45.6
44.5	44 30	0.7009	0.7133	0.9827	1.4267	1.4020	1.0176	45 30	45.5
44.6	44 36	0.7022	0.7120	0.9861	1.4242	1.4044	1.0141	45 24	45.4
44.7	44 42	0.7034	0.7108	0.9896	1.4217	1.4069	1.0105	45 18	45.3
44.8	44 48	0.7046	0.7096	0.9930	1.4192	1.4093	1.0070	45 12	45.2
44.9	44 54	0.7059	0.7083	0.9965	1.4167	1.4118	1.0035	45 06	45.1
45.0°	45°00′	0.7071	0.7071	1.0000	1.4142	1.4142	1.0000	45°00′	45.0°
		cos θ	sin θ	cot θ	sec θ	csc θ	tan θ	θ deg-min	θ deg

Table II Trigonometric Functions—Radians or Real Numbers

↓→	sin	cos	tan	cot	sec	csc
.00	.0000	1.0000	.0000	–	1.000	–
.01	.0100	1.0000	.0100	99.997	1.000	100.00
.02	.0200	.9998	.0200	49.993	1.000	50.00
.03	.0300	.9996	.0300	33.323	1.000	33.34
.04	.0400	.9992	.0400	24.987	1.001	25.01
.05	.0500	.9988	.0500	19.983	1.001	20.01
.06	.0600	.9982	.0601	16.647	1.002	16.68
.07	.0699	.9976	.0701	14.262	1.002	14.30
.08	.0799	.9968	.0802	12.473	1.003	12.51
.09	.0899	.9960	.0902	11.081	1.004	11.13
.10	.0998	.9950	.1003	9.967	1.005	10.02
.11	.1098	.9940	.1104	9.054	1.006	9.109
.12	.1197	.9928	.1206	8.293	1.007	8.353
.13	.1296	.9916	.1307	7.649	1.009	7.714
.14	.1395	.9902	.1409	7.096	1.010	7.166
.15	.1494	.9888	.1511	6.617	1.011	6.692
.16	.1593	.9872	.1614	6.197	1.013	6.277
.17	.1692	.9856	.1717	5.826	1.015	5.911
.18	.1790	.9838	.1820	5.495	1.016	5.586
.19	.1889	.9820	.1923	5.200	1.018	5.295
.20	.1987	.9801	.2027	4.933	1.020	5.033
.21	.2085	.9780	.2131	4.692	1.022	4.797
.22	.2182	.9759	.2236	4.472	1.025	4.582
.23	.2280	.9737	.2341	4.271	1.027	4.386
.24	.2377	.9713	.2447	4.086	1.030	4.207
.25	.2474	.9689	.2553	3.916	1.032	4.042
.26	.2571	.9664	.2660	3.759	1.035	3.890
.27	.2667	.9638	.2768	3.613	1.038	3.749
.28	.2764	.9611	.2876	3.478	1.041	3.619
.29	.2860	.9582	.2984	3.351	1.044	3.497
.30	.2955	.9553	.3093	3.233	1.047	3.384
.31	.3051	.9523	.3203	3.122	1.050	3.278
.32	.3146	.9492	.3314	3.018	1.053	3.179
.33	.3240	.9460	.3425	2.920	1.057	3.086
.34	.3335	.9428	.3537	2.827	1.061	2.999
.35	.3429	.9394	.3650	2.740	1.065	2.916
.36	.3523	.9359	.3764	2.657	1.068	2.839
.37	.3616	.9323	.3879	2.578	1.073	2.765
.38	.3709	.9287	.3994	2.504	1.077	2.696
.39	.3802	.9249	.4111	2.433	1.081	2.630
	sin	cos	tan	cot	sec	csc

Table II Trigonometric Functions—Radians or Real Numbers (continued)

↓→	sin	cos	tan	cot	sec	csc
.40	.3894	.9211	.4228	2.365	1.086	2.568
.41	.3986	.9171	.4346	2.301	1.090	2.509
.42	.4078	.9131	.4466	2.239	1.095	2.452
.43	.4169	.9090	.4586	2.180	1.100	2.399
.44	.4259	.9048	.4708	2.124	1.105	2.348
.45	.4350	.9004	.4831	2.070	1.111	2.299
.46	.4439	.8961	.4954	2.018	1.116	2.253
.47	.4529	.8916	.5080	1.969	1.122	2.208
.48	.4618	.8870	.5206	1.921	1.127	2.166
.49	.4706	.8823	.5334	1.875	1.133	2.125
.50	.4794	.8776	.5463	1.830	1.139	2.086
.51	.4882	.8727	.5594	1.788	1.146	2.048
.52	.4969	.8678	.5726	1.747	1.152	2.013
.53	.5055	.8628	.5859	1.707	1.159	1.978
.54	.5141	.8577	.5994	1.668	1.166	1.945
.55	.5227	.8525	.6131	1.631	1.173	1.913
.56	.5312	.8473	.6269	1.595	1.180	1.883
.57	.5396	.8419	.6410	1.560	1.188	1.853
.58	.5480	.8365	.6552	1.526	1.196	1.825
.59	.5564	.8309	.6696	1.494	1.203	1.797
.60	.5646	.8253	.6841	1.462	1.212	1.771
.61	.5729	.8196	.6989	1.431	1.220	1.746
.62	.5810	.8139	.7139	1.401	1.229	1.721
.63	.5891	.8080	.7291	1.372	1.238	1.697
.64	.5972	.8021	.7445	1.343	1.247	1.674
.65	.6052	.7961	.7602	1.315	1.256	1.652
.66	.6131	.7900	.7761	1.288	1.266	1.631
.67	.6210	.7838	.7923	1.262	1.276	1.610
.68	.6288	.7776	.8087	1.237	1.286	1.590
.69	.6365	.7712	.8253	1.212	1.297	1.571
.70	.6442	.7648	.8423	1.187	1.307	1.552
.71	.6518	.7584	.8595	1.163	1.319	1.534
.72	.6594	.7518	.8771	1.140	1.330	1.517
.73	.6669	.7452	.8949	1.117	1.342	1.500
.74	.6743	.7385	.9131	1.095	1.354	1.483
.75	.6816	.7317	.9316	1.073	1.367	1.467
.76	.6889	.7248	.9505	1.052	1.380	1.452
.77	.6961	.7179	.9697	1.031	1.393	1.437
.78	.7033	.7109	.9893	1.011	1.407	1.422
.79	.7104	.7038	1.009	.9908	1.421	1.408
	sin	cos	tan	cot	sec	csc

Table II Trigonometric Functions—Radians or Real Numbers (continued)

↓→	sin	cos	tan	cot	sec	csc
.80	.7174	.6967	1.030	.9712	1.435	1.394
.81	.7243	.6895	1.050	.9520	1.450	1.381
.82	.7311	.6822	1.072	.9331	1.466	1.368
.83	.7379	.6749	1.093	.9146	1.482	1.355
.84	.7446	.6675	1.116	.8964	1.498	1.343
.85	.7513	.6600	1.138	.8785	1.515	1.331
.86	.7578	.6524	1.162	.8609	1.533	1.320
.87	.7643	.6448	1.185	.8437	1.551	1.308
.88	.7707	.6372	1.210	.8267	1.569	1.297
.89	.7771	.6294	1.235	.8100	1.589	1.287
.90	.7833	.6216	1.260	.7936	1.609	1.277
.91	.7895	.6137	1.286	.7774	1.629	1.267
.92	.7956	.6058	1.313	.7615	1.651	1.257
.93	.8016	.5978	1.341	.7458	1.673	1.247
.94	.8076	.5898	1.369	.7303	1.696	1.238
.95	.8134	.5817	1.398	.7151	1.719	1.229
.96	.8192	.5735	1.428	.7001	1.744	1.221
.97	.8249	.5653	1.459	.6853	1.769	1.212
.98	.8305	.5570	1.491	.6707	1.795	1.204
.99	.8360	.5487	1.524	.6563	1.823	1.196
1.00	.8415	.5403	1.557	.6421	1.851	1.188
1.01	.8468	.5319	1.592	.6281	1.880	1.181
1.02	.8521	.5234	1.628	.6142	1.911	1.174
1.03	.8573	.5148	1.665	.6005	1.942	1.166
1.04	.8624	.5062	1.704	.5870	1.975	1.160
1.05	.8674	.4976	1.743	.5736	2.010	1.153
1.06	.8724	.4889	1.784	.5604	2.046	1.146
1.07	.8772	.4801	1.827	.5473	2.083	1.140
1.08	.8820	.4713	1.871	.5344	2.122	1.134
1.09	.8866	.4625	1.917	.5216	2.162	1.128
1.10	.8912	.4536	1.965	.5090	2.205	1.122
1.11	.8957	.4447	2.014	.4964	2.249	1.116
1.12	.9001	.4357	2.066	.4840	2.295	1.111
1.13	.9044	.4267	2.120	.4718	2.344	1.106
1.14	.9086	.4176	2.176	.4596	2.395	1.101
1.15	.9128	.4085	2.234	.4475	2.448	1.096
1.16	.9168	.3993	2.296	.4356	2.504	1.091
1.17	.9208	.3902	2.360	.4237	2.563	1.086
1.18	.9246	.3809	2.427	.4120	2.625	1.082
1.19	.9284	.3717	2.498	.4003	2.691	1.077
	sin	cos	tan	cot	sec	csc

Table II Trigonometric Functions—Radians or Real Numbers (continued)

↓→	sin	cos	tan	cot	sec	csc
1.20	.9320	.3624	2.572	.3888	2.760	1.073
1.21	.9356	.3530	2.650	.3773	2.833	1.069
1.22	.9391	.3436	2.733	.3659	2.910	1.065
1.23	.9425	.3342	2.820	.3546	2.992	1.061
1.24	.9458	.3248	2.912	.3434	3.079	1.057
1.25	.9490	.3153	3.010	.3323	3.171	1.054
1.26	.9521	.3058	3.113	.3212	3.270	1.050
1.27	.9551	.2963	3.224	.3102	3.375	1.047
1.28	.9580	.2867	3.341	.2993	3.488	1.044
1.29	.9608	.2771	3.467	.2884	3.609	1.041
1.30	.9636	.2675	3.602	.2776	3.738	1.038
1.31	.9662	.2579	3.747	.2669	3.878	1.035
1.32	.9687	.2482	3.903	.2562	4.029	1.032
1.33	.9711	.2385	4.072	.2456	4.193	1.030
1.34	.9735	.2288	4.256	.2350	4.372	1.027
1.35	.9757	.2190	4.455	.2245	4.566	1.025
1.36	.9779	.2092	4.673	.2140	4.779	1.023
1.37	.9799	.1994	4.913	.2035	5.014	1.021
1.38	.9819	.1896	5.177	.1931	5.273	1.018
1.39	.9837	.1798	5.471	.1828	5.561	1.017
1.40	.9854	.1700	5.798	.1725	5.883	1.015
1.41	.9871	.1601	6.165	.1622	6.246	1.013
1.42	.9887	.1502	6.581	.1519	6.657	1.011
1.43	.9901	.1403	7.055	.1417	7.126	1.010
1.44	.9915	.1304	7.602	.1315	7.667	1.009
1.45	.9927	.1205	8.238	.1214	8.299	1.007
1.46	.9939	.1106	8.989	.1113	9.044	1.006
1.47	.9949	.1006	9.887	.1011	9.938	1.005
1.48	.9959	.0907	10.983	.0910	11.029	1.004
1.49	.9967	.0807	12.350	.0810	12.390	1.003
1.50	.9975	.0707	14.101	.0709	14.137	1.003
1.51	.9982	.0608	16.428	.0609	16.458	1.002
1.52	.9987	.0508	19.670	.0508	19.695	1.001
1.53	.9992	.0408	24.498	.0408	24.519	1.001
1.54	.9995	.0308	32.461	.0308	32.476	1.000
1.55	.9998	.0208	48.078	.0208	48.089	1.000
1.56	.9999	.0108	92.620	.0108	92.626	1.000
1.57	1.0000	.0008	1,255.8	.0008	1,255.8	1.000
1.58	1.0000	−.0092	−108.65	−.0092	−108.65	1.000
1.59	.9998	−.0192	−52.067	−.0192	−52.08	1.000
1.60	.9996	−.0292	−34.233	−.0292	−34.25	1.000
	sin	cos	tan	cot	sec	csc

Table III Common Logarithms

Common Logarithms

x	0	1	2	3	4	5	6	7	8	9
1.0	.0000	.0043	.0086	.0128	.0170	.0212	.0253	.0294	.0334	.0374
1.1	.0414	.0453	.0492	.0531	.0569	.0607	.0645	.0682	.0719	.0755
1.2	.0792	.0828	.0864	.0899	.0934	.0969	.1004	.1038	.1072	.1106
1.3	.1139	.1173	.1206	.1239	.1271	.1303	.1335	.1367	.1399	.1430
1.4	.1461	.1492	.1523	.1553	.1584	.1614	.1644	.1673	.1703	.1732
1.5	.1761	.1790	.1818	.1847	.1875	.1903	.1931	.1959	.1987	.2014
1.6	.2041	.2068	.2095	.2122	.2148	.2175	.2201	.2227	.2253	.2279
1.7	.2304	.2330	.2355	.2380	.2405	.2430	.2455	.2480	.2504	.2529
1.8	.2553	.2577	.2601	.2625	.2648	.2672	.2695	.2718	.2742	.2765
1.9	.2788	.2810	.2833	.2856	.2878	.2900	.2923	.2945	.2967	.2989
2.0	.3010	.3032	.3054	.3075	.3096	.3118	.3139	.3160	.3181	.3201
2.1	.3222	.3243	.3263	.3284	.3304	.3324	.3345	.3365	.3385	.3404
2.2	.3424	.3444	.3464	.3483	.3502	.3522	.3541	.3560	.3579	.3598
2.3	.3617	.3636	.3655	.3674	.3692	.3711	.3729	.3747	.3766	.3784
2.4	.3802	.3820	.3838	.3856	.3874	.3892	.3909	.3927	.3945	.3962
2.5	.3979	.3997	.4014	.4031	.4048	.4065	.4082	.4099	.4116	.4133
2.6	.4150	.4166	.4183	.4200	.4216	.4232	.4249	.4265	.4281	.4298
2.7	.4314	.4330	.4346	.4362	.4378	.4393	.4409	.4425	.4440	.4456
2.8	.4472	.4487	.4502	.4518	.4533	.4548	.4564	.4579	.4594	.4609
2.9	.4624	.4639	.4654	.4669	.4683	.4698	.4713	.4728	.4742	.4757
3.0	.4771	.4786	.4800	.4814	.4829	.4843	.4857	.4871	.4886	.4900
3.1	.4914	.4928	.4942	.4955	.4969	.4983	.4997	.5011	.5024	.5038
3.2	.5051	.5065	.5079	.5092	.5105	.5119	.5132	.5145	.5159	.5172
3.3	.5185	.5198	.5211	.5224	.5237	.5250	.5263	.5276	.5289	.5302
3.4	.5315	.5328	.5340	.5353	.5366	.5378	.5391	.5403	.5416	.5428
3.5	.5441	.5453	.5465	.5478	.5490	.5502	.5514	.5527	.5539	.5551
3.6	.5563	.5575	.5587	.5599	.5611	.5623	.5635	.5647	.5658	.5670
3.7	.5682	.5694	.5705	.5717	.5729	.5740	.5752	.5763	.5775	.5786
3.8	.5798	.5809	.5821	.5832	.5843	.5855	.5866	.5877	.5888	.5899
3.9	.5911	.5922	.5933	.5944	.5955	.5966	.5977	.5988	.5999	.6010
4.0	.6021	.6031	.6042	.6053	.6064	.6075	.6085	.6096	.6107	.6117
4.1	.6128	.6138	.6149	.6160	.6170	.6180	.6191	.6201	.6212	.6222
4.2	.6232	.6243	.6253	.6263	.6274	.6284	.6294	.6304	.6314	.6325
4.3	.6335	.6345	.6355	.6365	.6375	.6385	.6395	.6405	.6415	.6425
4.4	.6435	.6444	.6454	.6464	.6474	.6484	.6493	.6503	.6513	.6522
4.5	.6532	.6542	.6551	.6561	.6571	.6580	.6590	.6599	.6609	.6618
4.6	.6628	.6637	.6646	.6656	.6665	.6675	.6684	.6693	.6702	.6712
4.7	.6721	.6730	.6739	.6749	.6758	.6767	.6776	.6785	.6794	.6803
4.8	.6812	.6821	.6830	.6839	.6848	.6857	.6866	.6875	.6884	.6893
4.9	.6902	.6911	.6920	.6928	.6937	.6946	.6955	.6964	.6972	.6981
5.0	.6990	.6998	.7007	.7016	.7024	.7033	.7042	.7050	.7059	.7067
5.1	.7076	.7084	.7093	.7101	.7110	.7118	.7126	.7135	.7143	.7152
5.2	.7160	.7168	.7177	.7185	.7193	.7202	.7210	.7218	.7226	.2735
5.3	.7243	.7251	.7259	.7267	.7275	.7284	.7292	.7300	.7308	.7316
5.4	.7324	.7332	.7340	.7348	.7356	.7364	.7372	.7380	.7388	.7396
x	0	1	2	3	4	5	6	7	8	9

Table III Common Logarithms (continued)

x	0	1	2	3	4	5	6	7	8	9
5.5	.7404	.7412	.7419	.7427	.7435	.7443	.7451	.7459	.7466	.7474
5.6	.7482	.7490	.7497	.7505	.7513	.7520	.7528	.7536	.7543	.7551
5.7	.7559	.7566	.7574	.7582	.7589	.7597	.7604	.7612	.7619	.7627
5.8	.7634	.7642	.7649	.7657	.7664	.7672	.7679	.7686	.7694	.7701
5.9	.7709	.7716	.7723	.7731	.7738	.7745	.7752	.7760	.7767	.7774
6.0	.7782	.7789	.7796	.7803	.7810	.7818	.7825	.7832	.7839	.7846
6.1	.7853	.7860	.7868	.7875	.7882	.7889	.7896	.7903	.7910	.7917
6.2	.7924	.7931	.7938	.7945	.7952	.7959	.7966	.7973	.7980	.7987
6.3	.7993	.8000	.8007	.8014	.8021	.8028	.8035	.8041	.8048	.8055
6.4	.8062	.8069	.8075	.8082	.8089	.8096	.8102	.8109	.8116	.8122
6.5	.8129	.8136	.8142	.8149	.8156	.8162	.8169	.8176	.8182	.8189
6.6	.8195	.8202	.8209	.8215	.8222	.8228	.8235	.8241	.8248	.8254
6.7	.8261	.8267	.8274	.8280	.8287	.8293	.8299	.8306	.8312	.8319
6.8	.8325	.8331	.8338	.8344	.8351	.8357	.8363	.8370	.8376	.8382
6.9	.8388	.8395	.8401	.8407	.8414	.8420	.8426	.8432	.8439	.8445
7.0	.8451	.8457	.8463	.8470	.8476	.8482	.8488	.8494	.8500	.8506
7.1	.8513	.8519	.8525	.8531	.8537	.8543	.8549	.8555	.8561	.8567
7.2	.8573	.8579	.8585	.8591	.8597	.8603	.8609	.8615	.8621	.8627
7.3	.8633	.8639	.8645	.8651	.8657	.8663	.8669	.8675	.8681	.8686
7.4	.8692	.8698	.8704	.8710	.8716	.8722	.8727	.8733	.8739	.8745
7.5	.8751	.8756	.8762	.8768	.8774	.8779	.8785	.8791	.8797	.8802
7.6	.8808	.8814	.8820	.8825	.8831	.8837	.8842	.8848	.8854	.8859
7.7	.8865	.8871	.8876	.8882	.8887	.8893	.8899	.8904	.8910	.8915
7.8	.8921	.8927	.8932	.8938	.8943	.8949	.8954	.8960	.8965	.8971
7.9	.8976	.8982	.8987	.8993	.8998	.9004	.9009	.9015	.9020	.9025
8.0	.9031	.9036	.9042	.9047	.9053	.9058	.9063	.9069	.9074	.9079
8.1	.9085	.9090	.9096	.9101	.9106	.9112	.9117	.9122	.9128	.9133
8.2	.9138	.9143	.9149	.9154	.9159	.9165	.9170	.9175	.9180	.9186
8.3	.9191	.9196	.9201	.9206	.9212	.9217	.9222	.9227	.9232	.9238
8.4	.9243	.9248	.9253	.9258	.9263	.9269	.9274	.9279	.9284	.9289
8.5	.9294	.9299	.9304	.9309	.9315	.9320	.9325	.9330	.9335	.9340
8.6	.9345	.9350	.9355	.9360	.9365	.9370	.9375	.9380	.9385	.9390
8.7	.9395	.9400	.9405	.9410	.9415	.9420	.9425	.9430	.9435	.9440
8.8	.9445	.9450	.9455	.9460	.9465	.9469	.9474	.9479	.9484	.9489
8.9	.9494	.9499	.9504	.9509	.9513	.9518	.9523	.9528	.9533	.9538
9.0	.9542	.9547	.9552	.9557	.9562	.9566	.9571	.9576	.9581	.9586
9.1	.9590	.9595	.9600	.9605	.9609	.9614	.9619	.9624	.9628	.9633
9.2	.9638	.9643	.9647	.9652	.9657	.9661	.9666	.9671	.9675	.9680
9.3	.9685	.9689	.9694	.9699	.9703	.9708	.9713	.9717	.9722	.9727
9.4	.9731	.9736	.9741	.9745	.9750	.9754	.9759	.9763	.9768	.9773
9.5	.9777	.9782	.9786	.9791	.9795	.9800	.9805	.9809	.9814	.9818
9.6	.9823	.9827	.9832	.9836	.9841	.9845	.9850	.9854	.9859	.9863
9.7	.9868	.9872	.9877	.9881	.9886	.9890	.9894	.9899	.9903	.9908
9.8	.9912	.9917	.9921	.9926	.9930	.9934	.9939	.9943	.9948	.9952
9.9	.9956	.9961	.9965	.9969	.9974	.9978	.9983	.9987	.9991	.9996
x	0	1	2	3	4	5	6	7	8	9

Table IV Four-Place Logarithms of Trigonometric Functions—Angle x in Degrees

Attach −10 to Logarithms Obtained from this Table

x	L sin x	L cos x	L tan x	L cot x	L sec x	L csc x	
0°00′	No value	10.0000	No value	No value	10.0000	No value	90°00′
10′	7.4637	.0000	7.4637	12.5363	.0000	12.5363	50′
20′	.7648	.0000	.7648	.2352	.0000	.2352	40′
30′	7.9408	.0000	7.9409	12.0591	.0000	12.0592	30′
40′	8.0658	.0000	8.0658	11.9342	.0000	11.9342	20′
50′	.1627	10.0000	.1627	.8373	.0000	.8373	10′
1°00′	8.2419	9.9999	8.2419	11.7581	10.0001	11.7581	89°00′
10′	.3088	.9999	.3089	.6911	.0001	.6912	50′
20′	.3668	.9999	.3669	.6331	.0001	.6332	40′
30′	.4179	.9999	.4181	.5819	.0001	.5821	30′
40′	.4637	.9998	.4638	.5362	.0002	.5363	20′
50′	.5050	.9998	.5053	.4947	.0002	.4950	10′
2°00′	8.5428	9.9997	8.5431	11.4569	10.0003	11.4572	88°00′
10′	.5776	.9997	.5779	.4221	.0003	.4224	50′
20′	.6097	.9996	.6101	.3899	.0004	.3903	40′
30′	.6397	.9996	.6401	.3599	.0004	.3603	30′
40′	.6677	.9995	.6682	.3318	.0005	.3323	20′
50′	.6940	.9995	.6945	.3055	.0005	.3060	10′
3°00′	8.7188	9.9994	8.7194	11.2806	10.0006	11.2812	87°00′
10′	.7423	.9993	.7429	.2571	.0007	.2577	50′
20′	.7645	.9993	.7652	.2348	.0007	.2355	40′
30′	.7857	.9992	.7865	.2135	.0008	.2143	30′
40′	.8059	.9991	.8067	.1933	.0009	.1941	20′
50′	.8251	.9990	.8261	.1739	.0010	.1749	10′
4°00′	8.8436	9.9989	8.8446	11.1554	10.0011	11.1564	86°00′
10′	.8613	.9989	.8624	.1376	.0011	.1387	50′
20′	.8783	.9988	.8795	.1205	.0012	.1217	40′
30′	.8946	.9987	.8960	.1040	.0013	.1054	30′
40′	.9104	.9986	.9118	.0882	.0014	.0896	20′
50′	.9256	.9985	.9272	.0728	.0015	.0744	10′
5°00′	8.9403	9.9983	8.9420	11.0580	10.0017	11.0597	85°00′
10′	.9545	.9982	.9563	.0437	.0018	.0455	50′
20′	.9682	.9981	.9701	.0299	.0019	.0318	40′
30′	.9816	.9980	.9836	.0164	.0020	.0184	30′
40′	8.9945	.9979	8.9966	11.0034	.0021	11.0055	20′
50′	9.0070	.9977	9.0093	10.9907	.0023	10.9930	10′
6°00′	9.0192	9.9976	9.0216	10.9784	10.0024	10.9808	84°00′
10′	.0311	.9975	.0336	.9664	.0025	.9689	50′
20′	.0426	.9973	.0453	.9547	.0027	.9574	40′
30′	.0539	.9972	.0567	.9433	.0028	.9461	30′
40′	.0648	.9971	.0678	.9322	.0029	.9352	20′
50′	.0755	.9969	.0786	.9214	.0031	.9245	10′
7°00′	9.0859	9.9968	9.0891	10.9109	10.0032	10.9141	83°00′
	L cos x	L sin x	L cot x	L tan x	L csc x	L sec x	x

Table IV Four-Place Logarithms of Trigonometric Functions—Angle *x* in Degrees (continued)

Attach −10 to Logarithms Obtained from this Table

x	L sin *x*	L cos *x*	L tan *x*	L cot *x*	L sec *x*	L csc *x*	
7°00′	9.0859	9.9968	9.0891	10.9109	10.0032	10.9141	83°00′
10′	.0961	.9966	.0995	.9005	.0034	.9039	50′
20′	.1060	.9964	.1096	.8904	.0036	.8940	40′
30′	.1157	.9963	.1194	.8806	.0037	.8843	30′
40′	.1252	.9961	.1291	.8709	.0039	.8748	20′
50′	.1345	.9959	.1385	.8615	.0041	.8655	10′
8°00′	9.1436	9.9958	9.1478	10.8522	10.0042	10.8564	82°00′
10′	.1525	.9956	.1569	.8431	.0044	.8475	50′
20′	.1612	.9954	.1658	.8342	.0046	.8388	40′
30′	.1697	.9952	.1745	.8255	.0048	.8303	30′
40′	.1781	.9950	.1831	.8169	.0050	.8219	20′
50′	.1863	.9948	.1915	.8085	.0052	.8137	10′
9°00′	9.1943	9.9946	9.1997	10.8003	10.0054	10.8057	81°00′
10′	.2022	.9944	.2078	.7922	.0056	.7978	50′
20′	.2100	.9942	.2158	.7842	.0058	.7900	40′
30′	.2176	.9940	.2236	.7764	.0060	.7824	30′
40′	.2251	.9938	.2313	.7687	.0062	.7749	20′
50′	.2324	.9936	.2389	.7611	.0064	.7676	10′
10°00′	9.2397	9.9934	9.2463	10.7537	10.0066	10.7603	80°00′
10′	.2468	.9931	.2536	.7464	.0069	.7532	50′
20′	.2538	.9929	.2609	.7391	.0071	.7462	40′
30′	.2606	.9927	.2680	.7320	.0073	.7394	30′
40′	.2674	.9924	.2750	.7250	.0076	.7326	20′
50′	.2740	.9922	.2819	.7181	.0078	.7260	10′
11°00′	9.2806	9.9919	9.2887	10.7113	10.0081	10.7194	79°00′
10′	.2870	.9917	.2953	.7047	.0083	.7130	50′
20′	.2934	.9914	.3020	.6980	.0086	.7066	40′
30′	.2997	.9912	.3085	.6915	.0088	.7003	30′
40′	.3058	.9909	.3149	.6851	.0091	.6942	20′
50′	.3119	.9907	.3212	.6788	.0093	.6881	10′
12°00′	9.3179	9.9904	9.3275	10.6725	10.0096	10.6821	78°00′
10′	.3238	.9901	.3336	.6664	.0099	.6762	50′
20′	.3296	.9899	.3397	.6603	.0101	.6704	40′
30′	.3353	.9896	.3458	.6542	.0104	.6647	30′
40′	.3410	.9893	.3517	.6483	.0107	.6590	20′
50′	.3466	.9890	.3576	.6424	.0110	.6534	10′
13°00′	9.3521	9.9887	9.3634	10.6366	10.0113	10.6479	77°00′
10′	.3575	.9884	.3691	.6309	.0116	.6425	50′
20′	.3629	.9881	.3748	.6252	.0119	.6371	40′
30′	.3682	.9878	.3804	.6196	.0122	.6318	30′
40′	.3734	.9875	.3859	.6141	.0125	.6266	20′
50′	.3786	.9872	.3914	.6086	.0128	.6214	10′
14°00′	9.3837	9.9869	9.3968	10.6032	10.0131	10.6163	76°00′
	L cos *x*	L sin *x*	L cot *x*	L tan *x*	L csc *x*	L sec *x*	*x*

Table IV Four-Place Logarithms of Trigonometric Functions—Angle x in Degrees (continued)

Attach −10 to Logarithms Obtained from this Table

x	L sin x	L cos x	L tan x	L cot x	L sec x	L csc x	
14°00′	9.3837	9.9869	9.3968	10.6032	10.0131	10.6163	76°00′
10′	.3887	.9866	.4021	.5979	.0134	.6113	50′
20′	.3937	.9863	.4074	.5926	.0137	.6063	40′
30′	.3986	.9859	.4127	.5873	.0141	.6014	30′
40′	.4035	.9856	.4178	.5822	.0144	.5965	20′
50′	.4083	.9853	.4230	.5770	.0147	.5917	10′
15°00′	9.4130	9.9849	9.4281	10.5719	10.0151	10.5870	75°00′
10′	.4177	.9846	.4331	.5669	.0154	.5823	50′
20′	.4223	.9843	.4381	.5619	.0157	.5777	40′
30′	.4269	.9839	.4430	.5570	.0161	.5731	30′
40′	.4314	.9836	.4479	.5521	.0164	.5686	20′
50′	.4359	.9832	.4527	.5473	.0168	.5641	10′
16°00′	9.4403	9.9828	9.4575	10.5425	10.0172	10.5597	74°00′
10′	.4447	.9825	.4622	.5378	.0175	.5553	50′
20′	.4491	.9821	.4669	.5331	.0179	.5509	40′
30′	.4533	.9817	.4716	.5284	.0183	.5467	30′
40′	.4576	.9814	.4762	.5238	.0186	.5424	20′
50′	.4618	.9810	.4808	.5192	.0190	.5382	10′
17°00′	9.4659	9.9806	9.4853	10.5147	10.0194	10.5341	73°00′
10′	.4700	.9802	.4898	.5102	.0198	.5300	50′
20′	.4741	.9798	.4943	.5057	.0202	.5259	40′
30′	.4781	.9794	.4987	.5013	.0206	.5219	30′
40′	.4821	.9790	.5031	.4969	.0210	.5179	20′
50′	.4861	.9786	.5075	.4925	.0214	.5139	10′
18°00′	9.4900	9.9782	9.5118	10.4882	10.0218	10.5100	72°00′
10′	.4939	.9778	.5161	.4839	.0222	.5061	50′
20′	.4977	.9774	.5203	.4797	.0226	.5023	40′
30′	.5015	.9770	.5245	.4755	.0230	.4985	30′
40′	.5052	.9765	.5287	.4713	.0235	.4948	20′
50′	.5090	.9761	.5329	.4671	.0239	.4910	10′
19°00′	9.5126	9.9757	9.5370	10.4630	10.0243	10.4874	71°00′
10′	.5163	.9752	.5411	.4589	.0248	.4837	50′
20′	.5199	.9748	.5451	.4549	.0252	.4801	40′
30′	.5235	.9743	.5491	.4509	.0257	.4765	30′
40′	.5270	.9739	.5531	.4469	.0261	.4730	20′
50′	.5306	.9734	.5571	.4429	.0266	.4694	10′
20°00′	9.5341	9.9730	9.5611	10.4389	10.0270	10.4659	70°00′
10′	.5375	.9725	.5650	.4350	.0275	.4625	50′
20′	.5409	.9721	.5689	.4311	.0279	.4591	40′
30′	.5443	.9716	.5727	.4273	.0284	.4557	30′
40′	.5477	.9711	.5766	.4234	.0289	.4523	20′
50′	.5510	.9706	.5804	.4196	.0294	.4490	10′
21°00′	9.5543	9.9702	9.5842	10.4158	10.0298	10.4457	69°00′
	L cos x	L sin x	L cot x	L tan x	L csc x	L sec x	x

Table IV Four-Place Logarithms of Trigonometric Functions—Angle x in Degrees (continued)

Attach −10 to Logarithms Obtained from this Table

x	L sin x	L cos x	L tan x	L cot x	L sec x	L csc x	
21°00′	9.5543	9.9702	9.5842	10.4158	10.0298	10.4457	69°00′
10′	.5576	.9697	.5879	.4121	.0303	.4424	50′
20′	.5609	.9692	.5917	.4083	.0308	.4391	40′
30′	.5641	.9687	.5954	.4046	.0313	.4359	30′
40′	.5673	.9682	.5991	.4009	.0318	.4327	20′
50′	.5704	.9677	.6028	.3972	.0323	.4296	10′
22°00′	9.5736	9.9672	9.6064	10.3936	10.0328	10.4264	68°00′
10′	.5767	.9667	.6100	.3900	.0333	.4233	50′
20′	.5798	.9661	.6136	.3864	.0339	.4202	40′
30′	.5828	.9656	.6172	.3828	.0344	.4172	30′
40′	.5859	.9651	.6208	.3792	.0349	.4141	20′
50′	.5889	.9646	.6243	.3757	.0354	.4111	10′
23°00′	9.5919	9.9640	9.6279	10.3721	10.0360	10.4081	67°00′
10′	.5948	.9635	.6314	.3686	.0365	.4052	50′
20′	.5978	.9629	.6348	.3652	.0371	.4022	40′
30′	.6007	.9624	.6383	.3617	.0376	.3993	30′
40′	.6036	.9618	.6417	.3583	.0382	.3964	20′
50′	.6065	.9613	.6452	.3548	.0387	.3935	10′
24°00′	9.6093	9.9607	9.6486	10.3514	10.0393	10.3907	66°00′
10′	.6121	.9602	.6520	.3480	.0398	.3879	50′
20′	.6149	.9596	.6553	.3447	.0404	.3851	40′
30′	.6177	.9590	.6587	.3413	.0410	.3823	30′
40′	.6205	.9584	.6620	.3380	.0416	.3795	20′
50′	.6232	.9579	.6654	.3346	.0421	.3768	10′
25°00′	9.6259	9.9573	9.6687	10.3313	10.0427	10.3741	65°00′
10′	.6286	.9567	.6720	.3280	.0433	.3714	50′
20′	.6313	.9561	.6752	.3248	.0439	.3687	40′
30′	.6340	.9555	.6785	.3215	.0445	.3660	30′
40′	.6366	.9549	.6817	.3183	.0451	.3634	20′
50′	.6392	.9543	.6850	.3150	.0457	.3608	10′
26°00′	9.6418	9.9537	9.6882	10.3118	10.0463	10.3582	64°00′
10′	.6444	.9530	.6914	.3086	.0470	.3556	50′
20′	.6470	.9524	.6946	.3054	.0476	.3530	40′
30′	.6495	.9518	.6977	.3023	.0482	.3505	30′
40′	.6521	.9512	.7009	.2991	.0488	.3479	20′
50′	.6546	.9505	.7040	.2960	.0495	.3454	10′
27° 0′	9.6570	9.9499	9.7072	10.2928	10.0501	10.3430	63° 00′
10′	.6595	.9492	.7103	.2897	.0508	.3405	50′
20′	.6620	.9486	.7134	.2866	.0514	.3380	40′
30′	.6644	.9479	.7165	.2835	.0521	.3356	30′
40′	.6668	.9473	.7196	.2804	.0527	.3332	20′
50′	.6692	.9466	.7226	.2774	.0534	.3308	10′
28° 00′	9.6716	9.9459	9.7257	10.2743	10.0541	10.3284	62° 00′
	L cos x	L sin x	L cot x	L tan x	L csc x	L sec x	x

Table IV Four-Place Logarithms of Trigonometric Functions—Angle x in Degrees (continued)

Attach—10 to Logarithms Obtained from this Table

x	L sin x	L cos x	L tan x	L cot x	L sec x	L csc x	
28° 00′	9.6716	9.9459	9.7257	10.2743	10.0541	10.3284	62° 00′
10′	.6740	.9453	.7287	.2713	.0547	.3260	50′
20′	.6763	.9446	.7317	.2683	.0554	.3237	40′
30′	.6787	.9439	.7348	.2652	.0561	.3213	30′
40′	.6810	.9432	.7378	.2622	.0568	.3190	20′
50′	.6833	.9425	.7408	.2592	.0575	.3167	10′
29° 00′	9.6856	9.9418	9.7438	10.2562	10.0582	10.3144	61° 00′
10′	.6878	.9411	.7467	.2533	.0589	.3122	50′
20′	.6901	.9404	.7497	.2503	.0596	.3099	40′
30′	.6923	.9397	.7526	.2474	.0603	.3077	30′
40′	.6946	.9390	.7556	.2444	.0610	.3054	20′
50′	.6968	.9383	.7585	.2415	.0617	.3032	10′
30° 00′	9.6990	9.9375	9.7614	10.2386	10.0625	10.3010	60° 00′
10′	.7012	.9368	.7644	.2356	.0632	.2988	50′
20′	.7033	.9361	.7673	.2327	.0639	.2967	40′
30′	.7055	.9353	.7701	.2299	.0647	.2945	30′
40′	.7076	.9346	.7730	.2270	0654	.2924	20′
50′	.7097	.9338	.7759	.2241	.0662	.2903	10′
31° 00′	9.7118	9.9331	9.7788	10.2212	10.0669	10.2882	59° 00′
10′	.7139	.9323	.7816	.2184	.0677	.2861	50′
20′	.7160	.9315	.7845	.2155	.0685	.2840	40′
30′	.7181	.9308	.7873	.2127	.0692	.2819	30′
40′	.7201	.9300	.7902	.2098	.0700	.2799	20′
50′	.7222	.9292	.7930	.2070	.0708	.2778	10′
32° 00′	9.7242	9.9284	9.7958	10.2042	10.0716	10.2758	58° 00′
10′	.7262	.9276	.7986	.2014	.0724	.2738	50′
20′	.7282	.9268	.8014	.1986	.0732	.2718	40′
30′	.7302	.9260	.8042	.1958	0740	.2698	30′
40′	.7322	.9252	.8070	.1930	.0748	.2678	20′
50′	.7342	.9244	.8097	.1903	.0756	.2658	10′
33° 00′	9.7361	9.9236	9.8125	10.1875	10.0764	10.2639	57° 00′
10′	.7380	.9228	.8153	.1847	.0772	.2620	50′
20′	.7400	.9219	.8180	.1820	.0781	.2600	40′
30′	.7419	.9211	.8208	.1792	.0789	.2581	30′
40′	.7438	.9203	.8235	.1765	.0797	.2562	20′
50′	.7457	.9194	.8263	.1737	.0806	.2543	10′
34° 00′	9.7476	9.9186	9.8290	10.1710	10.0814	10.2524	56° 00′
10′	.7494	.9177	.8317	.1683	.0823	.2506	50′
20′	.7513	.9169	.8344	.1656	.0831	.2487	40′
30′	.7531	9160	.8371	.1629	.0840	.2469	30′
40′	.7550	.9151	.8398	.1602	.0849	.2450	20′
50′	.7568	.9142	.8425	.1575	.0858	.2432	10′
35° 00′	9.7586	9.9134	9.8452	10.1548	10.0866	10.2414	55° 00′
	L cos x	L sin x	L cot x	L tan x	L csc x	L sec x	x

Table IV Four-Place Logarithms of Trigonometric Functions—Angle x in Degrees (continued)

Attach —10 to Logarithms Obtained from this Table

x	L sin x	L cos x	L tan x	L cot x	L sec x	L csc x	
35° 00′	9.7586	9.9134	9.8452	10.1548	10.0866	10.2414	55° 00′
10′	.7604	.9125	.8479	.1521	.0875	.2396	50′
20′	.7622	.9116	.8506	.1494	.0884	.2378	40′
30′	.7640	.9107	.8533	.1467	.0893	.2360	30′
40′	.7657	.9098	.8559	.1441	.0902	.2343	20′
50′	7675	.9089	.8586	.1414	.0911	.2325	10′
36° 00′	9.7692	9.9080	9.8613	10.1387	10.0920	10.2308	54° 00′
10′	.7710	.9070	.8639	.1361	.0930	.2290	50′
20′	.7727	.9061	.8666	.1334	.0939	.2273	40′
30′	.7744	.9052	.8692	.1308	.0948	.2256	30′
40′	.7761	.9042	.8718	.1282	.0958	.2239	20′
50′	.7778	.9033	.8745	.1255	.0967	.2222	10′
37° 00′	9.7795	9.9023	9.8771	10.1229	10.0977	10.2205	53° 00′
10′	.7811	.9014	.8797	.1203	.0986	.2189	50′
20′	.7828	.9004	.8824	.1176	.0996	.2172	40′
30′	.7844	.8995	.8850	.1150	.1005	.2156	30′
40′	.7861	.8985	.8876	.1124	.1015	.2139	20′
50′	.7877	.8975	.8902	.1098	.1025	.2123	10′
38° 00′	9.7893	9.8965	9.8928	10.1072	10.1035	10.2107	52° 00′
10′	.7910	.8955	.8954	.1046	.1045	.2090	50′
20′	.7926	.8945	.8980	.1020	.1055	.2074	40′
30′	.7941	.8935	.9006	.0994	.1065	.2059	30′
40′	.7957	.8925	.9032	.0968	.1075	.2043	20′
50′	.7973	.8915	.9058	.0942	.1085	.2027	10′
39° 00′	9.7989	9.8905	9.9084	10.0916	10.1095	10.2011	51° 00′
10′	.8004	.8895	.9110	.0890	.1105	.1996	50′
20′	.8020	.8884	.9135	.0865	.1116	.1980	40′
30′	.8035	.8874	.9161	.0839	.1126	.1965	30′
40′	.8050	.8864	.9187	.0813	.1136	.1950	20′
50′	.8066	.8853	.9212	.0788	.1147	.1934	10′
40° 00′	9.8081	9.8843	9.9238	10.0762	10.1157	10.1919	50° 00′
10′	.8096	.8832	.9264	.0736	.1168	.1904	50′
20′	.8111	.8821	.9289	.0711	.1179	.1889	40′
30′	.8125	.8810	.9315	.0685	.1190	.1875	30′
40′	.8140	.8800	.9341	.0659	.1200	.1860	20′
50′	.8155	.8789	.9366	.0634	.1211	.1845	10′
41° 00′	9.8169	9.8778	9.9392	10.0608	10.1222	10.1831	49° 00′
10′	.8184	.8767	.9417	.0583	.1233	.1816	50′
20′	.8198	.8756	.9443	.0557	.1244	.1802	40′
30′	.8213	.8745	.9468	.0532	.1255	.1787	30′
40′	.8227	.8733	.9494	.0506	.1267	.1773	20′
50′	.8241	.8722	.9519	.0481	.1278	.1759	10′
42° 00′	9.8255	9.8711	9.9544	10.0456	10.1289	10.1745	48° 00′
	L cos x	L sin x	L cot x	L tan x	L csc x	L sec x	x

Table IV Four-Place Logarithms of Trigonometric Functions—Angle x in Degrees (continued)

Attach -10 to Logarithms Obtained from this Table

x	L sin x	L cos x	L tan x	L cot x	L sec x	L csc x	
42° 00′	9.8255	9.8711	9.9544	10.0456	10.1289	10.1745	48° 00′
10′	.8269	.8699	.9570	.0430	.1301	.1731	50′
20′	.8283	.8688	.9595	.0405	.1312	.1717	40′
30′	.8297	.8676	.9621	.0379	.1324	.1703	30′
40′	.8311	.8665	.9646	.0354	.1335	.1689	20′
50′	.8324	.8653	.9671	.0329	.1347	.1676	10′
43° 00′	9.8338	9.8641	9.9697	10.0303	10.1359	10.1662	47° 00′
10′	.8351	.8629	.9722	.0278	.1371	.1649	50′
20′	.8365	.8618	.9747	.0253	.1382	.1635	40′
30′	.8378	.8606	.9772	.0228	.1394	.1622	30′
40′	.8391	.8594	.9798	.0202	.1406	.1609	20′
50′	.8405	.8582	.9823	.0177	.1418	.1595	10′
44° 00′	9.8418	9.8569	9.9848	10.0152	10.1431	10.1582	46° 00′
10′	.8431	.8557	.9874	.0126	.1443	.1569	50′
20′	.8444	.8545	.9899	.0101	.1455	.1556	40′
30′	.8457	.8532	.9924	.0076	.1468	.1543	30′
40′	.8469	.8520	.9949	.0051	.1480	.1531	20′
50′	.8482	.8507	9.9975	.0025	.1493	.1518	10′
45° 00′	9.8495	9.8495	10.0000	10.0000	10.1505	10.1505	45° 00′
	L cos x	L sin x	L cot x	L tan x	L csc x	L sec x	x

Index